Service life estimation and extension of civil engineering structures

Related titles:

Strengthening and rehabilitation of civil infrastructures using fibre-reinforced polymer (FRP) composites
(ISBN 978-1-84569-448-7)
The book discusses the mechanical and in-service properties, the relevant manufacturing techniques and aspects related to externally bonded FRP composites to strengthen/rehabilitate/retrofit civil engineering structural materials. The book concentrates upon: the mechanical properties of the FRP materials used; the analysis and design of strengthening/rehabilitating/retrofitting beams and columns manufactured from reinforced concrete (RC), metallic and masonry materials; failure modes of strengthening systems; site preparation of the two adherend materials; durability issues; quality control, maintenance and repair of structural systems; case studies.

Structural health monitoring of civil infrastructure systems
(ISBN 978-1-84569-392-3)
Structural health monitoring (SHM) has been defined as the use of in-situ, non-destructive sensing and analysis of structural characteristics, including the structural response, for detecting changes that may indicate damage or degradation. Essentially, an SHM system must have the ability to collect, validate and make accessible, operational data on the basis of which decisions related to service life management can be made. This important book provides a comprehensive review of the topic of structural health monitoring of civil infrastructure systems including aspects associated with sensors, data collection, data interpretation, application and specification.

Durability of composites for civil structural applications
(ISBN 978-1-84569-035-9)
This comprehensive book on durability of FRP composites will make it easier for the practising civil engineer and designer to use these materials on a routine basis. It addresses the current lack, or inaccessibility, of data related to the durability of these materials, which is proving to be one of the major challenges to the widespread acceptance and implementation of FRP composites in civil infrastructure. The book should help further the acceptance of composites for civil structural applications, by providing this source for engineers, decision makers, and students involved in architectural engineering, construction and materials, disaster reduction, environmental engineering, maritime structural technology, transportation engineering and urban planning.

Details of these and other Woodhead Publishing materials books can be obtained by:

- visiting our web site at www.woodheadpublishing.com
- contacting Customer Services (e-mail: sales@woodheadpublishing.com; fax: +44 (0) 1223 832819; tel.: +44 (0) 1223 499140 ext. 130; address: Woodhead Publishing Limited, 80 High Street, Sawston, Cambridge CB22 3HJ, UK)

If you would like to receive information on forthcoming titles, please send your address details to: Francis Dodds (address, tel. and fax as above; e-mail: francis.dodds@woodheadpublishing.com). Please confirm which subject areas you are interested in.

Service life estimation and extension of civil engineering structures

Edited by
Vistasp M. Karbhari and Luke S. Lee

WOODHEAD PUBLISHING

Oxford Cambridge Philadelphia New Delhi

Published by Woodhead Publishing Limited, 80 High Street, Sawston, Cambridge CB22 3HJ, UK
www.woodheadpublishing.com

Woodhead Publishing, 1518 Walnut Street, Suite 1100, Philadelphia, PA 19102-3406, USA

Woodhead Publishing India Private Limited, G-2, Vardaan House, 7/28 Ansari Road, Daryaganj, New Delhi – 110002, India

First published 2011, Woodhead Publishing Limited

British Library Cataloguing in Publication Data
A catalogue record for this book is available from the British Library.

ISBN 978-0-08-101496-7 (print)
ISBN 978-0-85709-092-8 (online)

The publisher's policy is to use permanent paper from mills that operate a sustainable forestry policy, and which has been manufactured from pulp which is processed using acid-free and elemental chlorine-free practices. Furthermore, the publisher ensures that the text paper and cover board used have met acceptable environmental accreditation standards.

Typeset by Toppan Best-set Premedia Limited, Hong Kong

Contents

Contributor contact details

(* = main contact)

Editors

Professor Vistasp M. Karbhari
University of Alabama in Huntsville
366 Shelbie King Hall
Huntsville, AL 35899
USA

E-mail: vistasp.karbhari@uah.edu

Professor Luke S. Lee
Department of Civil Engineering
University of the Pacific
3601 Pacific Avenue
Stockton
CA 95219
USA

E-mail: LLee4@pacific.edu

Chapter 1

Professor L. C. Hollaway
University of Surrey
Guildford
Surrey
GU2 7XH
UK

E-mail: L.Hollaway@surrey.ac.uk

Chapter 2

Professor Khaled Soudki
Department of Civil Engineering
University of Waterloo
Waterloo
Ontario
N2L 3G1
Canada

E-mail: soudki@uwaterloo.ca

Chapter 3

Assistant professor Rebecca A. Atadero
Colorado State University
Department of Civil and Environmental Engineering
Campus Delivery 1372
Fort Collins
CO 80523
USA

E-mail: Rebecca.Atadero@colostate.edu

Chapter 4

Professor Luke S. Lee – see Editors

Chapter 5

Professor Dan M. Frangopol*
Department of Civil and Environmental Engineering
ATLSS Engineering Research Center
Lehigh University
117 ATLSS Dr.
Bethlehem
PA 18015-4729
USA

E-mail: dan.frangopol@lehigh.edu

Sunyong Kim
Graduate Research Assistant
Department of Civil and Environmental Engineering
ATLSS Engineering Research Center
Lehigh University
117 ATLSS Dr.
Bethlehem
PA 18015-4729
USA

E-mail: suk206@lehigh.edu

Chapter 6

Professor Robert E. Melchers
Centre for Infrastructure Performance and Reliability
The University of Newcastle
University Drive, Callaghan
NSW 2308
Australia

E-mail: rob.melchers@newcastle.edu.au

Chapter 7

Professor Yongtao Dong*
Department of Civil and Environmental Engineering
University of Alaska Fairbanks
P. O. Box 755900
Fairbanks
AK 99775-5900
USA

E-mail: ydong2@alaska.edu

Dr Farhad Ansari
Department of Civil and Materials Engineering
University of Illinois at Chicago
842 W. Taylor St.
Chicago
IL 60607-7023
USA

E-mail: fansari@uic.edu

Chapter 8

Dr Charles Sikorsky
California Department of Transportation
Division of Engineering Services
1801 30th Street MS 9-2/SI
Sacramento
CA 95816
USA

Email: charles_sikorsky@dot.ca.gov

Professor Vistasp M. Karbhari –
see Editors

Chapter 9

Professor Vistasp M. Karbhari –
see Editors

Chapter 10

Professor Mohammad Najafi
Department of Civil Engineering
The University of Texas at Arlington
Box 19308
428 Nedderman Hall
Arlington
TX 76019
USA

E-mail: najafi@uta.edu

Introduction

The infrastructure of constructed facilities for the transportation and housing of people, goods, and services, which grew exponentially over the latter half of the last century is rapidly approaching a critical period, with increasing signs of deterioration and reduced functionality. Deficiencies in the existing bridge inventory, for example, range from those related to wear, environmental deterioration, and aging of structural components, to increased traffic demands and changing patterns, and from insufficient detailing at the time of construction/original design, and the use of substandard materials in initial construction, to inadequate maintenance and rehabilitation measures taken through the life of the structure. These deficiencies are not isolated to just bridge and other transportation-related structures alone, but are endemic to the built environment, ranging from residential housing and industrial/commercial structures to pipelines used for the distribution of water and sewage. In addition, explosive growth of populations in urban areas across the globe has resulted in substantially increased needs for services, putting severe stress on the existing life-lines and often resulting in their use at levels greatly over those for which they were designed. This, again, results in a decrease in functional life expectancy and the need for expansion, either through rebuilding or through rehabilitation.

While engineers and the owners of structural systems often talk in terms of service life, the term is, in itself, ill defined. While it is a critical consideration in design and planning, it is simultaneously a somewhat nebulous consideration in the overall management system in that there is unlikely to be a specific end date planned for the structure – rather it is expected that maintenance, albeit on an irregular schedule, will result in continuous extension of useful life, at least to the point when the structure is either demolished due to a complete change in needs for that area (such as with changes in functionality of buildings) or replaced with a newer structure (such as in the case of bridges with more lanes or higher load capacity). Although service life is often defined in terms of deterioration rates, and time to structural or functional deficiency, it is also affected by methods of assessment of assets in the inventory. In most cases the assessment is not single functioned but is based on trade-offs between aspects related to cost, risk, and convenience. The schema is further complicated by the knowledge that most civil structures are built with extremely high factors of safety and redundancy; hence, benign neglect will not cause catastrophic failure immediately, resulting in needed maintenance actions being deferred. Until recently, there was also very little attention paid to the financial aspects accruing from functional deficiency, such as due to congestion resulting from bridges with posted speed limits or lower than the required number of lanes, thereby not emphasizing the economic effects of such deficiencies on society.

In contrast to consumer goods and even automobiles, which have limited expectations of service life, the public generically has expectations of very long service-lives for infrastructure facilities. While the expected design life of a bridge might well be 50 years, the general expectation is that it would 'last forever', or at least that it could be maintained at intervals to ensure continued service. There is, however, an increasing focus on both the determination of useful remaining service life of structures and of designing new structural systems for well-defined periods of functionality. This is aided by the development and use of newer and more durable materials, and methods of construction, as well as the development and use of techniques of analyses that include aspects of reliability and stochastic mechanics. Rather than being an art, service life estimation and decisions regarding extension of service life are increasingly being made on the basis of solid scientific, engineering and economic principles, melding risk management with design. The development of integrative tools, such as those involved in Structural Health Monitoring (SHM), add to the palette of available methods for monitoring the health of structures and of making assessments regarding the need for maintenance and/or rehabilitation in real-time rather than on the basis of emergencies and/or pre-determined periods of neglect.

Vistasp M. Karbhari

Part I

Using fibre reinforced polymer (FRP) composites to extend the service life of civil engineering structures

1
Key issues in the use of fibre reinforced polymer (FRP) composites in the rehabilitation and retrofitting of concrete structures

L. C. HOLLAWAY, University of Surrey, UK

Abstract: Fibre reinforced polymer (FRP) composites were first used in the building industry during the late 1960s to construct all-composite buildings; also at this time, the construction industry proposed that rebars for reinforcing concrete should be made from composite materials and prestressing tendons for concrete beams should also be manufactured from FRP composites. However, it was not until the mid-1980s that a few research/design teams throughout the world seriously investigated the use of composites to prestress concrete beams and to utilise the material in conjunction with the more conventional materials. FRP composites were seen at that time as materials with a high strength and stiffness and to have a high corrosion resistance; they could be tailored to any design requirement but their high cost was a drawback to their use. In addition, the attitude of the civil engineering industry generally was against the materials but in the late 1980s, composite materials started to have their first major successes in the field of flexural and shear strengthening and seismic retrofitting of degraded concrete structures. This initial thrust has now been extended to areas such as confining concrete columns and strengthening beam/column joints, but in spite of this FRP composites have not yet been fully accepted by all areas of the civil engineering industry. In the strengthening and seismic retrofitting area of civil engineering, the cost of the FRP material is only a relatively small percentage of the overall cost for the work, and because of its physical properties, its fabrication on site can be undertaken much more speedily than if a more conventional material were to be used. Its low weight has an economic benefit in decreasing the erection time and reducing any costly closure period of the highway.

This chapter discusses the mechanical and in-service properties of advanced polymer composite materials and their components, their long term loading characteristics and the manufacturing techniques available for use, specifically in terms of rehabilitation and retrofitting of reinforced and prestressed concrete structures. It considers the preparation of the surfaces of the concrete, and of the polymer composite plate, and suggests ways of joining the plate to the adherend. Finally, it anticipates future trends in the field of upgrading structural members.

Key words: rehabilitation, retrofitting, fibre reinforced polymer, speed.

1.1 Introduction

The rehabilitation and retrofitting of reinforced concrete (RC) and prestressed concrete (PC) structures utilising advanced polymer composites (in the construction industry these materials are referred to as fibre reinforced polymer (FRP) composites) are now well established techniques. Over the last three decades, externally-bonded FRP composite strengthening has been shown to be particularly attractive and it has many advantages, including the light weight and strength of the FRP laminates. Furthermore, the ease of handling and installation techniques without heavy equipment include a simple application in overhead installations where the viscosity of the epoxy adhesive is sufficient to support the self-weight of the FRP strips during curing. (However, the author would not recommend this procedure except under very special circumstances.) The strips can cross other FRP laminate strips, with adjustments in the thickness of the epoxy adhesive layer. In addition, the fabrication technologies for the production of FRP composites have been revolutionised by sophisticated manufacturing techniques. These methods have enabled polymer composite materials to produce good quality laminates with minimal voids and accurate fibre alignment.

This chapter will discuss: (i) The composite materials and their component parts used in the civil infrastructure, (ii) the mechanical, in-service and physical properties of composite materials, (iii) the different manufacturing methods for the FRP composites specifically for the techniques of rehabilitation and retrofitting, and the influence that the various methods have on the FRP composites' mechanical and in-service properties, (iv) the rehabilitation of RC and PC flexural and shear structural members and the failure modes of the upgraded structural systems, and (v) the retrofitting of RC columns.

1.2 Key issues affecting composite use in construction

For 40 years following the Second World War, the construction industry suffered from a lack of investment, and by the early 1990s there was a serious decline in that industry in both the United States and Britain. The construction industry was regarded as low technology, low skill and labour intensive compared with most other industries. Although economic factors were the main cause of this decline, which meant that methods and materials had changed little over the 40-year period, other underlying factors were believed to include the industry's widespread over-reliance on the use of litigation and arbitration to settle disputes and claims regarding 'value for money'. The concern over the lack of investment, performance and productivity of the UK construction industry led the UK Government to commission Sir Michael Latham (Latham, 1993) to investigate the procurement and contractual relations within, and the structure of, the construction industry. It was reported that disputes and conflicts had taken their toll on moral and team spirit and that defensive attitudes were commonplace. Following this report, Latham lead a year-long enquiry with the aim of ending 'the culture of conflict and inefficiency that dogs Britain's largest industry'. The second Latham Report, which was published in 1994, concluded that the traditional methods of procurement and contract management and its adversarial culture caused inefficiency and ineffectiveness within the industry. The Latham Reports set the agenda for reform and gave the industry targets. Whilst these reports were widely welcomed, the implementation of their recommendations was perceived to be slow and, as a result, the UK Government established the Construction Task Force, led by Sir John Egan (1998). The Egan Report, 'Rethinking Construction', (commissioned by John Prescott, the deputy Prime Minister at the time) was published in 1998 and acknowledged that its foundation was the Latham Report.

The report concluded that there was 'growing dissatisfaction with construction among both private and public sector clients. Projects are widely seen as unpredictable in terms of delivery on time, within budget and to the standards of quality expected'. The Report saw a need for a change of style, culture and process in the construction industry. To this end, it identified five areas that needed to be in place to secure improvement in the industry, namely, committed leadership, focus on the customer, integrated processes and teams, quality-driven agenda, and commitment to people. Within these areas, the need for partnering and sharing of risk between contractor and client was identified.

The findings and conclusions of the Latham and Egan Reports are reflected in the 'Strategic Forum for Construction Report' (2001); these findings are now being implemented by the Government.

These discords within the construction industry were responsible for the lack of investment in research and development and for the practical methods of operation and the materials used in the construction industry changing little over the 40-year period after the Second World War. These problems resulted in a complete lack of interest by the plastics/composites industry in the civil engineering sphere, where the possibility of utilising new materials and therefore market opportunities were not visible to potential investors. By contrast, the use of composites technology in aircraft applications and the use of fibre reinforced polymers in boat hulls and car bodies were developing rapidly. Thus, the technological revolution in materials and processing in other sectors of the manufacturing industry had largely by-passed the construction industry. However, civil engineering construction is a large sector in the economy of the UK with an output of over

£60 bn at the beginning of the 21st century; this represented 8% of GDP. There are few with the temerity to try to reform such a large industry. There are even fewer with the ability to succeed.

It should be said that initially (during the 1970s), composites made with the higher performance fibres were too expensive to make much impact in civil engineering beyond niche applications. By the middle to late 1980s, there were a few forward looking research and consulting engineers who saw the potential of FRP composites for structural use in civil engineering. Furthermore, in the early 1990s, as the defence markets waned, increased importance was placed by the fibre and the FRP manufacturers on the cost reductions for the continued growth of the FRP industry; they could also see a potentially large market in construction. As the cost of FRP materials continued to decrease and the need for aggressive infrastructure renewal became increasingly evident in the developed world, pressure mounted for the use of these materials in the civil infrastructure.

From the late 1980s, and continuing to the present, a vast number of RC structures were/are in urgent need of repair and strengthening, due to either a change in use or structural degradation. Material deterioration in concrete structures typically involves corrosion of the internal steel reinforcement due to long-term chloride ingress or carbonation of the concrete. Many concrete structures were built prior to the introduction of modern design codes, and hence did not meet modern design requirements. Furthermore, RC structures built between the Second World War and the 1980s were often designed with little attention to durability issues, and have thus suffered severe structural degradation. Moreover, concrete structures are susceptible to hazard events, and to design or construction errors. Material deterioration is particularly likely in aggressive marine or industrial environments, and in cold regions where de-icing salts are used. Prestressed concrete members are susceptible to steel strand fatigue, and may require strengthening to prevent further loss of prestress. Strengthening is often required due to structural modifications, such as the formation of openings in a concrete slab. Structural assessment may highlight inadequate reinforcement within the concrete, when compared with today's more stringent requirements. Of particular concern was the seismic performance of structures originally designed for gravity loads only. Seismic upgrade of these structures can have profound economic and social implications. RC buildings that were not designed for seismic loads can have inadequate ductility and a lack of robustness. The need for economically viable seismic retrofit systems has driven research in both the USA and Japan into FRP strengthening.

The practice of bonding carbon/epoxy composites to reinforced concrete beams to increase their flexural capacity was first reported in the mid-1980s, when Germany and Switzerland replaced steel with FRP plates. The FRP composite material began to be viewed as a promising improvement in externally bonded repairs (Meier, 1987; Meier and Kaiser, 1991; Rostasy and Budelman, 1992). Kaiser (1989), load-tested carbon FRP composites and showed the validity of the strain compatibility method (i.e. classical approach for RC sections) in the analysis of repaired members. In the United States, Ritchie *et al.* (1991) and Saadatmanesh and Ehsani (1991a, b) studied the static behaviour of RC beams with externally-bonded glass FRP plates and developed analytical methods also based on strain compatibility. In 1992, Triantafillou and Plevris (1992) added concepts of fracture mechanics to this classical method. Berset (1992) investigated the use of externally-bonded composites to strengthen RC beams in shear. Plevris and Triantafillou (1994) developed an analytical model for predicting the creep and shrinkage behaviour of RC members strengthened with various types of FRP plates. In Saudi Arabia, Sharif *et al.* (1994), using both Roberts' theory and strain compatibility, developed a theoretical algorithm for predicting the flexural strength and the plate separation load of repaired beams. At the turn of the 20th century, and into 21st, FRP plate bonding was widely recognized as a cost-effective technique when compared with the utilisation of steel plates (Karbhari and Seible, 2000).

Steel was considered to have a number of disadvantages such as:

- Uncertainty regarding its durability and corrosion effects.
- Situations where chloride-contaminated concrete has to be removed prior to bonding.
- The need for plates to be subjected to careful surface preparation, including the application of resistant priming systems.
- Difficulty of shaping and fixing complex profiles.
- The weight of plate making transport difficult.
- The elaborate and expensive falsework that is required to maintain steelwork in position during bonding.

Flexural and shear failures are the two main failure criteria of RC beams. The former mode is a ductile failure, whereas the latter criterion is brittle in nature. When the RC beam is upgraded using an externally bonded FRP composite plate, its flexural failure mode exhibits a much reduced ductility over that of the original RC beam, but it still shows a ductile behaviour. Therefore, a strengthened beam must have a shear capacity which is greater than that of its flexural capacity; this may entail upgrading the beam in shear as well. Both types of upgrading will be discussed later in this chapter.

In 1996 the Civil Engineering Research Foundation (CERF) in the USA, recommended the use of high-performance materials and systems in construction, citing potentially substantial cost savings due to lower volumes of materials needed, reduced maintenance and longer lifetimes (Civil Engineering Research Foundation, Executive Report 93-5011, 1993).

The benefits of using FRP composite materials can be listed as follows:

- The fibres can be introduced in a variety of positions, volume fractions and directions in the matrix to obtain maximum efficiency, allowing the composites to be tailor-made to suit the required shape and specification.
- The resulting materials have high strength and stiffness in the fibre direction at a fraction of the weight of steel.
- The transportation and handling the composite materials are easy.
- Less false work is required than with steel plates, and FRP composites can be used in areas of limited access.
- Continuous lengths of FRP can be readily produced which, because of their low bending stiffness (thickness between 1.0 and 1.5 mm), can be delivered to site in rolls.
- CFRP (carbon fibre reinforced polymer) and AFRP (aramid fibre reinforced polymer) composites exhibit excellent fatigue and creep properties and require less energy per kg to manufacture and to transport to the site than that for steel.

The drawbacks to the use of FRP are

- The intolerance to uneven bonding surfaces, which may cause peeling of the plate away from the concrete surface.
- The possibility of brittle failure modes.
- Risk of fire.
- Risk of vandalism.
- Risk of accidental damage unless the strengthening is protected. (A particular concern for bridges over roads is the risk of soffit reinforcement being hit by over-height vehicles.)
- High material costs.

The relative economics of the fibre composites and other strengthening systems depend upon circumstances. It will be necessary to compare them both in the short and the long term. The latter may be difficult to quantify, as the life-time behaviour can only be estimated. The costs that should be taken into account are:

- *Costs of access and possession time.*
- *Costs of materials* – FRP composites will generally be more expensive than conventional materials. However, the cheaper transportation of the FRP material from factory to site and its installation savings can offset the higher material costs, which rarely exceed 20% of the overall project costs. Furthermore, when traffic management costs are included, the use of FRP provides a cost saving in the region of 17.5% over steel.
- *Costs of closure by highway work* – High closure costs are often incurred by highway works. Railtrack UK have estimated that strengthening with

FRP materials is approximately 30% cheaper than the equivalent strengthening using steel plates.

- *Whole-life costing* – Whole-life costing is an important part of decision making about repair and strengthening of the structural member. BS DD ENV 1504 Part 9 (1997) has listed amongst the factors to be considered:
  - (i) The number and cost of repair cycles acceptable during the design life of the concrete structure.
  - (ii) The cost of the alternative protection or repair options, including future maintenance and access costs.

The whole-life cost of a repair or strengthening solution is the sum of the initial installation cost and the future maintenance costs over the remaining life of the structure.

The ROBUST project (Hollaway and Leeming, 1999) showed that FRP is more economical than steel in construction. Plate thicknesses, which are less for polymer composites than for steel, may be important from an aesthetic viewpoint. In applications where corrosion, length of the required strengthening plate, and handling of it on site are of greater significance (for example, bridge rehabilitation), fibre composites become a more attractive alternative.

An alternative technique to externally bonded FRP laminates is the Near Surface Mounted (NSM) reinforcement; this method is becoming a serious competitor to the bonding of rigid composite plates. The technique will be discussed later in the chapter.

It should be noted that, in the past, clients have related the construction industry to one which utilises bulk materials, namely, concrete and steels. The clients' confidence in FRP technology has been and will further be realised when they see the benefits of FRP in terms of speed of installation, durability of the material and cost effectiveness, often through the results of case studies. Reduced labour, shut-down costs and site constraints usually offset the material cost of FRP, which is of the order of 13% of the total cost of the upgrading, thus making FRP strengthening systems very competitive when compared to traditional strengthening techniques, such as steel plate bonding and section enlargement. From a structural perspective, materials and fabrication methods are developing rapidly, *viz.* the wet lay-up, pultrusion and the pre-impregnated (prepreg) methods, together with the new approaches, the cold melt prepreg, the NSM reinforcement and the structural re-pointing systems. Currently, FRP composites are gaining considerable worldwide interest and growing acceptance in the construction industry. The expansion and the range of potential applications and such issues as the long-term durability and material compatibility are based upon an understanding of the behaviour of the material over the life-span of the structure, the design procedures of the upgrading technologies and the implementation of new manufacturing techniques.

1.3 Basic fibre/matrix components and composites

Polymers consist of long chains or networks, built by the repeated linkage of small reactive molecules. Each of the simple units is called a mer, hence a polymer is the result of the joining together of many identical units. When the chain gets long enough (*viz.* a molecular weight above 5000), new properties appear such as hardness, stiffness and mechanical strength, which have no relation to the original liquids or gases that comprise their primary molecules. Thus, depending on the detailed chemistry and the spatial arrangement of the chains and their cross-links, the following can be produced: plastics, rubbers, adhesives, coatings and fibres. The first production of a thermosetting polymer commenced in 1909; the resin was a phenolic and when this was combined with a compatible curing agent, the cured polymer ('Bakelite') was formed. The first commercial production of glass fibre filaments was in 1935 (Hollaway, 1994; Schwartz, 1997), but it was not until the 1940s that composites technology was used in aircraft applications and in radomes during the Second World War. By the 1950s, glass fibre reinforced polymer (GFRP) boat hulls and GFRP car bodies were developed; the polymers used in the early days of composites were made from unsaturated polyester resins. However, under certain conditions of exposure, glass fibres are sensitive to alkaline environments and moisture attack and, in addition, creep affects glass fibres more than any other types of fibres (Barbero, 1999). Further discussions regarding these two important topics with respect to upgrading RC structures will be given later in this chapter. The higher performance carbon fibre, using the precursor polyacrylonitrile (PAN) was invented and developed by Watt, Phillips and Johnson at the former Royal Aircraft Establishment (now QinetiQ) at Farnborough, UK (Watt *et al.*, 1966).

With a clearer understanding of the chemistry and physics of polymers, it became possible to combine them with fibres in order to produce a large range of hitherto unknown materials which are referred to as 'advanced composites'. Carbon fibre reinforced polymers (CFRPs) were developed during the 1960s for specialised applications. Unlike glass fibre, carbon fibre is an electrical conductor and hence galvanic corrosion could take place if fibres are placed in direct contact with metals, (e.g. when upgrading metallic structural components, Miller *et al.*, 2001); however, this effect is irrelevant when CFRP is bonded to RC members. Creep and relaxation of carbon fibres are practically zero (Balazs and Borosnyoi, 2001).

Aramid fibres were first developed in 1965 and likewise do not creep or fatigue under load. They have anisotropic mechanical properties, with higher strength and modulus of elasticity values in their longitudinal direction compared to their transverse direction.

Currently, the two main polymers used for the upgrading of RC structural members are vinylesters and the epoxies. The fibres used with these polymers are glass, aramid and carbon. These man-made fibres (and composites made with them) have stability, combined with outstanding physical and mechanical properties; fibre properties will be discussed in Section 1.4.3. Furthermore, combined with a low specific weight, these attributes make the composites ideal engineering materials, able to outperform many of the other materials used in civil engineering. However, there are complications; composites are manufactured by various methods, manually, semi-automated and automated. These methods will be briefly described in Section 1.5.2 and it will be shown that the properties of the composite will be influenced by their manufacturing method. Composites can be fabricated with a pronounced grain and are most efficient when constructed with the fibres running in the principal stress directions; they can also be readily made having quasi-isotropic or anisotropic properties.

1.3.1 The upgrading of structural members

Arguably, one of the most important issues currently for the civil/structural engineer worldwide is the repair of deteriorated, damaged and substandard concrete structural members and concrete structures. As explained before, a considerable volume of the civil infrastructure of the developed countries of the world was completed in the middle period of the last century and is now over 60 years old. Furthermore, structures that were built after the Second World War had little attention paid to their durability and consequently, many of these structures are currently structurally deficient and are in need of strengthening to allow their continued use. In addition, during the second half of the last century, civil engineers in the USA and in Japan had inadequate knowledge of seismic design and, as a result, in 1995 the Hyogoken-Nanbu earthquake caused a great disaster to the city of Kobe, Japan; this is one of many serious disasters caused by seismic actions. In response to this tragedy, independent efforts, coordinated by various organizations such as the Japan Building Disaster Prevention Association (JBDPA) and the American Concrete Institute (ACI), led to implementing appropriate provisions. In 1999, the JBDPA published the 'Seismic Retrofitting Guideline' (JBDPA, 1999) and this was followed by guidelines by Fukuyama *et al.* (2003). In Europe during the latter part of the last century, nearly 84 000 reinforced and prestressed concrete bridges required maintenance, repair and strengthening, with an annual budget of £215 M, excluding traffic management cost (Leeming and Darby, 1999). Moreover, upgrading the US civil engineering infrastructure during the latter part of the last century has been estimated to have cost $20 trillion (NSF, 1993).

Within the scope of 'strengthening' concrete, it is essential to differentiate between the terms *repair, rehabilitation, strengthening* and *retrofitting*. These terms are often erroneously interchanged but they do refer to four different structural upgrading procedures. The term *repair* to a structure implies the filling of a crack in a concrete member by the injection of a resin polymer into the crack. The term *rehabilitation* refers to the improvement of the functional deficiency of a structure such as a severely degraded structural component which would require the addition of a structural member (in our case, a FRP composite component) to return it to its original structural form. The term *strengthening* of a structure is specific to those cases where the addition or application of a structural component (again in the current case, the FRP composite component) would enhance the existing designed performance level. The term *retrofit* is specifically used to relate to the upgrading of the part of the structure damaged during a seismic event; this could involve the use of composite jackets for the confinement of a damaged column. Within the last decade, repairing, rehabilitating, strengthening, and retrofitting existing concrete structures by utilising advanced FRP composites has been shown, throughout the world, to be a powerful, cost-effective and viable alternative solution to the use of steel; this latter material has now largely been replaced by FRP composites. However, there are drawbacks – the reduction in ductility that is experienced in the strengthened member, and also the fact that the plastic range that the section previously encountered is no longer a flat plateau due to the presence of the FRP composite upgrading. It is worthy of note that this loss in ductility has lead various design guidelines of FRP composite external strengthening to prohibit or discourage the application of moment redistribution from strengthened cross-sections (fib, 2001), leading to onerous conditions for such strengthening, particularly when the original design was based on moment redistribution. One of the problems with a section strengthened with FRP composites is its relatively unpredictable ultimate mode of failure (e.g. peeling, delamination); failure criteria will be considered in Section 1.8.4.

A structure will deteriorate with time, due to the service conditions to which it is subjected. This deterioration will continue during its service life and if it is unchecked the structure will then become non-functional. The main reasons for this deterioration are:

- *Corrosion* – this form of deterioration is the most common mechanism of structural degradation, particularly where a member is exposed to an aggressive environment, such as the de-icing salts used on highways. Corrosion can lead to a loss of member cross-section, and consequently a reduction in the structural capacity of the member.

- *Fatigue* – this form of deterioration is caused by a loading event. It can also result from hazard events, such as impact (e.g. 'bridge bashing' by over-height vehicles), vandalism, fire, blast loading or inappropriate structural alterations during maintenance. A single event may not be structurally significant, but multiple events could cause significant cumulative degradation to a structure.
- *Design or construction errors* – due to poor design, construction, workmanship and management, or the use of inferior materials.

This chapter will not be considering crack repair but there are specifications that provide requirements for the repair of cracks in concrete by injecting epoxy resin into the affected region of the RC beam. These are ACI Committee (503.7-07) (2007) and ASTM C881, Type I and Type IV (2002).

1.4 The component parts of fibre reinforced polymer (FRP) materials used for rehabilitation of reinforced concrete (RC) and prestressed concrete (PC) structural members

1.4.1 Introduction

It has been stated in Section 1.3 that the principal constituents of the composite are the fibre and the matrix (the polymer); but of equal importance is the interface between the fibre and matrix. This last component is an anisotropic transition region with a graduation of properties. The interface is required to provide adequate chemical and physical bonding stability between the fibre and the matrix in order to maximise the coupling between the two phases and thus allow stresses to be dispersed through the matrix, and thence to be transferred to the reinforcement.

1.4.2 Polymers used in the rehabilitation of structural members

The polymer material is formed from two components, the resin and curing agent (hardener). In civil engineering construction there are two types of curing procedures, one using a cold-cure resin and the other using a hot-cure resin. The cold-cure system is generally used for the fabrication of the composite and for adhesives used on site, and the hot-curing system is used for manufacturing composite components in a factory under controlled conditions. The more refined production of composites and curing procedures, particularly the hot-cured polymers, has produced compaction of the two components, lower voids and higher stiffness and strength values.

Table 1.1 Typical mechanical properties of the three thermosetting polymers used to rehabilitate engineering

Material	Specific strength	Ultimate tensile strength (MPa)	Modulus of elasticity in tension (GPa)	Coefficient of linear expansion (10^{-6}/°C)
Polyester	1.28	45–90	2.5–4.0	100–110
Vinylester	1.07	90	4.0	80
Epoxy	1.03	90–110	3.5	45–65

There are three types of thermosetting matrices that can be used in the rehabilitation of structures; these are polyester, vinylester and epoxy, all of which are thermosetting resins and can be reinforced with fibres to fabricate the fibre/matrix composite. These polymers are usually made from liquid or semi-solid precursors (resins) which cure irreversibly and on completion of the cure process, a hard solid polymer is formed by chemical cross-linking which is a tightly bound three-dimensional network of polymer chains; these materials cannot be reformed into other shapes. Currently, polyesters are used only occasionally to rehabilitate RC structural members; vinylester polymers have replaced them. Table 1.1 gives typical mechanical properties for the three thermosetting polymers.

1.4.3 Fibres used in the rehabilitation of structural members

The fibres used for rehabilitation in conjunction with the previously listed polymers are glass fibre, aramid fibre and carbon fibre.

Glass fibres

Glass fibres are manufactured by the rapid drawing of molten glass (temperature 1200°C – 1400°C) through platinum bushings, thus forming filaments, and 200 of these filaments form a strand. The technique has been discussed in Hollaway (2009). There are four main types of glass that can be used for fibre production; these are E-glass, S-glass, A-glass and C-glass. However, there are other glass fibres that have been developed to provide specific in-service properties. For instance, AR-glass is an alkali-resistant glass and is used for the reinforcement of glass fibre reinforced cement (GFRC). If alkali-resistant glass composite is used in a site environment or in humid conditions, it might degrade in strength and toughness but this will be over a long period of time. E-glass is an alumino-borosilicate glass, and has low alkali content, of the order of 2%. S-glass is stronger and stiffer

Table 1.2 Typical mechanical properties of four types of glass fibres

Glass fibre type	Elastic modulus (GPa)	Tensile strength (MPa)	Ultimate strain (%)
E	72.0	2400	3.3
AR	80.0	3030	3.8
S-2	88.0	4600	5.2
C	72	2200	3.1

than E-glass and has a greater resistance to corrosion. C-glass has been developed to resist chemical attack, in particular by acids. However, the strength of all the glass fibres depends strongly on the surface quality of the fibre itself and all the well-known types of glass fibre are affected by the influence of aggressive media. Table 1.2 gives typical mechanical properties for four types of glass fibre. Glass fibre is a complex subject and McCrum (1971) has listed important facts relating to the material, a few of the relevant ones being:

(i) Glass fibres are much stronger than the bulk material. This is because the molecules of the fibre have been orientated into one direction by virtue of the manufacturing technique.
(ii) Newly-made glass filaments, if contaminated by dust or water vapour, immediately lose half their original strength.
(iii) If glass fibres are loaded in tension, their strength decreases as the load increases.
(iv) When a glass fibre is held under a constant load well below its instantaneous ultimate strength, it will eventually fail. This phenomenon is known as static failure or stress corrosion; its occurrence depends upon the environmental conditions into which it is placed; water vapour is particularly serious.

The reason for the phenomenon of item (iv) above is that the surface of glass contains sub-microscopic voids, which act as stress concentrations; these are known as 'Griffiths cracks'. A state of high stress at the tip of the crack, in conjunction with the influence of corrosive water vapour is sufficient to propagate the crack until fracture occurs. The fibres can be protected from these influences by the application of a size to their surfaces at the end of the manufacturing procedure. During the 1960s, AR-glass was developed to resist both acid and alkali attack. However, until recently it has not been possible to use AR glass to reinforce polymer composites because the surface treatment applied to the fibre degraded with exposure over time, leading to a loss of performance. This difficulty has now been overcome and AR-glass fibres are now available for use with normal and chemically-resistant resins, mainly unsaturated polyesters, vinylesters and epoxies, for the manufacture of highly resistant and durable GFRP composite structures. The commercial name of the new reinforcement is ARcoteX™ and it was introduced by Saint-Gobain Vetrotex (Almenara and Thornburrow, 2004).

The main properties of the glass fibres are given in Table 1.2.

Aramid fibres

Aramid fibres are prepared from liquid crystal polymers, which in the liquid state retain some measure of crystalline molecular orientation and order. The fibres may be prepared by extruding a hot concentrated solution of the polymer through a spinneret, followed by some stretching, solvent removal and heat-treatment stages. The solvents are very strong chemicals, such as 100% sulphuric acid, and they produce fibres that are generally resistant to less aggressive media. The main advantage of aramid fibres is their very low density, giving high values of tensile specific strength and stiffness, combined with good toughness and, as a result of this latter characteristic, they are frequently used in applications where impact resistance is important. However, the bonds between the molecules are relatively weak hydrogen bonds and, consequently, the fibres have low shear modulus, poor transverse properties and low axial compression strength. These effects are reflected in the resulting compressive properties of composites where, in bending and compression, the response is non-linear with relatively low ultimate strengths. As a result, they are frequently used in composite hybrid construction with other fibres, such as carbon, with the aramid providing toughness characteristics to the composite. Composite properties that rely upon the transverse and compressive properties of the fibre would also be expected to be lower than other fibre composites, for example bolt bearing properties. Aramid fibres are difficult to cut, both before and after laminating. Table 1.3 provides typical mechanical properties of aramid fibres.

Table 1.3 Typical tensile strength properties of Kevlar 49 aramid fibres

Material	
Unimpregnated twisted yarn textile test – ASTM D885	
ultimate tensile failure (MPa)	2760
Impregnated strand test – ASTM D2343	
ultimate tensile failure (MPa)	3620
Tensile modulus (GPa)	125
Elongation to break (%)	2.5
Density (g/cm^3)	1.44

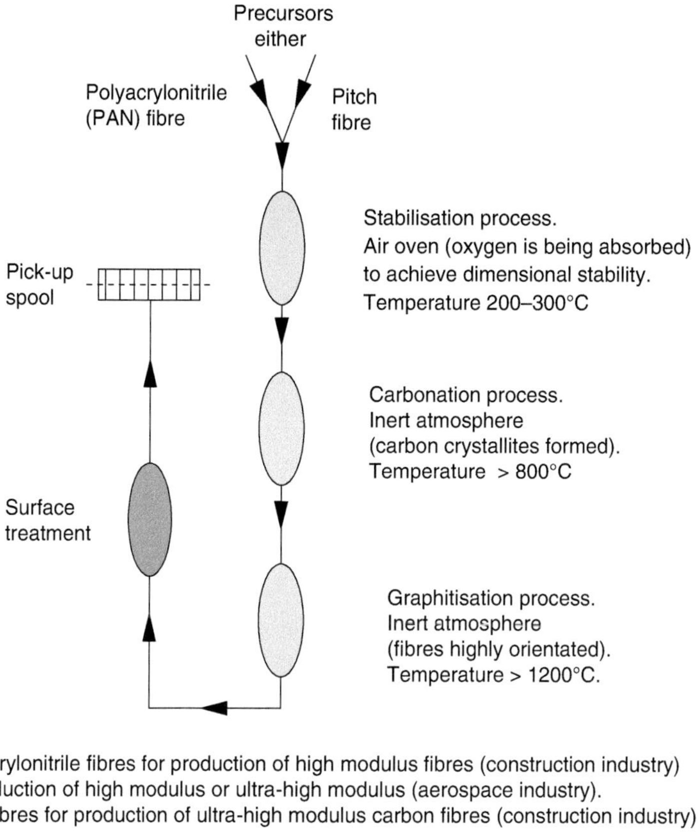

1.1 Diagrammatic representation of the manufacture of carbon fibres.

Table 1.4 Typical mechanical properties of carbon fibres

Fibre	Fibre type	Elastic modulus (GPa)	Tensile strength (MPa)	Ultimate strain (%)
Pan based fibre Hysol Grafil Apollo	HM[1]	300	5200	1.73
	UHM[2]	450	3500	0.77
	HS[3]	260	5020	2.00
Pan based fibre BASF Celion	G-40-700	300	4960	1.66
	Gy 80	572	1860	0.33
Pan based fibre Torayca	T300	234	3530	1.51
Pitch based fibres Hysol Union Carbide	T-300	227.5	2758.0	1.76
	T-500	241.3	3447.5	1.79
	T-600	241.3	4137.0	1.80
	T-700	248.2	4550.7	1.81

[1] High modulus (Intermediate modulus in USA and Asia).
[2] Ultra high modulus (High modulus in USA and Asia).
[3] High strain.

Carbon fibres

Carbon fibres are produced by passing the precursor fibre (either polyacrylonitrile (PAN) or pitch fibre precursor) through an air oven at 220°C to 230°C, where oxygen is absorbed; the oxidised fibres are then passed into a chamber with an inert atmosphere at temperatures from 400°C to 1600°C, for carbonisation of the fibres. As the raw fibre emerges from the carbonising furnace, it tends to pick up surface charges from rubbing contact with guides, rollers, etc.; therefore, a size to provide a protective coating on the surface of the fibre is applied. There are many different polymer solution sizes that can be applied, but a low molecular weight epoxy resin is generally used. In addition, a surface treatment is applied to the carbon fibre in order to improve adhesion to the matrix; the treatment is essentially an oxidation process. This latter process cleans the surface of the fibre and produces an etching on the surface; in addition, it provides reactive oxide sites by chemical reaction. The manufacture of carbon fibre is shown diagrammatically in Fig. 1.1.

Typically, the diameters of carbon fibre filaments are 5 and 8 μm and they are combined into tows containing 5000–12 000 filaments. A common size of untwisted carbon fibre 'tow' is known as '12 K filaments' (containing 12 000 filaments); these are sold as:

(i) High modulus (H-M) fibres (known as intermediate modulus fibres in some parts of the world, including the USA, stiffness 250–350 GPa). The stiffness of the H-M carbon fibre (manufactured by using polyacrylonitrile (PAN) fibre precursor) used in upgrading concrete structural members is generally about 300 GPa.
(ii) Ultra high modulus (UH-M) fibres, manufactured using the pitch fibre for the civil engineering industry or the PAN fibre for the aircraft industry (known in some parts of the world including the USA, as high modulus fibres, stiffness 300–1000 GPa). These are generally used to upgrade steel structures.

In addition to strong and high stiffness fibres, rehabilitated concrete structures require in-service resistance to high temperature and aggressive environmental conditions; carbon fibres, in general, are not affected by moisture, solvents, bases and weak acids. Table 1.4 provides typical mechanical properties of some types of carbon fibres.

1.5 Advanced fibre/polymer composites

1.5.1 Introduction

Advanced fibre/polymer composites are formed through the physical combination of two or more materials; these are the fibres, the matrix and any fillers/additives that may be required to give the component specific physical properties. The combination is along the following themes:

- Optimisation of the polymer composite formulation.
- Characterisation and control of the matrix microstructure and the fibre–matrix interface.
- Optimisation of the composite forming processes.
- Evaluation of the composites performance.
- Non-destructive characterisation of the composite.

The manufacturing technologies allow optimisation and control of the structure of the composite, e.g. fibre–matrix interactions, the fibre/volume ratio, degree of cure and fibre arrangement. The chosen method will also make it possible to optimise the formation processes in terms of economics, productivity, product performance, quality and reproducibility. However, the manufacturing technology eventually chosen for a specific upgrading procedure will be highly dependent upon the site conditions and the degree of difficulty of applying the specific fabricating method; the chosen procedure will clearly influence the quality of the composite.

1.5.2 Manufacturing techniques of FRP composites for upgrading reinforced concrete and prestressed concrete structural systems

Introduction

There are a number of techniques available for the manufacture of advanced polymer composites used in the external upgrading of RC structural beams; the method finally chosen will depend upon the site condition and the structure to be upgraded. There are three major manufacturing methods: (i) the manual production process, (ii) the semi-automated process, and (iii) the automated process. Each method will have an effect upon the quality, performance and therefore characteristics of the final composite. The automated fabrication methods have a high degree of production control, composite compaction and complete cure compared to the manual fabricated techniques undertaken on site for the upgrading of structures, and therefore the automated process will have higher values of strength and stiffness compared to those of other techniques.

The various manufacturing processes given here are all relevant to the upgrading of RC and PC structural members in flexure and shear. These processes have been discussed and described in many publications, including Hollaway and Head (2001). Consequently, only basic descriptions of these methods and an indication where they are used in the rehabilitation/retrofit techniques in civil engineering will be given.

Manual technique

The three manual methods for the manufacture of advanced polymer composites that may be used to upgrade RC and PC structures are a variation on the general 'wet lay-up' method. The three techniques are:

- The Replark method.
- The Dupont method.
- The Tonen Forca method.

These techniques are essentially the same and consist of *in-situ* impregnation by a resin into dry fibres in the form of sheets or fabrics. These are laid onto, or wrapped around, the structural member and are generally of widths varying between 150 mm and 1500 mm. The general procedure for the manual methods is:

(i) A first layer of the resin plus curing agent is applied to the prepared surface of the structural member; this acts as an adhesive as well as the matrix of the composite.
(ii) A fibre sheet, cut to size, is adhered to the concrete using the first laminating layer of resin; it is pressed down onto the surface using a roller to expel entrapped air between the fibres and the polymer.
(iii) A second fibre sheet and impregnating polymer are applied.
(iv) Application of polymer and sheets are repeated in the case of multiple plies.
(v) The system is then allowed to cure. As this procedure is a site operation, a cold-cure polymer is used and the cure temperature will be that of the ambient temperature.

The Replark method (Replark, 1999) was developed by the Mitsubishi Chemical Corporation and is a carbon fibre sheet which is impregnated with a cold-cure epoxy resin (Epotherm) and the fibres are Mitsubishi manufactured fibres, unidirectionally orientated. The sheet has a paper backing, which serves to keep the fibres in position; the paper backing is removed when the sheet is placed in position onto the structure. The cold-cured epoxy resin is cured at ambient temperature.

The Dupont method is a system using Kevlar aramid fibres, which is marketed as a repair system for concrete structures. The application of the material to the surface to be rehabilitated is similar to that for Replark.

The Tonen Forca method uses a unidirectional carbon fibre sheet which is impregnated with a cold-cure epoxy resin and is marketed in the UK by Kyokuto Boeki Kaisha Ltd. The system was originally developed by Mitsubishi Chemical Corporation and is therefore similar to the Replark system.

Semi-automated processes

The three semi-automated methods that can be used for the manufacture of advanced polymer composites to upgrade structural RC and PC structural members are:

(i) The cold-melt factory-made pre-impregnated fibre (prepreg), site-cured. The cold-melt factory-made pre-impregnated fibre (prepreg) to be site-cured is made in the factory and frozen to –20°C until required; the prepreg is then removed from the freezer and raised in temperature to the site value. The prepreg in this state is about 6% polymerised and is 'floppy' in nature. The curing procedure is then undertaken and is in conjunction with a compatible film adhesive. The two component parts are placed onto or wrapped around the prepared structural member and are cured in one operation under an elevated temperature of 65°C applied for 16 hours or 80°C applied for 4 hours. A vacuum-assisted pressure of 1 bar is applied for simultaneous compaction of the composite and the film adhesive before and during the curing procedure (Hollaway *et al.*, 2006). It is likely that this method will be used increasingly for strengthening/stiffening degraded structural members. In the UK, the manufacturing specialist in the production of cold-melt factory made pre-impregnated fibre composites for the construction industry is ACG in Derbyshire.

(ii) The site filament winding method. The site filament winding method was invented in Japan in the early 1990s, when the Ohbayashi system for bridge retrofits was developed. The idea was then extended into the USA by XXsys Technologies, Inc., San Diego, California, for seismic retrofitting and strength restoration of concrete columns using continuous carbon fibre. The specialised technology associated with the technique is based upon filament winding but uses partially cured prepregs of carbon fibre tows and a hot-cured polymer wound around damaged structural columns; this procedure forms a carbon fibre jacket. The partially cured carbon fibre prepreg is manufactured under factory controlled conditions. It is then continuously wound on site, around the columns to be upgraded using a robot. This results in good quality control and rapid installation. The XXsys carbon fibre composite jackets are installed with a fully automated machine called a Robo-Wrapper™ and a portable oven is used for controlled elevated temperature curing. Figure 1.2 illustrates the system. An advantage of this automated process is that good quality control is provided and, as a consequence, a composite of high strength-to-weight ratio is pro-

1.2 The XXsys carbon fibre jacket site filament winding technique (© XXsys Technologies Inc., San Diego, Ca, USA).

duced. The carbon fibre jacket which is eventually formed around the column increases the shear capacity of the column, confines the concrete, and greatly enhances its ductility in the flexural plastic hinge region. The system is expensive and has not been used routinely in the United States or Japan for the FRP rehabilitation and retrofitting of concrete structures.

(iii) The resin infusion under flexible tooling (RIFT) process. The resin infusion under flexible tooling (RIFT) process is one in which the composite is manufactured by an automated process but requires a substantial amount of manual input. Dry fibres are preformed in a mould in the fabrication shop and are sent to the site. The preform is then attached to the structure and a resin supply is channelled into it. The prepreg and resin supply is then enveloped in a vacuum bagging system. As the resin flows through the dry fibre preform it forms both the composite material and the adhesive bond between the CFRP and the structure. The process provides high fibre volume fraction composites, in the order of 55%, that have high strength and stiffness values. One disadvantage with the infusion process is that a considerable amount of the resin is wasted during the production procedure. With the increase in legislation to limit styrene emissions (mainly from polyester and vinyl ester resin systems into the work place) this process has been the key factor in promoting new technology in the manufacture of fibre reinforced plastics composites. Styrene emissions can be reduced by the development of resin systems with low styrene emission, improved ventilation and air filtering systems, and closed moulding techniques.

The automated process

The two automated processes that are available to the construction industry for upgrading a RC or PC structural member are:

- The pultrusion technique to form a FRP rigid plate.
- The factory-made and cured prepreg.

(i) The pultrusion technique is used quite extensively in the construction industry and a large percentage of this output is used in association with upgrading bridge and building structures. The upgrading FRP plates, for bonding to the soffit (for flexural resistance) or to the sides (for shear resistance) of the beam, or round or other geometric shapes for the NSM rods, are manufactured in a factory using a hot-cure resin. The dies forming the component operate at temperatures between 120°C and 135°C and the component is pulled from the die rather than being pushed through it, as in conventional extrusion. The plates produced will generally be fully cured but this does depend upon their size; the large sizes may require post-curing. Care must be taken to ensure that there is complete wetting of the fibres during the production stage, thus preventing voids forming in the composite. Wetting of the fibres takes the form of passing the fibres through a resin bath before they enter the heated die or by injecting resin through ports in the heated die. In addition, the fibres must be well distributed throughout the plate; it is not usual to have fibre weight fractions (f.w.f) greater than 60%. For rehabilitation of structures, a roughened surface is required on the side of the pultruded unit to be bonded to the structural member for increased bonding capability. This roughened surface can be provided by using a peel-ply on the bonding side. However, if this system is used it is advisable to have a peel ply fabricated onto both sides of the plate to maintain symmetry within the pultruded plate. A peel-ply is a layer of nylon or polyester* fabric incorporated onto the surface of the composite during manufacture. The peel-ply is stripped from the pultruded surface immediately prior to bonding to the adherent, thus providing a clean, textured surface to the composite unit. Furthermore, to provide resistance to hostile environments, a resin-rich exterior surface to the pultruded section can be fabricated using a surface veil, which is also incorporated into the structural component at the time of manufacture. Carbon, aramid and glass fibres and epoxy, vinylester and polyester polymer materials have all been used for the production of pultruded units for the plate bonding. Epoxy resins are probably the most difficult

* Polyester fabrics use *saturated* polyesters which are thermoplastic – unlike the unsaturated thermosetting polyesters used in the FRP matrices.

to pultrude but do have the advantage of low shrinkage during polymerisation (3–4%). Vinylesters have a shrinkage value of 6–10% and polyesters have a large shrinkage during polymerisation (12–19%). Further information may be obtained from Starr (2000) and Hollaway and Head (2001).

Some of the companies that manufacture pultruded sections and provide details of geometries and mechanical properties of pultruded profiles are Fibreforce Composites (UK), Fiberline (Denmark), Sika (Switzerland), Creative Pultrusions Inc. (USA), Fyfe Co. LLC (USA), Strongwell Corporation (USA), Bedford Inc. (USA), Shandong Qingyun Maoyuan FRP Products Co.Ltd (China). Contact details for these companies are to be found within the References (Section 1.14).

(ii) Factory-made and factory-cured prepreg

The factory made and factory cured prepreg is manufactured as a rigid plate. It is then transported to the site and bonded onto the degraded beam using a cold-cured adhesive; the bonding procedure is similar to that for pultruded plate. The method of manufacture of the prepreg in the factory is the same as that for the semi-automated prepreg process except that it is cured in the factory under controlled conditions.

The company that provides details of the factory-made and factory-cured prepreg is Advanced Composite Group (ACG), UK (see References, Section 1.14, for contract details)

Manufacturers of FRP products have traditionally provided design guidance for the use of pre-cured FRP strengthening plates, but more recently, professional organisations have developed general purpose design guides for the use of pre-cured and bonded FRP strengthening products for concrete structures. The design guidelines currently available are TR 55, 2nd Ed, (Concrete Society, 2000); fib, (2001); ACI, Committee 440-2R-08 (2008); CSA-S806-02, (2002).

The near surface mounted bars

An alternative method to the externally bonded techniques is the near surface mounted (NSM) technique. The technique is based upon the concept of placing pultruded glass or carbon FRP rods or strips into pre-cut grooves in the soffit of flexural beams or into the sides of the beams if the cover concrete is not sufficiently thick; the grooves are then filled with an adhesive, which is usually a high-viscosity epoxy or cement paste. Figure 1.3 shows the positioning of the NSM bars in the cross-section of a RC beam. The NSM rods concept is not new as the method was used in Europe in the 1940s (Asplund, 1949) and by Blaschko and Zilch (1999). The technique offers several advantages over the externally bonded rigid FRP plate method. These are:

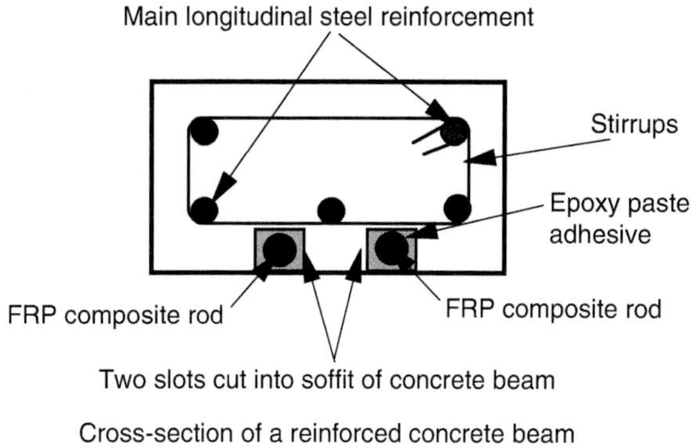

1.3 Typical positioning of NSM bars in the cross-section of a RC beam.

- No surface preparation work (other than grooving); this implies minimum installation time compared to FRP plate bonding.
- Customised grooving tools allow site operatives to cut the appropriate grooves in one pass, and the choice of high viscosity epoxy adhesive paste and groove-filling material allows the two components to be readily gunned into the groove.
- The technique is attractive when strengthening in the negative moment region.
- A significant decrease in the probability of damage from fire, acts of vandalism, mechanical damage and ageing effects is apparent when compared with the external plate bonding method.
- The technique is particularly useful in seismic retrofit of RC column–beam joints, providing either additional strength or ductility when moving the failure zone from the column to the beam (Prota *et al.*, 2001).
- The embedded system provides protection from the external environment.
- Such systems require minimal intervention.

The disadvantages of the technique are:

- The possibility of debonding at the ends of the rods or plates
- The process bonds the rods or plates into the weakest part of the RC beams (the cover concrete).
- There is a possibility of sawing through the shear link rods or cutting the longitudinal steel reinforcement.
- Durability of GFRP composite rods, which may be in close proximity to a high-alkali concrete component is questionable.

Test results of De Lorenzis *et al.* (2002) and El-Hacha and Rizkalla (2004), have indicated that NSM reinforcement can significantly increase the flexural capacity of RC elements. As is the case with externally bonded laminates, the bond may be the limiting factor of the efficiency of this technology. El-Hacha and Rizkalla (2004) strengthened RC Tee-beams with identical CFRP composite strips used in both the NSM reinforcement and the externally bonded reinforcement techniques, and concluded that an increase in strength of 4.8 times greater was obtained in the former case compared with the latter. This was due to early debonding failure of the external FRP composite as opposed to the tensile rupture of the NSM reinforcement strips. There have been cases where debonding failures have been experienced by the NSM rods; this failure mechanism can be prevented by providing a larger bond length or by anchoring the NSM bars in the beam flange. A design recommendation is given in Rizkalla *et al.* (2003) and a state-of-the-art review on NSM reinforcement can be found in De Lorenzis and Teng (2007). The first UK application of NSM bars was in 2002 (Farmer, 2003). Farmer (2004) discusses the UK experience and the development of best practice for strengthening concrete elements using NSM reinforcement. In that paper, hopes were expressed that within four years there would be a set of design and construction guidelines for utilisation of this technique.

Design codes for NSM systems are currently being developed.

1.5.3 Important in-service properties of the advanced fibre/polymer composite for service life estimation

High performance fibre/thermosetting polymer composites are designed to provide specific properties in highly demanding environments. These composites usually possess high dimensional stability at elevated temperatures and excellent thermal resistance, low water absorption, good chemical resistance, high mechanical strength, excellent stiffness and high compressive strength. However, the degree to which individual composites perform in practice is highly dependent upon their manufacturing techniques and a short discussion will be given here about those physical and mechanical properties which are particularly important for the service life estimation of rehabilitated and retrofitted FRP composite materials, namely, the polymerisation of the resin, the glass transition temperature of the polymer, the coefficient of thermal expansion of the composite, chemical resistance, permeability of the composite and mechanical properties.

Polymerisation of the resin

When the composite unit is manufactured by an automated factory production method under a controlled temperature and humidity environment the hot-cured resin is used (examples of these systems are the pultrusion and the cold-melt prepregs). If, however, the composite material is manually fabricated in the field (examples of these are the wet lay-up and the XXsys Technologies systems), or when the adhesive polymer is employed on site, the cold-cure systems would generally be used under ambient temperature conditions for the fabrication and the curing of the polymer and the adhesive. The curing procedure for polymers is important because the long-term performance properties of the composite and its durability will be influenced by the degree of cure of the resin, particularly for site-cured systems. The operating temperatures of the automated manufacturing processes will generally vary between 120°C and 135°C; the actual value will depend upon which system is being used. Consequently, the glass transition temperature (T_g) of the polymer (see next section) will be much higher for the hot-cured systems than for the cold-cured ones. It is therefore advisable that all cold cured polymers used in civil engineering, after their initial cure, should be post-cured at an elevated temperature for a certain length of time; the higher the post-cure temperature the shorter will be the post-cure period. However, the curing procedure used for the combined cold-cured composite polymer and adhesive (the wet lay-up systems) and the adhesive used for bonding rigid composite plates to concrete, should be applied with caution as an elevated temperature applied to the cold-cure adhesive, if it is near the T_g of the polymer, can soften and cause an increase in viscosity response, a reduction in mechanical performance and an increase in susceptibility to moisture absorption. The hot-cured resins used in the above automated manufacturing methods for composites are not generally post-cured due to the factory controlled method of production and the high temperature value of manufacture; the site temperature into which they will be placed is unlikely to reach their T_g value.

The glass transition temperature

The glass transition temperature (T_g) of an amorphous or an amorphous/crystalline material (the epoxy polymer used in upgrading a structure comes under the second description) is the temperature at which it changes from (or to) a brittle or vitreous state to (or from) a plastic state; this change takes place over a temperature range and the mid point of that range is taken as the T_g. On a decreasing temperature, the state of the polymer is reversed. Figure 1.4 illustrates the relationship between the volume of sample and between stiffness of sample against temperature. The T_g depends upon the detailed chemical structure of the polymer, which in turn depends upon the cross-linking of their molecules for strength. The numerical value of the T_g may be quoted with slightly different values depending upon the testing technique used; these methods are dynamic mechanical thermal analysis (DMTA) and differential scanning calorimetry (DSC). The

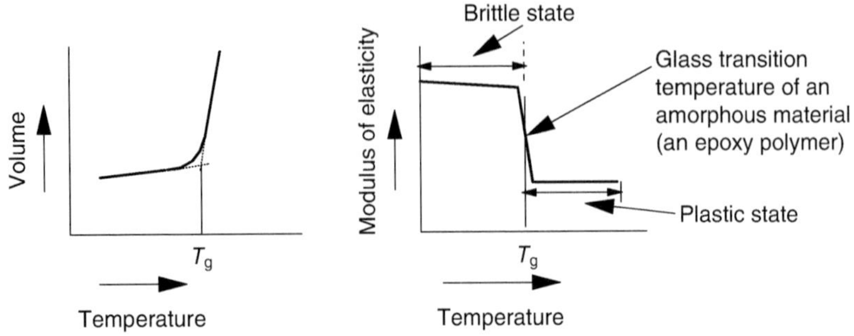

1.4 Relationship between the volume of sample and between stiffness of sample against temperature. T_g is measured by Differential Scanning Calorimetry (DSC) or by Dynamic Mechanical Thermal Analysis (DMTA). These methods will give slightly different values.

drastic changes that any polymer will experience at the T_g will cause serious reductions in their mechanical/physical properties, viz. modulus of elasticity, strength and hardness. It will be clear that polymer composite structural members should not be exposed to temperatures above the T_g value of the polymer material. The values of the T_g will generally be about 15°C–20°C above the curing temperature.

The coefficient of thermal expansion

The coefficient of thermal expansion of polymers is much higher than that of the fibre component of the fibre/matrix composite; the value of the former is of the order of 100×10^{-6}/°C. The equivalent values for glass, aramid and carbon fibre components are in the region of 10×10^{-6}/°C, -2×10^{-6}/°C, and -0.9×10^{-6}/°C respectively, depending upon the type of specific fibre. The thermal expansion of the polymer is stabilised by the fibre to a value near to that of the materials used in civil engineering such as concrete and steel. The actual value will depend upon the type of fibre, the fibre array and the fibre volume fraction of the composite. The value of the coefficient of thermal expansion will vary with the temperature and with the temperature range into which the composite is placed. The differential thermal expansion of the composite and of the RC member may cause stresses to develop at the bond line during any large swing in temperature (Hamilton and Dolan, 2000). However, this should not cause a problem with the upgrading of the structure as the post-cure temperature will not generally exceed 50°C; the elasticity of the adhesive should take up any movement between the two components. Furthermore, the site temperature variations for flexural upgrading of bridge beams will not be great because the soffit of the beams will be protected from the direct rays of the sun.

Chemical resistance

The resistance of the polymer to chemical attack, whether this is from the natural or from a chemical environment, depends upon its chemical composition and the bonding in the monomer. As all glass fibres are very susceptible to alkaline environments, which is not the case with carbon or aramid fibres, it is necessary to choose the polymer carefully; the matrix component of the FRP composite protects the fibre.

Permeability

Polymers with a high degree of cross-linking will generally have low permeability, thus gases and other small particles will not readily permeate through them. Haque and Shamsuzzoha, (2003), Liu *et al.* (2005) and Hackman and Hollaway (2006) have shown that the ingress of moisture will permeate through polymers over time, particularly if the polymer (or composite) is permanently immersed in water or salt solution, or is exposed to de-icing salt solutions. Carbon fibres do not absorb liquids and are subsequently resistant to all forms of ingress from alkalis or solvents (Balazs and Borosnyoi, 2001). Aramid fibres do suffer reduction in tensile strength when exposed to an alkaline environment (Balazs and Borosnyoi, 2001), and therefore the long-term properties of the protecting matrix used to form CFRP and AFRP composites are important to the overall properties of the composite. Most glass fibres have limited solubility in water but they are very dependent upon the pH value of any liquid that may percolate through the polymer. Chloride ions will attack and dissolve the surface of E-glass fibre. Moisture is readily adsorbed and can exacerbate microscopic cracks and surface defects in the fibre, thus reducing the tensile strength of the glass fibre. These fibres have high surface area to weight and the increased surface makes them much more susceptible to chemical attack. Again, it is the properties of the protecting polymer that are important to the overall properties of the composite.

Providing the rehabilitation is not exposed to moisture continuously, such as below the surface level of the liquid, the composite will continuously be in the wet/dry condition, either in the normal environment or within a splash zone; these exposures should not have a disastrous effect upon its durability. However, care should be exercised when the upgrading is on the soffit of a highway bridge; this situation in many parts of the world will be exposed to spray from salt solutions during the winter period.

Durability of externally applied polymer composites used to upgrade structural members

The civil engineering construction environments which cause major concern for the durability of the FRP composites are:

Moisture and aqueous solutions (alkaline environments)

The polymer has a vital role in protecting the fibre in FRP composites. It has been shown that the presence of moisture significantly affects the mechanical properties of polymer matrices and therefore will affect the FRP composite, particularly when it is immersed in water at high temperatures (Apicella *et al.*, 1982, 1983; Sheard *et al.* 1997). The mechanical properties of FRP composites are generally less sensitive to moisture compared to the polymer, but the composite is affected by fibre debonding and by delamination when exposed for some time to moisture. Plasticisation, a physical ageing of thermosetting polymers due to water sorption and the loss of low molecular weight substances, will lead to moisture-related degradation of the fibres, matrix and interface, a lowering of the T_g, and hence reductions in longitudinal and transverse tensile strengths, compressive strength and shear strength. Jacobs and Jones (1993) also showed that, as a result of an increase in relative humidity, a decrease in T_g resulted. Karbhari *et al.* (1998) found that, when testing short-term (1 year) environmental effects of immersing uniaxial, biaxial and triaxial E-glass/vinylester composites in deionized water, their tensile strengths were significantly affected, and to a lesser extent their tensile moduli. It is therefore vitally necessary to use the appropriate epoxies or vinylesters when the composite is known to be in contact with moisture during its life. However, it should be mentioned that during the service life of polymer composites used to rehabilitate structural members, the environment in which these structures are exposed is rarely as hostile as those described above. It will be realised that all civil engineering structural materials will degrade to a greater or lesser degree, and FRP composites do have an excellent resistance to hostile environments and are superior to most other civil engineering materials.

The difficulties in acquiring data relating to the long-term durability properties of FRP material are:

- The length of time involved in gathering the relevant information on site.
- The many different polymers on the market, each referred to by their generic name; some of these will have been modified by chemists over the years to improve one of their mechanical or physical performance properties; these modifications may have altered their durability performance.

- The difficulty of relating the results of accelerated tests to field tests.
- The possibility of changing the characteristics of the polymer during the process of undertaking the accelerated tests.

To gain an understanding of the long-term durability of polymer composite structural members built in the natural environment, it is desirable to monitor these members for any changes in their stiffness characteristics or any signs of degradation. This is the most relevant durability test to be applied to new structures. It is emphasised that to undertake accelerated tests in a laboratory, particularly if these tests involve elevated temperature testing, or to test over short periods of time followed by extrapolation of results, can lead to erroneous conclusions. These latter tests, by their nature, are focused on one specific resin system and one environment application, thus making extrapolations to other systems and environments difficult.

An example of long-term durability investigative work is that of ISIS, Canada Research Network of Centres of Excellent (Mufti *et al.*, 2005a, b, c). Although not directly related to the external plate-bonding technique, ISIS are investigating five concrete structures reinforced with GFRP rebars to provide information on the reliability of GFRP materials exposed to a concrete environment. So far, some of the tests have been running for some ten years. Further discussions on these tests may be found in Hollaway (2007).

Thermal effects

The thermal effects relevant to FRP materials used in rehabilitation systems are coefficient of thermal expansion and thermal conductivity. It is unlikely that either of these properties will cause any problem during the lifetime of a rehabilitated structure as the bonded composite is generally in the shadows of the structure itself. If any surface of this composite is exposed to direct sunlight causing a rise in its surface temperature above 60°C, an increase in stress value may develop at the bond line. Table 1.1 gives typical values of the coefficient of thermal expansion of polymers used in construction; a typical value of concrete is $10 \times 10^{-6}/°C$.

Fire

The performance of FRP-strengthened structures in fire is of great concern when considering applications of FRP in buildings. Fire resistance is not one single property; it may be defined by a number of fire reaction properties (Mouritz, 2006), including:

- Time to ignition.
- The minimum percentage of oxygen required to sustain flaming.

- Heat release rate.
- Flame spread rate.
- Smoke density and smoke toxicity.
- Burn-through resistance.
- Structural integrity retention.

In many situations, concern about fire resistance can severely restrict potential applications of FRP. The fundamental problem is that FRP composites are inherently combustible and they may release toxic fumes; this situation is particularly significant in buildings, tunnels and confined spaces. Furthermore, typical polymer resins for FRP composites in civil engineering applications have T_g values between 60°C and 80°C. Higher external temperatures than these values will cause the polymer to soften and degrade resulting in a loss of its mechanical properties. The high percentage of fibres in the composite will act as a barrier to the fire and cause a slowing down of its progress, but the bond between the FRP composite and the concrete surface will be the component of the strengthening system most affected by the high temperature, as a large percentage of the bond strength will almost certainly be lost just above 100°C. Generally, the design for the rehabilitation of bridges is such that if the upgrading composite plate is destroyed, the bridge will not collapse but will be closed for structural inspection.

There are fire protection coatings or fire-resistant organic resins that can be used to reduce the effects of fire. These are: (i) flame retardant polymers, (ii) intumescent coatings, (iii) thermal barriers. Further information on these systems may be obtained from Sorathia and Beck (1996), Kandola and Horrocks (1998), Sorathia *et al.* (2003) and Mouritz and Gibson (2006).

Bisby *et al.* (2005) and Chowdhury (2005) have undertaken full-scale fire tests on FRP-strengthened columns and beams; all the beams were insulated on the outside of the FRP with either a gypsum-based or a cementitious spray-applied fire protection. The authors concluded that FRP-strengthened structures can provide adequate performance in a fire if protected. Performance-based fire safety design methods (procedures that take into account fire protection schemes) are recommended as the best potential design approach for FRP-strengthened concrete structures.

Ultra-violet radiation

Carbon and glass fibres are not affected by ultraviolet light but aramid fibres, when exposed to this light, change colour and their strength is reduced. However, when the fibres are encapsulated in polymer resin, this degeneration occurs only near the outer surface of the composite and there is little effect on its overall mechanical properties. The manufacturers of resins should be able to give advice on the UV resistance. If a protection is required, UV absorbers can be added to the polymer at the time of manufacture or a cementitious or other over-coating can be used.

The mechanisms of durability are considered in Karbhari (2007).

Long-term loading of structural systems

The long term loading conditions which may have an added durability effect on the material are:

(i) Fatigue.
(ii) Creep.

(i) *Fatigue of FRP composites.* Fatigue is the degrading or failure of the mechanical properties of a composite material after repeated application of stress. In addition, due to the anisotropic nature of the material, a complex failure mechanism and extensive damage can be caused. Under tensile loads, unidirectional continuous fibre composites have good fatigue properties, which are essentially linear to failure when loaded parallel to the longitudinal fibre. However, if the composite contains off axes plies, various damage mechanisms can occur under loading and these will redistribute the internal load and cause the stress–strain response to become non-linear. Generally, polymeric composites experience progressive fatigue degradation due to failure of the fibres, fibre-stacking sequence and type of fatigue loading. Furthermore, a reduction in the strength and stiffness of advanced composites often result from fatigue damage and the actual effect on its properties will be highly dependent upon the composite design, manufacturing procedure and nature of the loading. For instance, under a tension–tension fatigue load, a directional fibre array at 0° to the major axis will show very little strength reduction until immediately before failure, whereas under this loading regime, a randomly orientated fibre array composite will show a gradual strength reduction until failure. The fibre/matrix interface characteristics, such as the fibre surface treatment, the fibre sizing and the presence of interface modifiers, have an effect upon the fatigue performance of FRP composites. Generally it can be argued that any improvements of the fibre/matrix adhesion brings about modification of one or more of the above mentioned parameters; this results in enhancement of the fatigue performance of a broad range of composite materials. However, this overall improvement has not been adequately quantified and very few analytical models exist that relate certain interface characteristics with one or more fatigue parameters. It is also known that environmental ageing has a detrimental effect on the fatigue performance, due primarily to the deterioration of the properties of constituent materials, as well as to interfacial weakening.

The conditions on which fatigue behaviour will depend are:

- The resin type and proportion.
- The reinforcement type, orientation and proportions.
- The nature of the fatigue loading (mean stress and amplitude).
- The frequency of the fatigue loading.
- The ambient temperature and environment.
- The effect of the interface between the fibre and the matrix.

For FRP composite bridge structures subject to cyclic loading, fatigue becomes an important limit state that needs to be considered by the designer. Analytical and experimental investigations have been undertaken for some 45 years by the aerospace, marine, and mechanical engineering industries. Over this period, fatigue design data have been generated for a wide range of composite material systems under axial and flexural fatigue loading, as well as environmental conditions. However, in the civil engineering industry, fatigue of FRP composites has only been actively investigated for some 20 years – Demers (1998a, b) has discussed fatigue strength degradation and fatigue life diagrams of E-glass FRP composites and carbon FRP composites. Moreover, there have been only a few studies published on the fatigue loading behaviour of beams strengthened with CFRP composites, *viz.* Karbhari *et al.* (2000), Hefferman and Erki (2004) and Gheorghiu *et al.* (2004a). Lopez *et al.* (2001) undertook investigations on a combination of fatigue loadings and low temperatures, and Gheorghiu *et al.* (2004b) on a combination of fatigue and harsh water exposure; this latter work showed a variation in results, depending on the method of fabrication of the CFRP composite upgrade on to the concrete substrate. Furthermore, there are virtually no data dealing with the fatigue response of beams subjected to different fatigue load levels and number of cycles. Gheorghiu *et al.* (2004c), during their fatigue testing of externally strengthened beams with CFRP composites, showed a consistent response independently of the number of fatigue cycles. The specimens maintained most of their initial stiffness until the end of the test. Initially, the beams showed an important deflection increase up to 105 cycles, then the response stabilized, with deflections being asympototic to the maximum values corresponding to the load amplitudes. The beams were then loaded to rupture. The authors showed that the number of fatigue cycles had no significant effect on the load–deflection response of the beams. The total deflection, including the residual deflection due to cycling, was not significantly different from that of the control beams; the ultimate load reached by the beams did not appear to be influenced by fatigue cycling. Aidoo *et al.* (2004) studied the effect of a one-dimensional FRP composite rehabilitation system on the flexural fatigue performance of RC bridge T-beam girders, with and without bonded FRP reinforcement on their tensile surfaces; the beams were tested with a concentrated load at mid-span, under constant amplitude cyclic loading. The authors stated that the fatigue behaviour of such rehabilitated beams is controlled by the fatigue behaviour of the reinforcing steel. The fatigue life of such a beam can be increased by the application of an FRP composite plate, which relieves some of the stress carried by the steel. However, the increase in fatigue life of the beam is influenced by the quality of the bond between the carbon FRP and concrete substrate. Debonding is initiated at the centre of the beam and progresses towards the supports; it is driven partially by crack distribution and shear distribution.

(ii) *Creep characteristics of FRP composites.* The polymer component of an advanced composite is a visco-elastic material and therefore the FRP composite material will creep, but as the carbon, glass or aramid fibres have very low creep characteristics they have a stabilizing influence on the advanced composite material and the final creep value will be dependent upon the fibre volume fraction and the orientation of the fibres. Linear visco-elastic theory is well documented in Nutting (1921), Lockett, (1972) and Ward (1971). The deformational behaviour of composites is dependent upon the way in which the polymer composite is subjected to, and the value of, the applied load, and hence strain; these latter values will differ under different short-term conditions of testing. To enable the creep characteristics of polymer composite materials to be realistically established, long-term creep tests are generally required under field conditions. However, this method is time-consuming and will provide information under only one loading arrangement and the environmental conditions at one location. Creep tests produce curves of elongation against time at different stress levels, and although they are not able to produce data that may be converted directly into stress–strain curves, constant time sections through families of such creep curves have been used to produce isochronous stress–strain curves. Likewise, it is possible for data to be taken over a few days or weeks and an extrapolation to be made beyond the test duration, via a mathematical equation. If this procedure is adopted, it does imply that the continuous creep process will obey the same law as existed at the commencement of the test. The creep data are modelled by a power law function and, when plotted on a logarithmic axis (log strain/log time), a linear graph results; this method should yield a confident extrapolated value up to twice the test duration. Kontou (2005) studied the tensile creep behaviour of unidirectional glass-fibre/epoxy polymer composites at three different temperature levels (*viz.* 298K, 333K and 353K); the tests were performed on off-axes of 0°, 15°, 30° and 60°. The creep strain was negligible at room temperature, but, as would be expected, it was considerable at the higher temperatures. It was considered that the creep strain was composed of elastic, visco-elastic and visco-plastic parts. The visco-plastic part was calculated through a functional form and it was assumed that the visco-plastic response of polymer composites arises mainly from the matrix visco-plasticity. The author claims that the model predictions in terms of creep compliances were satisfactory; however, the experimental results did not compare favourably.

To estimate the creep characteristics of polymer composites, a time–temperature superposition principle (TTSP) technique could be used. This method utilises the accelerating effect of increased temperature on the creep behaviour of polymer composites. A simple creep equation can model a family of creep curves at different test temperatures. Identical specimens are placed under the same stress level, but are exposed to different temperatures, and the creep compliance data are plotted on a logarithmic scale against a logarithmic time scale. Assuming time–temperature correspondence, a master curve can be plotted by shifting the higher temperature curves parallel to the same axes, and the amount of shift to achieve the superposition of the data gives the reduced time factor for each test temperature. Figure 1.5 illustrates this technique. Likewise, the applied stress–time superposition principle (TSSP) can be applied for polymer composite materials: Figure 1.6 shows a typical creep stress–time relationship. The time–temperature and stress–time superposition principles are discussed more fully in Aklonis and MacKnight (1983) and in Cessna (1971), respectively.

Maksimov and Plume (2004) explored the possibilities of predicting the creep in a unidirectionally aligned glass fibre–polymer, an

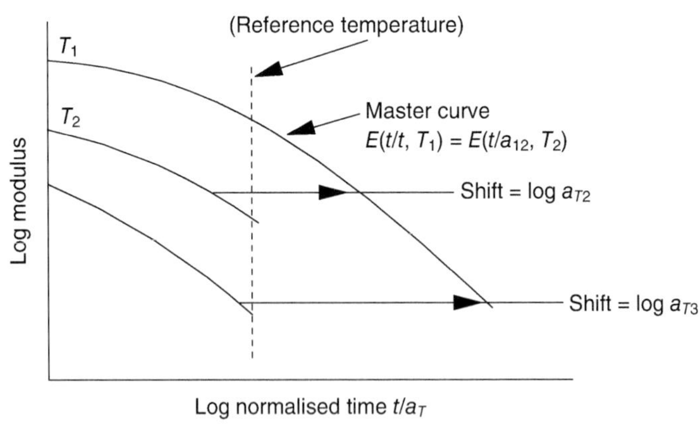

1.5 Typical schematic graph for extrapolation procedure for TTSP.

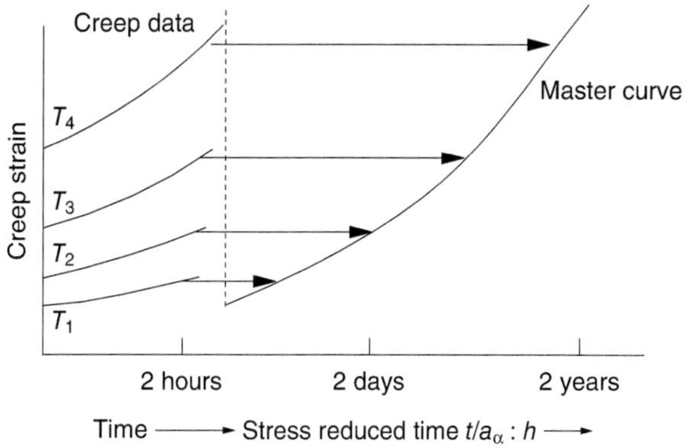

1.6 Stress time superposition principle (TSSP).

organoplastic, a carbon fibre–polymer, and a boroplastic composite material; all tests were either in tension in the direction of the fiber reinforcement or in the transverse direction. The results of the creep of these composite materials were obtained by the method of temperature–time analogy, with the aid of functions characterising the matrix's viscoelastic compliance; these results were compared with long term creep tests up to three years. In addition, long-term duration tests (up to five years) were undertaken on the polymers. The mutually independent components of the viscoelastic compliance tensor under conditions of creep were calculated with the aid of the Laplace transformation and an earlier verified variant of determining the mean elastic characteristics of a composite material, whereupon the originals of the sought functions were obtained by a numerical inverse (Laplace transformation).

There seems to be little reference in the literature to research work undertaken on the creep of the upgraded structural beams with FRP composites.

1.6 The strength and stiffness properties of advanced fibre/polymer composites for service life extension

Structural composites can have a wide spectrum of mechanical properties and these properties will be dependent upon the following four items:

(i) The relative proportions of fibre and matrix materials (the fibre/matrix volume or weight ratio).
(ii) The method of manufacture of the composite.

Table 1.5 Typical mechanical properties of long, directionally aligned fibre/matrix composites (fibre weight fraction 60%) manufactured by the pultrusion process (the matrix material is epoxy)

Fibre material used in composites	Specific weight	Tensile strength (MPa)	Tensile modulus (GPa)	Flexural strength (MPa)	Flexural modulus (GPa)
E-glass	1.9	1250	40	1450	40
Aramid 49	1.45	1300–2000	70–130	–	–
Carbon fibre (PAN)	1.6	2000–2600	150–250	≃1800	≃150.0
Carbon fibre (PITCH)	1.8	1000–1500	260–500	–	–

Table 1.6 Typical mechanical properties of glass fibre composites (fibre weight fraction 60%) manufactured by different fabrication methods

Method of manufacture	Tensile strength (MPa)	Tensile modulus (GPa)	Flexural strength (MPa)	Flexural modulus (GPa)
Wet lay-up	350	4–7	110–550	3–7
RTM	140–200	4–10	200–300	8–15
Filament winding	550–1380	30–50	700–1800	30–48
Pultrusion	1250	40	1450	40

(iii) The mechanical properties of the component parts (*viz.* carbon fibre, glass fibre or aramid, and the polymer).
(iv) The fibre orientations within the polymer matrix (which can take the form of unidirectional, bi-directional, various off-axis directions and randomly orientated arrays).

Table 1.5 illustrates typical properties of composites manufactured by the pultrusion method using long directionally aligned fibre reinforcements of glass, aramid and carbon with a fibre/matrix ratio by weight of 60%. Table 1.6 shows typical mechanical properties of glass fibre composites manufactured by different techniques; the effect that the methods of fabrication have on the properties can be seen.

1.7 Methods for attaching rigid plates to RC structural beams for their rehabilitation

1.7.1 Introduction

The main method for attaching FRP rigid plates to the soffit of RC structural beams is via an adhesive polymer. It is vitally important that the surfaces of the adherends are clean and free of any contaminants, such as oil or loose dust, etc. Recently, another fastening method has been introduced into the construction industry, known as the mechanical fastening procedure (MF-FRP), whereby a powder-actuated fastener and mechanical anchors are used. This latter method is a promising one, particularly where speed of installation and immediate use of the structure is imperative and the strengthening is intended for a limited period only; for instance, when it is known that the structure is to be replaced in the foreseeable future.

1.7.2 Surface preparation of adherents

A clean adherend surface is a necessary condition for adhesion but it is not a sufficient condition for bond durability. Most structural adhesives are the result of the formation of chemical bonds between the adherent surface atoms and the compounds constituting the adhesive. These chemical links are the load transfer mechanism between the adherents. Solvent degreasing is important, because it removes contaminants that inhibit the formation of the chemical bonds (Kinloch, 1987). Consequently, it is necessary to pre-treat the substrate of the adherents to enable the required surface properties to be achieved.

Concrete substrates are prepared by grit blasting; in the UK, 'Turbobead' Grade 7 angular chilled iron grit (Guyson, 1989), of nominal 0.18 mm particle size, is generally used. The grit is applied at a blasting pressure of 80 lb/in^2. With this operation, the surface cement layer (the laitance) must be removed, providing a uniform exposure of the underlying aggregate. Before the adhesive polymer is applied to the surface, all traces of dust must be removed by air jet or similar, and then it must be solvent cleaned. The performance of the adhesive joint is directly related to the successful application of the pre-treatment and this, in turn, depends upon the quality of the surface characteristics of the substrate in terms of topography and chemistry.

For *polymer composites*, two of the most widely used preparation techniques for FRP materials are abrasion followed by solvent cleaning, and the peel-ply method. Abrasion removes weak surface layers and contamination and increases the apparent surface energy and the rate of spreading of the adhesive. Although the *degree* of abrasion prior to bonding is known to affect subsequent bond strength and durability, the strength of bonded FRP joints depends on the roughness of the surfaces and the level of contaminants present. Consequently, a contamination-free surface is the factor of vital importance, but joint strength may be reduced by exaggerated surface roughness due to the entrapment of air.

Abrasion of the FRP surface can be carried out using sand or silicon carbide (SiC) paper, Scotchbrite cloth or pumice. SiC paper followed by cleaning is a convenient method and will give high strength joints. Light grit blasting is an alternative technique but it is vitally important that the surfaces are not over abraded, thus avoiding fibre damage to the carbon and glass fibre reinforced composites. The technique is not normally recommended for the preparation of FRP surfaces.

Peel-ply composite layers can be applied to glass and carbon pultruded composites and to fibre prepregs at the time of the manufacturing procedures. A peel-ply is a layer of nylon or polyester fabric incorporated onto the surface of the composite. It is stripped from the pultruded surface immediately prior to bonding to the adherend thus providing a clean, textured surface to the FRP composite unit. The success of this procedure is dependent upon (i) the clean removal of the peel-ply without plucking the fibres from the FRP composite matrix, and (ii) the complete removal of any remnants of the peel-ply from the FRP composite matrix. To achieve this, most peel-plies are coated with a release agent to ensure that their removal does not damage the underlying plies.

1.7.3 Structural adhesives

The two-part adhesive polymer

Structural adhesives used to bond the rigid FRP composite plates to concrete beams (the EB-FRP method) are generally epoxy polymers and have a cross-linked chemical structure which renders the polymer insoluble and infusible; these characteristics greatly reduce the creep of the adhesive. As with the laminating resins, there are two types of adhesives, the cold- and the hot-cured polymers. Epoxy polymers have mechanical and physical properties which make them ideal candidate for bonding dissimilar adherents; some typical mechanical properties of epoxy adhesives are given in Table 1.7. The physical properties of the epoxies are:

- A high, cured cohesive strength (the concrete of the rehabilitated concrete structure will generally fail first).
- A low shrinkage (residual bond-line strain in cured joints is reduced).
- A low creep and good strength retention under sustained load.
- Good wetting properties for concrete substrates.
- May be formulated to give a long pot life (this will allow the bonding operation of the FRP plate to be less rushed and more thoroughly undertaken).
- Adhesive polymer toughness may be increased by blending, filling or co-polymerising it with a tough polymer to improve impact and low temperature performance; this will cause a reduction in strength of the adhesive and in this situation it is necessary to compromise between strength and stiffness.

Table 1.7 Typical mechanical properties of cold-cured epoxy adhesive

Property	Epoxy adhesive polymer+
Tensile strength (MPa)	20–40
Tensile modulus (GPa)	1–10
Tensile failure strain (%)	1–4
Shear strength (MPa)	15–35
Shear modulus (GPa)	0.5–2.0
Shear failure strain (%)	5–50
Glass transition temperature (°C)	35–90*
Poissons Ratio	0.3–0.4
Thermal expansion (10^{-6}/°C)	30–70

+If post cured, otherwise the value will be dependent upon the environmental temperature at the time of curing.
* For design purposes, actual properties must be obtained from the manufacturer.

- Low creep and superior strength retention possible under sustained load.
- Able to accommodate irregular or thick bond-lines (necessary in the case of concrete soffits).
- With the correctly formulated polymers, possible to obtain operating temperatures in the range –30°C to +60°C for site service life. In the case where the FRP materials are to be bonded to the top surface of a bridge deck that is to receive hot bituminous surfacing, an adhesive with a high T_g must be used (Concrete Society 2000, 2003); this can be achieved by using a hot-cured adhesive.

Further information on adhesive bonding may be obtained from Mays and Hutchinson (1992), and on the structural use of adhesives from the Institution of Structural Engineers, UK (1999).

The essential requirements for structural adhesives are:

- The selection of a suitable adhesive polymer.
- A well-prepared adherend surface; this is a very important aspect in adhesive bonding, it might be necessary to use a coupling agent on the surface of the concrete substrate.
- An appropriate joint design.
- Adequate quality control at the fabrication stage.
- Protecting the joint from unacceptably hostile conditions in service.
- Post-bonding quality assurance.

The requirements for a satisfactory joint performance are:

- Good contact between the adhesive and the substrate.
- Absence of weak layers in the joint.

Table 1.8 Key performance issues (Hutchinson *et al.*, 2004) (by kind permission of Woodhead Publishing)

Item	Key performance issues
Substrate (parent material)	• Can the surface be prepared such that a sufficient level of adhesion is obtained? • Is the resistance to delaminating of the substrate sufficient for load sharing with the reinforcement?
Adhesive layer	• Can the adhesive transfer the load between the substrate and reinforcement for the design lifetime? • Will the adhesive remain stuck to the substrate for the design lifetime?
Composite (reinforcement)	• Can the adhesive transfer the load between the substrate and the reinforcement for the design lifetime? • Will the reinforcement area be sufficient to provide the required strengthening?

- The adhesive possessing appropriate mechanical properties.
- It might be desirable to prime cracked or porous surfaces of the concrete to generate suitable conditions for the application of the adhesive, but the epoxy should be tolerant to any alkalinity of the substrate and will adhere to concrete surfaces satisfactorily.

Hutchinson *et al.* (2004) have given in Table 1.8 the key performance QA/AC assessment of the material properties of the flexural reinforcement, the adhesive layer and the substrate that are required to design and install to the original design intent. They have stated that one of the critical issues with the substrate is defining what quality is required to obtain a sufficient level of adhesion. Lee *et al.* (1998) have provided 'A Best Practice Guide' for the assessment of adherend surface quality and guidance on the definitions of surface quality.

Hutchinson and Quinn (1999) have discussed the principal requirements for the formation of a satisfactory and durable bonding composite material system. Clarke (1996) and Hutchinson (1997) have provided design guidance on joining composite materials.

Adhesive films

Epoxy adhesive films are applied as solids and generally have no pot-life restrictions; they are environmentally friendly. When the cure heat is applied, the adhesive cures in place and any 'squeeze out' is avoided. Films are available in thicknesses of 2 to 8 mils, and both performance and processing characteristics are reproducible from one production process to another. The hot-cured epoxy adhesive films are a more robust adhesive compared to the cold-cured two-part adhesive polymer, and they exhibit high moisture and chemical resistance. They have fast cure rates at moderately elevated temperatures but the actual values should be provided by the manufacturer. In addition, they have long storage stability, even at ambient temperatures, although refrigeration is preferred. The adhesive film must be compatible with the polymer matrix of the FRP composite; this is an equivalent requirement to that for the two-part cold-cure adhesive.

The power-actuated fastening 'pins' method for fastening FRP strips onto the soffit of concrete beams

The power-actuated fastening (MF-FRP) 'pins' method, which attaches FRP plates to RC concrete members using closely spaced steel MF-FRP 'pins' and a limited number of steel expansion anchors was developed by Lamanna *et al.* (2001a, b). It is a rapid procedure that uses simple hand tools and unskilled labour. Unlike the externally bonded FRP composite method, the current system does not require surface preparation and, after the installation of the strips, the strengthened structure is ready for immediate use. Furthermore, the multiple small fasteners closely spaced on the beam used in this method distribute the load uniformly over the FRP plate and reduce stress concentrations which can lead to premature failure. A similar method was developed by Rizzo (2005), using concrete wedge bolts and anchors instead of pins. Tests also showed that using a steel washer improved the performance of the connection by spreading the clamping load on a bigger surface of the FRP and transferring a portion of the load by friction.

Strongwell (Chat-field, Minnesota, USA) produce FRP strips which are 102 mm wide and 3.2 mm thick. The fibres in the strip are a combination of unidirectional carbon and E-glass fibres continuous strand mats; the matrix material is a vinylester system formulated for the pultrusion process. The system strip is known as SafStrip™. The fabrication method is either to predrill the strip and concrete, or the fasteners can be 'shot' directly through the FRP strip and into the concrete. If the predrilling system is used, the FRP strips are predrilled at the required spacing and are then used as templates for the location of the drill holes in the concrete and for inserting the fasteners. The fastener pins are embedded into the strip and the concrete using a Power Actuating Fastening gun, such as a Hilti Corp. DX A41 or the Hilti Corp. DX 460 power actuating fastening system. There have been a number of studies describing the details of two systems, including the washer and the fastener types, their lengths, diameter and embedment lengths (Bank *et al.* 2002a, b; Lamanna *et al.* 2001a, b; Bank, 2004, 2006/7).

1.8 Rehabilitation of RC flexural members

1.8.1 Introduction

In recent years, a significant number of design codes and specifications have been published by technical organisations that provide guidance for design with FRP materials for civil engineering. The key publications are listed in Section 1.15, Appendix. The Concrete Society Technical Report 55 (Concrete Society, 2000) was the first document to give authoritative, independent design guidance for externally-bonded fibre composites for strengthening concrete structures. It is important to mention that some areas in this document do require updating when compared with the ACI Committee (440.2R-08) (2008), the fib (2001) Task Group 9.3 and Cripps *et al.* (2002). Clarke *et al.* (2004) have suggested one important area in TR 55 that requires modification; this is the choice of stress–strain model for the steel. The authors state that it makes little difference whether the parabolic or rectangular stress block is used for modelling concrete under high compression; the idealised steel stress–strain model used may affect the calculations. The authors continue by saying that the trilinear stress–strain model of BS 5400-2 (2006) is conservative when calculating the flexural strength of the element but for longitudinal shear, a bilinear version results in higher shear stresses. Consequently, the trilinear model may underestimate longitudinal shear.

Section 1.8.3 of this chapter discusses the use of mechanical anchorages in situations where it is difficult to provide sufficient anchorage length to delay or to prevent the onset of debonding at the free end of the unstressed soffit plate upgrade. The existing equations in Concrete Society (2000) for calculating the resulting anchorage force are acceptable but some clarification is needed for situations when the steel reinforcement has yielded and when members are non-prismatic.

1.8.2 The most relevant manufacturing procedures for the flexural upgrading of RC structural members

Section 1.5.2 discussed the basic techniques for the flexural strengthening of beams, using unstressed FRP soffit plates, that can be used for RC or PC structural beams; other methods may be suitable in particular situations. It is interesting to note that the first application of FRP strengthening was to beams, using wet lay-up sheets or pre-cured plates bonded to the tension face of the beam with the fibre direction aligned to the beam axis; these two applications are still the most popular methods currently utilised. The effectiveness of flexural strengthening of RC beams with FRP is evident from the large database of experiments, reported by Smith and Teng (2002a, b).

(i) The method most used is the *precast factory-made composite plate* (produced either by pultrusion or from a factory-made prepreg). This plate is site-bonded to the degraded beam with a cold-cure epoxy adhesive polymer. The *initial* polymerisation of the cold-cure adhesive system is never 100% and therefore, for best practice, the adhesive should be post-cured to a temperature in the order of 55°C (caution should be taken over the value of this temperature (see Section 1.5.3). The T_g of the adhesive (see Section 1.5.3) is then about 65°C–70°C, which is a satisfactory range for most site conditions in Northern Europe. If the adhesive is not post cured, the T_g value will be only about 15°C above the ambient temperature of cure (but eventually after several months, depending upon site temperature and conditions, about 98% of polymerisation will have taken place and the T_g value will have reached its ultimate upper limit). During this natural post-curing period, however, it has to be assumed that the adhesive has not been damaged due to environmental or external influences. It should be said that most structures that have been upgraded utilising the rigid FRP plate and cold-cure adhesive polymers have not been post-cured; this is not good practice but is accepted for economic reasons.

(ii) The *wet lay-up process* is the next most used of the rehabilitation/retrofitting techniques. Here, the entire fabrication is undertaken on the degraded structure on site. This method is used mainly for the strengthening of columns (see Section 1.12) and upgrading beams where access is difficult for methods (i) and (iv) to be utilised. However, it should be noted that this method, if used for flexural upgrading, is sensitive to unevenness of the surface of the RC beam and this problem can lead to debonding (Meier, 1995). It is important that during the surface preparation of the concrete, all uneven concrete surfaces are smoothed. This prevents the FRP composite under tension during loading of the structure from extending across an imperfection in the concrete. This imperfection would result in interfacial peeling stresses being developed and debonding of the composite from the concrete adjacent to the adhesive/concrete interface.

(iii) The *NSM rods* technique is one of the newest methods for upgrading concrete structural members. The system can preclude delamination-type failures, which are frequently observed when utilising the externally bonded reinforcement. This technique becomes particularly attractive for flexural strengthening in the negative moment regions of slabs and decks, where external reinforcement would be subjected to mechanical and environmental damage and would require protective cover which could interfere with the presence of floor finishes. The feasibility of using NSM rods and strips has been demonstrated by many researchers, including, De Lorenzis, (2002), De Lorenzis and

Nani (2002) and Hassan (2002). Hassan and Rizkalla (2003) have stated that the test results show that the efficiency of NSM strips, defined as the ratio of the percentage increase in capacity to construction cost, was three times that of the external bonded strips.

(iv) The *cold-melt, factory-made, site-cured prepreg and film adhesive* is the newest of the production methods and is a more sophisticated upgrade system than the earlier ones and forms a better union with the concrete by virtue of the vacuum-assisted pressure and elevated temperature cure on site. The advantage of this technique is that it allows any irregular geometry of the beam (for instance, the cross-sectional geometry of the beam or the curvature of the beam in elevation or in plan) to be accommodated, and immediate use of the upgraded structure on completion of the fabrication. Dimitriou (2004) showed that this system gave twice the yield load of the steel rebars of a rectangular RC beam tested in four point bending before failure compared with that of the rigid pultruded plate bonded with cold-cured adhesive at failure. There was no complete collapse of the beam; the bonded prepreg retained the damaged part of the beam, although the values of the strength and stiffness had been reduced considerably. The disadvantage with this process, currently, is that it is about twice as expensive as the first method due mainly to the greater site work required.

(v) The *resin infusion under flexible tooling* (RIFT) process is not generally used to upgrade RC structural members.

It is important to check the state of the shear resistance in the beam after the FRP composite upgrade design has been calculated, to ascertain that the original shear design of the beam has not been exceeded.

1.8.3 Unstressed and pre-stressed FRP soffit plates

There are two main systems used to rehabilitate or retrofit FRP composite members to a RC structural member. These are (i) an unstressed FRP composite and (ii) an FRP plate prestressed before bonding to the degraded beam.

The unstressed FRP soffit plate

Within this system there are three techniques:

- *The adhesive bonding of a precast FRP plate* to the soffit of the beam. This procedure requires a two-part cold-cure adhesive and it is used with or without mechanical anchors at the free end of the composite plate. In most instances, peeling (at the free end) can be prevented by providing adequate anchorage length beyond that required for flexural strengthening. In instances where it is difficult to provide sufficient anchorage length, some form of mechanical/adhesive anchorage device is required.

- *Wet lay-up technique.* In this case it would be advisable to provide a wet lay-up U-shaped laminate at the free end of the upgrade to provide a good anchorage device.
- *Cold melt prepreg composite.* This system contains a built-in anchorage device.

From experimental studies it has been shown that mechanical anchors used with the precast plates bonded with cold-cure adhesive delay and can prevent the onset of debonding (Sharif *et al.*, 1994; Garden and Hollaway, 1998). However, Smith and Teng (2001) have demonstrated that the increase in the plate-end debonding load appears to be limited. Experimental work undertaken by Chahrour and Soudki (2005) revealed that end-anchored, partially-bonded CFRP strips significantly enhanced the ultimate capacity of the control beam and performed better than the fully-bonded strip with no end anchorage. These observations seem to point to the importance of end-anchorage in strengthening schemes. There are three types that might be used:

(i) Prefabricated or *in situ* (wet lay-up) FRP U-strips at the free end of the plates to prevent debonding (see Fig. 1.7).
(ii) Steel bolts applied at the free ends of the plate, but care must be taken that the bolts do not tear through the unidirectional fibres. If this system is used, it is usual to introduce at the free ends of the plate a 0°–90° fibre array GFRP block with the fibres at ±45° to the longitudinal direction of the plate. This GFRP block, with predrilled holes,

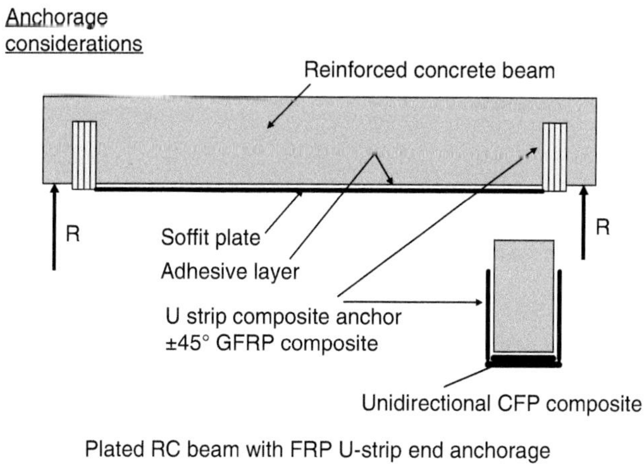

1.7 GFRP U-strips at the free ends of the strengthening CFRP plate to prevent debonding.

is bonded to the externally-bonded CFRP plate; the GFRP block is then used as a template to drill holes through the CFRP plate and the concrete. Hilti bolts or similar are then placed in position (through the GFRP block, CFRP plate and into the concrete) and struck to break a glass container to release the adhesive resin and curing agent for the completion of the bonding operation to the concrete.

(iii) Glass fibre anchor spikes have been used in the flexural strengthening of cantilever slabs (Teng *et al.*, 2001; Eshwar *et al.*, 2003). They have also been used as anchors for CFRP fabric-strengthened beams (Ekenel *et al.*, 2006).

A summary of different anchorage schemes and experimental investigations is given in Hollaway and Mays (1999), Van Gemert *et al.* (1999), Bank (2006/7) and Hollaway and Teng (2008).

The wet lay-up technique is generally used for strengthening RC columns but is also used in the flexural upgrading of RC beams. With this technique, there is probably more site work than the precast plate bonding method and the polymer used is the cold-cure one for the matrix of the composite, which also acts as the adhesive. Therefore, when this method is used, the recommendations of the manufacturer of the resin system regarding the site operation temperature must be adhered to.

The pre-stressed FRP soffit plate

Within this system there are three methods that are used to prestress the FRP plate. These are:

(i) to camber the RC beam, prior to the application of the bonded unstressed FRP soffit plate. On releasing the camber, the bonded plate becomes stressed. Anchor blocks as discussed above are employed (Saadatmanesh and Ehsani, Parts I & II, 1991a, b),

(ii) to tension the FRP plate against an independent external reaction frame. Anchor blocks are used and fitted, as stated above (Garden and Hollaway, 1999; Ferrier *et al.*, 2001; Hollaway, 2008), or

(iii) to tension the FRP plate against the RC beam to be strengthened. Anchor blocks are also used in this method (Nordin, 2004).

Each of these methods includes four phases to achieve the desired level of prestressing in the FRP plate.

- Firstly, the prestress is applied to the FRP plate with a hydraulic jack or equivalent device.
- Secondly, a cold-cure adhesive polymer is applied to the surface of the stressed composite plate and to the concrete beam; the composite plate makes contact with the soffit of the concrete beam.

- Thirdly, an anchoring system similar to the one described in the unstressed soffit plate section is applied to the free ends of the plate; the external prestressing force remains in place until the adhesive used to bond the FRP plate and the end blocks has polymerised.
- Fourthly, after the completion of polymerisation of the adhesive and when the anchors are fully supporting the plate, the FRP plates are cut near their ends and the prestress device is removed.

The level of prestress in the FRP is critical in order to guarantee the strengthening system does not fail near the anchorage zones when the prestress is released (Quantrill and Hollaway, 1998). As the bond between the un-stressed FRP plate and the concrete determines the design of externally-bonded reinforcement, the tensile strength of the FRP composite plate can be exploited only to about 50% of its ultimate value. The total prestressed FRP composite stress value then consists of the prestress value and the stress due to the subsequently applied external loads.

The advantages of this technique are:

- A more efficient use of the strengthening and serviceability capabilities of the expensive FRP composite material plate.
- A decrease in crack size and crack distance of the RC beam.
- The anchorages placed at the ends of the composite plate prevent delaminating and debonding effects.

Prior to external prestressing, all existing cracks must be epoxy-injected and spalls patched to ensure that the prestressing forces are distributed uniformly across the section of the member. El-Hacha *et al.* (2001) have discussed the characteristics of prestressed FRP composites, the techniques that have been developed, and field applications.

There are a few prestressing systems for epoxy-bonded external reinforcement; some are available on the market (Triantafillou *et al.*, 1992; Char *et al.*, 1994; Quantrill *et al.*, 1996; Garden *et al.*, 1998; Quantrill and Hollaway, 1998; Andrä and Maier, 1999; Sika, 2002; S&P, 2006; Yu *et al.*, 2008). However, the application of this method on the construction site is complex and time consuming, and at present it is not used extensively. Research work in this field has been undertaken, for instance, by Wight *et al.*, 2001; Wight and Erki, 2003; El-Hacha *et al.*, 2003; Piyong *et al.*, 2003).

1.8.4 Failure modes of FRP composites when upgrading reinforced concrete structures in flexure

Ten possible flexural failure modes have been identified as the cause of failure of RC beams bonded with FRP composite soffit plates (Ritchie *et al.*, 1991; Triantafillou and Plevris, 1992; Sharif *et al.*, 1994; Shahawy *et al.*,

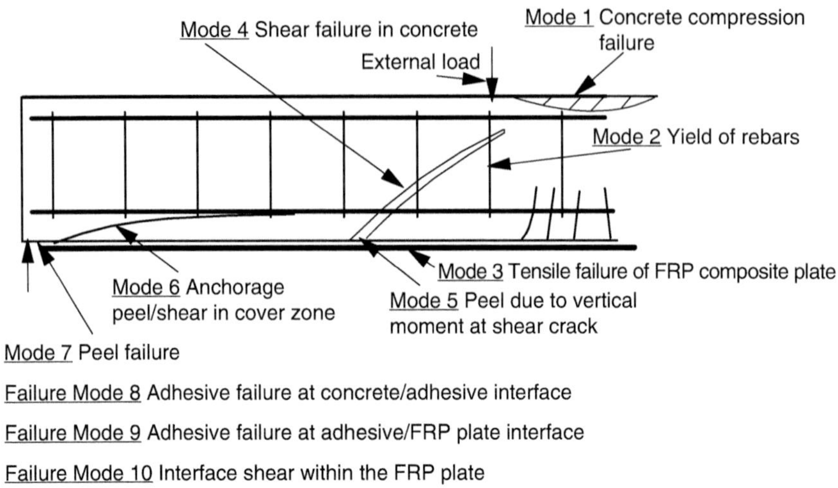

1.8 Typical failure modes in a RC structural beam upgraded with FRP composites (adapted from ICE Manual of Bridge Engineering 2nd ed. (Hollaway, p 514).

1996; Arduini and Nanni, 1997; Garden *et al.*, 1997; Grace *et al.*, 1998; Leeming and Darby, 1999; Nguyen *et al.*, 2001; Teng *et al.*, 2001). These are 'classical' failure modes, whereby full composite action is maintained between FRP and concrete until either *the concrete in compression crushes* or *the FRP ruptures in tension*. Figure 1.8 illustrates the ten possible failure modes.

Plate end debonding failure modes, *viz. the concrete cover separation mode* and *the plate-end interfacial debonding mode* result in the loss of composite action before the classical modes develop; these failures are undesirable as they are brittle in nature and occur at lower load levels than the classical failures. The intermediate crack-induced interfacial debonding failures, *viz. intermediate flexural crack-induced interfacial debonding* and *intermediate flexural shear crack-induced interfacial debonding*, are two further failure modes. The *plate end debonding failure modes* can be prevented by utilising anchoring devices, or adopting suitable factors of safety in the initial design. It will be clear that the *bond* between the FRP upgrade and the concrete has a profound impact on the failure mode of RC beams, and consequently on the ductility and the failure load of the strengthened member. The surface preparation of the substrate plays an important role in the bonding performance between the two dissimilar adherends and a well-performed preparation can reduce and sometimes prevent debonding failure modes. Central to the analysis and design of FRP for rehabilitation is the identification of all possible failure modes. There are a number of parameters that will influence the actual failure of a beam. These include:

- The amount of internal flexural and shear reinforcement.
- The material properties of the FRP composite plate.
- The adhesive properties and its thickness.

The ductility of a flexural member generally decreases as a result of strengthening, especially if the controlling failure mode is *debonding* or *FRP rupture*. To guarantee adequate ductility of a strengthened cross-section, the strain level of the internal steel reinforcement at ultimate should considerably exceed its yield strain, as indicated by available design recommendations, e.g. fib (Task Group 9.3) (EBR group) (2001) [commenced in 2001 and constantly updated] and ACI Committee (440.2R-08) (2008). ACI Committee (440.4R.04) (2004) also suggests that the lower ductility should be compensated with a higher reserve of strength through the use of a lower overall strength reduction factor.

1.8.5 Design of reinforced concrete beams rehabilitated by FRP composites

The design of FRP flexural strengthening of RC beam structural members is outside the scope of this chapter but information on this topic may be obtained from Teng *et al.* (2001, 2008) and the Appendix (Section 1.15) gives a list of FRP documents for the strengthening of RC structures.

1.9 Rehabilitation of PC flexural members

Compared to the flexural strengthening of RC members, there has been only a limited amount of strengthening work undertaken on PC flexural members. Only 10% of strengthened bridges in 2001 were prestressed (fib, 2001). The situation of PC structural members is complicated due to the long-term properties of creep, shrinkage and relaxation that are associated with the prestressing procedures; otherwise the verification and ultimate limit state design methodologies for rehabilitating PC flexural members is similar to that of the RC flexural members. The failure modes are similar to those of the RC members but the failure modes controlled by rupture of the prestressing tendons as well as limitations on cracking must be considered in the rehabilitation design.

1.10 Rehabilitation/strengthening of RC slabs

Numerous experimental investigations have demonstrated the effectiveness of strengthening slabs using FRP. The procedure for the flexural strengthening of slabs utilising FRP composite material is to employ strips or sheets of the material onto the tension face of the slab. Caution is required when sheets cover the entire slab as it is difficult and almost impossible to check on the quality of the bond between the sheet and the concrete. However, it is advisable to utilise wide strips rather than narrow ones as there is a reduced possibility of a debonding failure occurring. One-way slabs, which are simply supported, may be strengthened by bonding FRP strips along the longitudinal direction of the slab. Seim *et al.* (2001) demonstrated that the load capacity can be increased by up to 370%. A significantly higher load capacity and more uniform shear stresses are developed when the fabric covers the entire width of the slab. Mosallam and Mosalam (2003) report that the structural capacity of two-way slabs, when they were strengthened by FRP systems, increased by up to 200%. Concrete crushing was the most common form of mode failure, with localised debonding close to the ultimate load.

The failure modes of slabs are associated with intermediate crack-induced debonding as opposed to end plate debonding, which is the type found in RC beams.

Teng *et al.* (2002) have examined failure modes for slabs.

(i) One-way slabs
The following failure modes can occur:

- Flexural failure
- Intermediate crack-induced debonding
- Plate-end interfacial debonding
- Fibre anchor failure.

(ii) Two-way slabs
Teng *et al.* (2002) state that evaluation of the flexural strength of a two-way slab can be undertaken by the moment coefficient approach of BS 8110 (1997), where the maximum bending moments in both directions are found first. The reinforcement is then determined for each direction. Two modes of debonding failures can occur:

- Intermediate crack-induced debonding
- Plate-end interfacial debonding.

Teng *et al.* (2001) have discussed the strengthening of RC structural members with FRP composites and have also gives guidelines for the design of one-way and two-way slabs.

1.11 Shear strengthening of RC beams

Shear strengthening of a RC beam must be considered when the beam is deficient in shear capacity, whether this is the result of an increased external loading on the beam or as a result of insufficient shear capacity after flexural strengthening of the beam has been undertaken. It is vitally important to check its shear capacity after flexural upgrading. To provide shear resistance to a RC structural beam, FRP unidirectional composites can be bonded to the sides of the beam to strengthen it against shear failure. The design generally requires the principal fibre direction to be parallel to that of the maximum principal tensile stresses, but in practice the external FRP reinforcement (*viz.* the principal fibre direction) is invariably placed perpendicular to the member axis. In the cross-sectional plane, wet lay-up sheets (Section 1.5.2) or the cold-melt factory-made prepreg and film adhesive (see Section 1.5.2) can be (i) completely wrapped around and along the cross-section, (ii) wrapped on three sides (U-wrap), or (iii) bonded on two opposite sides of the beam; this method can also employ a rigid plate, bonded with cold-cure adhesive. The first pattern is the most efficient, and is typically adopted for shear strengthening columns; clearly, it is impractical for strengthening beams in the presence of an integral slab. In addition, the use of a continuous pattern may limit the migration of moisture and hence should be considered with caution. Shear strengthening on three sides of the beam, (item (ii)), is less efficient than item (i) due to the development of the bond length at the free end of the FRP composite. Strengthening on only the two side faces of the beam (item (iii)) is the least efficient scheme;

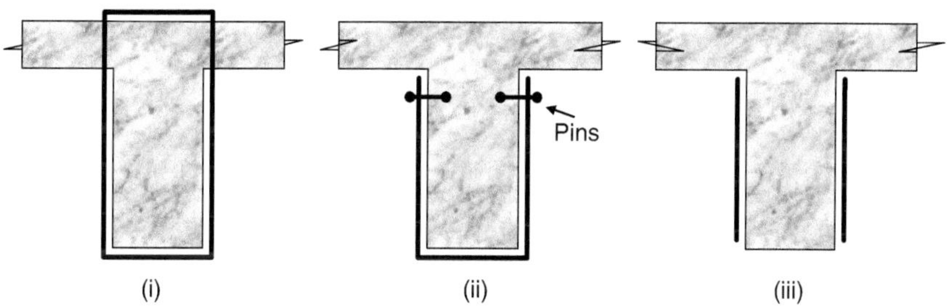

(a) Along the axis of the beam the strengthening system can either be continuous or discontinuous.
(b) The use of continuous pattern may limit the migration of moisture.
(c) System (i) is the most efficient but may be difficult to achieve. System (ii) is the most practical one and the debonding failure mechanism can be reduced by mechanical anchors at the free end of the wrap or by bonding the ends of the strips into core holes through the flange of the T-beam.
(d) The loss in ductility of FRP-strengthened cross-sections, and particularly the possiblity of brittle debonding failures, has led various design guidelines to prohibit or discourage the application of moment redistribution into or out from strengthened cross-section, leading to onerous conditions for this type of strengthening, particularly when the original design was based on moment redistribution.

1.9 Typical methods for shear strengthening of RC beams (adapted from Hollaway and Head, 2001).

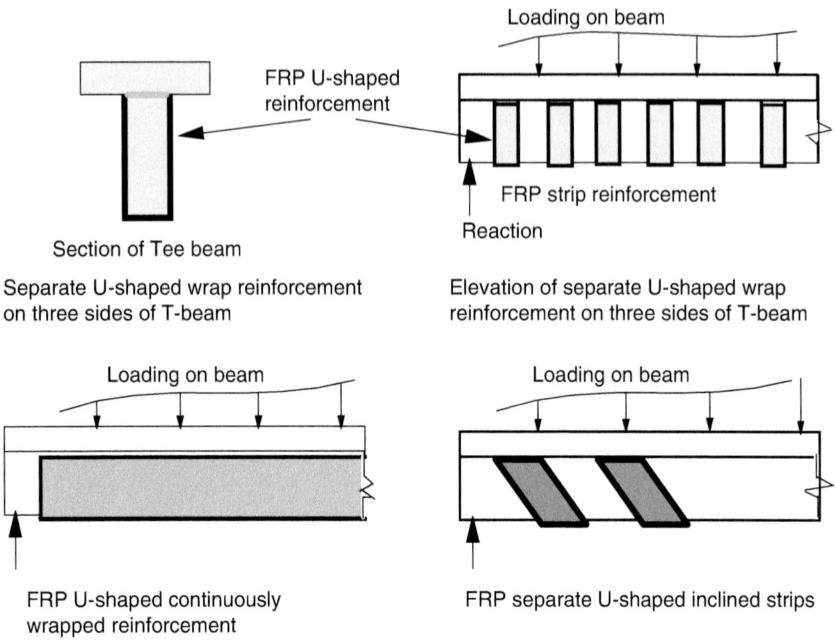

1.10 Methods for positioning U-shaped FRP shear wraps onto reinforced concrete T-beams.

this is due to the development of the bond length at the two free ends of the FRP composites. One technique to overcome the problem of item (iii) is to use prefabricated composite L-angles. These overcome the lower end of the bonded shear composite component as the bond length is developed in the soffit of the beam. These patterns are illustrated in Fig. 1.9, and Fig. 1.10 shows methods for the positioning of FRP wraps on to RC T-beams.

The bond between FRP composite and concrete plays a crucial role in the failure mode and the ultimate load of a shear-strengthened structural beam. The typical shear failure mode is associated with concrete diagonal tension and this causes debonding of the FRP laminate from the concrete and FRP fracture. Chen and Teng (2008) have discussed in detail these failure modes. If the load associated with these failure modes is large enough, the critical mode may shift to concrete diagonal compression, or to flexural failure; the latter is ductile compared with the shear failure and hence is more desirable.

1.12 Strengthening of RC columns

Reinforced concrete circular columns may require strengthening (i) in seismic regions, (ii) due to upgrading the columns or (iii) due to insufficient transverse reinforcement in the columns that will cause premature shear failure and brittle crushing of the concrete. Furthermore, a splice failure might occur if the longitudinal reinforcement is spliced at, or near, a potential hinge region. In these cases the RC columns can be strengthened by using FRP composite jackets where the fibres are orientated in the hoop direction. This technique is based on the fact that lateral confinement of concrete can substantially enhance the compressive strength and ductility of a column. No confinement will take place whilst the concrete column is in the elastic zone as the concrete undergoes little lateral expansion under low longitudinal loads and thus it does not react against the restraint of the jacket to produce confinement pressure. When the concrete column reaches its plastic zone, it rapidly expands and fully activates the jacket such that a small increase in longitudinal stress will cause a large (relative to the elastic zone) increase in lateral expansion. This expansion causes two actions to be initiated; firstly it deteriorates the internal structure of the concrete and secondly, it causes an increase in confining pressure with the fibres in the jacket exhibiting linear elastic behaviour until failure. A circular concrete column is uniformly confined by a FRP composite jacket and many models have been developed to provide information on the axial compressive strength and the stress–strain curve for these columns (Teng *et al.*, 2001; Lam and Teng, 2002); the last mentioned publication provides a large data base of tests reported in the literature.

There are a number of techniques available for strengthening existing RC columns under both static loads and seismic retrofit; in addition there are aspects specific to seismic retrofit which will be discussed separately. There are three methods which are generally used in practice. These are:

- Filament winding.
- Wrapping.
- Prefabricated-shell jacketing.

1.12.1 The site filament winding process

This has been described in Section 1.5.2 *Semi-automated processes (ii)*.

1.12.2 The wrapping method

This is a similar procedure to filament winding except that FRP wraps (either sheets or strips) are made of preformed layers of fibres and these are applied to the column; the resin impregnates these wraps by the wet lay-up technique. Sheets are easier to apply and will protect the column from the environment as they cover its entire surface area. Furthermore, the uniform 'confinement' provided by the composite sheets enhances both the strength and ductility of the concrete. The strips, on the other hand, allow more flexibility in regulating the amount of material used and this will lower the material cost, but the application process in both cases is manually orientated and therefore the cost of fabrication is about the same for both cases.

Although surface bonded, the FRP sheets overlap each other when wrapped around a column, thus developing the required strength without the possibility of surface delamination. In a seismic environment, each layer may have the fibres oriented in multiple directions to protect the column from reverse cyclic loading or other loadings due to seismic activity. In a non-seismic environment, to increase the shear strength of the column the 'wrap' bonding scheme is applied to the column in the hoop or horizontal direction with the fibres oriented at 90° to the column axis.

1.12.3 Prefabricated shell jacketing method

This is a prefabricated FRP shell made in two half circles to allow them to be placed around the column. The effectiveness of the confinement will be realised only if the two half circles of the jacket are bonded to the column or injected with an expansive and shrinkage-compensated cement, or some other like method.

As stated above for a circular column the confinement provided by the FRP composite is uniform and the confining pressure at rupture failure of the FRP is given by:

$$f_l = (2f_{frp}t) / d \qquad [1.1]$$

where f_l is the lateral pressure

f_{frp} is the tensile strength of the FRP in the hoop direction

t is the total thickness of the FRP composite

d is the diameter of the confined concrete column.

The compressive strength of FRP-confined concrete can be related to the lateral confining pressure as:

$$f_{cc}'/f_{co}' = 1 + k_1 (f_l/f_{co}') \qquad [1.2]$$

where f_{cc}' and f_{co}' are the compressive strengths of confined and the unconfined concrete, respectively, and k_1 is the confinement effectiveness coefficient.

Lam and Teng (2002) proposed a value of 2 for k_1. This was based on the analysis of a large database assembled from published literature.

In square and rectangular columns, the confining pressure provided by the FRP composite is much less effective than that developed in a circular

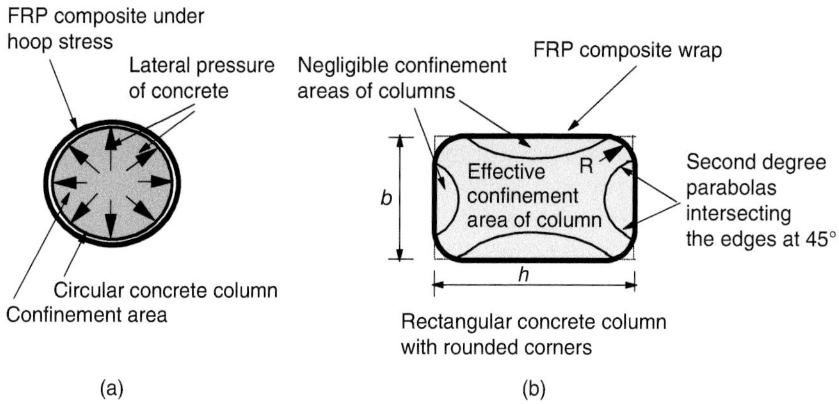

(a) Confined circular columns show an increase in compressive strength and ductility when exposed to uniaxial loads.
(b) Confinement of rectangular columns shows a moderate increase in axial strength but does not exhibit ductile properties due to stress concentrations at the corners. As the size of a rectangular column increases, the effectiveness of the FRP confinement decreases.
The FRP composite can be effectively used to enhance shear strength, to increase flexural ductility and to supplement concrete reniforcement detailing in seismic-related applications.

1.11 Effective confining pressure by FRP composite on a circular and rectangular RC column.

column. Mander *et al.* (1988) and Cusson and Paultre (1995) investigated steel confinement of rectangular cross-sections of concrete columns; these investigations illustrated an effective confinement area. Figure 1.11 shows the confinement areas for rectangular and circular columns. The corners of square and rectangular columns are rounded, firstly to enhance the confinement, and secondly to prevent the corners of the column from damaging the confining material and hence reducing its tensile failure strength. The rectangular concrete columns with FRP retrofit will have reduced stiffness and ductility benefits compared to those of the circular columns. The reduction in these properties is a function of the corner radii of the cross-section; as the values decrease, the mechanical properties of the retrofit also decrease. On the other hand, the internal steel reinforcement of the column limits the maximum value of the corner radii. Furthermore, the FRP composite wrap/shell must be applied over the entire column in order to increase flexural strength and to achieve uniform confinement. The use of discontinuous confining wraps leaves areas between them exposed and the failure modes will shift to those locations. Usually, the only exposed region in FRP wrap is between the cap beam at the bottom of the structure and the start of the RC column; this gap is a few centimetres long.

Although there are differences between the use of steel and FRP confinement, a general observation is that the confining pressure varies over the cross-section of the column. To improve on the confinement of rectangular (or square) columns, an effective way is to modify their shape by obtaining an elliptical (or circular if the columns are square) cross-section before wrapping the column with FRP sheets. Priestley and Seible (1995) were among the first researchers to investigate the modification of the cross-section of a column by adding concrete sections before confinement by FRP wraps/shell. Saadatmanesh *et al.* (1997) proposed an active confinement by the use of fast-curing cement to obtain the shape modification. Teng and Lam (2002) have illustrated methods of modifying a rectangular concrete column to an elliptical one by placing a FRP shell around the column and filling the space between the shell and column with concrete. When a rectangular column is confined by an FRP jacket with a shape modification applied, an elliptical jacket is less effective than a circular one but more effective than a rectangular column without modification. The behaviour is highly dependent upon the aspect ratio (a/b). If this ratio tends to unity, the behaviour is similar to that of the circular concrete column, but as the aspect ratio increases, the effectiveness of the FRP jacket confinement reduces. At a value of aspect ratio of about four, all benefits of confinement are virtually lost. A modified rectangular concrete column system is shown in Fig. 1.12.

Lam and Teng (2003) have introduced stress/strain models for characterising the behaviour of FRP-confined concrete; these are significantly different from that produced by Mander *et al.* (1988) for steel-confined concrete. It was found that the strength of a FRP-confined concrete column was lower than that of a steel-confined column for the same amount of hoop force. It is outside the scope of this chapter to provide detailed analysis of the assessment of existing and new strength models, failure modes, eccentri-

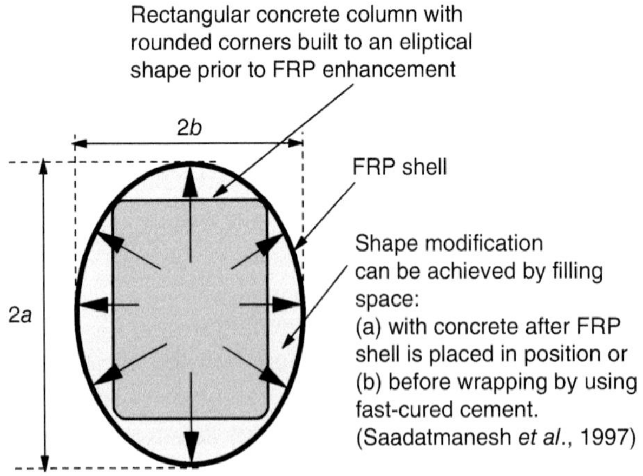

1.12 Effective confinement for a rectangular column by space modification.

cally loaded and typical behaviour of FRP confined RC columns; these have been discussed by Teng and Lam (2002).

1.12.4 Surface preparation for the bonding of FRP jackets to reinforced concrete columns

The process of preparing the surfaces of the concrete columns for bonding, and the bonding procedure, are undertaken in a similar way to that described for flexural strengthening (Section 1.7.2).

1.13 Conclusions and future trends

Advanced polymer composites have been available to the civil/structural engineer for rehabilitation, retrofitting and strengthening of RC structural members for more than two decades. During this short period, the composites' influence in the area of upgrading of structures has been rapid and successful; however, it is important for the engineer to be aware of:

(i) the environment into which the material is placed and how successful it will be to resist degradation over the longer term. The current problem associated with the durability of composites is the accessibility of information on the topic; it is sparse, not well documented, and not accessible to the civil engineer (Karbhari *et al.*, 2003). This problem can only be overcome by site monitoring of rehabilitated structures and this will provide long-term characteristics of the polymer composite in a specific environment. Ideally, this information should then be made available to the scientific community. However, there is a cost associated with monitoring and this tends to be a major obstacle to its use.

(ii) the many statements claiming success regarding the durability of FRP material. Practical evidence suggests that if designed and fabricated appropriately polymer composite systems can provide longer lifetimes and lower maintenance costs than conventional civil engineering structural materials. However, a high percentage of the current knowledge on durability is based on experimental accelerated laboratory testing of composite materials.

Engineers must be aware that the results derived from using composite materials coupons that have been exposed to accelerated tests (usually at elevated temperatures and immersed in concentrated salt solutions) to predict the long-term response of a site structural system need to be investigated and verified. In the past, published accelerated test results have degraded coupon specimens infinitely more quickly than those under degradation in the natural environment. The results of these accelerated tests have produced erroneous conclusions regarding the ability of FRP composites to withstand normal civil engineering environments. In allied investigations, the conclusions of Mufti *et al.* (2005a, b, c) have shown that FRP composites have excellent durability properties when exposed to severe environmental conditions. Further work needs to be undertaken in the area of accelerated testing in order to relate the laboratory test results to those obtained from site investigations.

(iii) the degree to which individual composites perform in practice, which is highly dependent upon the manufacturing techniques which determine their physical, mechanical and in-service properties, and the service life extension. This in turn is dependent upon the polymer and fibres used in the rehabilitation and retrofitting of the RC structure.

The manual, the semi-automated or the automated methods of manufacture, the cold-cure or hot-cure resins for bonding the FRP plate on to the RC concrete beam, and the need for post cure, will all have an influence upon the quality and performance and therefore upon the characteristics of the final composite. The cold-melt polymer composite prepreg is the newest technique to enter the rehabilitation scenario, and the most successful, but it is about twice as expensive as the pultrusion or wet lay-up methods presently used. Further effort is being devoted to reducing the manufacturing/site-fabrication costs, and generally improving the material.

(iv) the safe use of composites for the rehabilitation of buildings, which is dependent upon the polymer matrix, and probably to a greater extent upon the adhesive having a high fire resistance; a discussion on fire resistance is given in Section 1.5.3. A number of new fire-resistant polymers have the potential to improve the fire performance of civil engineering composites. Furthermore, research is being undertaken into additives which may be incorporated into the polymer to improve its fire resistance. One such additive is montmorillonite clay nanocomposites (Wang and Wilkie, 2006; Hackman, 2007). It should be said, however, that currently, the cost of incorporating nano-composites into the polymer of FRP composites is too expensive for the rehabilitation of structural members. Generally, the design for the rehabilitation of bridges is such that if the upgrading composite plate is destroyed by fire, the bridge continues to maintain its stability and will not collapse.

Special care should be taken when designing structural members that are to be upgraded using a cold-cure adhesive to bond a rigid composite plate, or when the cold-cured matrix and adhesive wet lay-up system are used. Care is particularly important when the system is immediately subjected to load before the matrix and/or adhesive are fully cured. The structural response and an estimation of the quality of the upgrade and its long-term response must be analysed. The effect of partially cured resins in this situation requires further investigation.

The methods of rehabilitating and retrofitting FRP composites to various structural components have been given. The design of these components is outside the scope of this chapter. Currently there are a number of design guides and aids available for the design engineer, and more are gradually entering the profession by research institutes and design bodies. The Appendix (Section 1.15) provides a list of design guides, codes of practice and specifications.

In Section 1.2, one or two drawbacks to the use of FRP were mentioned. The risk of fire and vandalism are probably the most difficult of the drawbacks to solve. However, both of these items can be solved, at a price, by protecting the surfaces of the material, but such measures would defeat the object and advantages of the utilisation of composites. Research is being undertaken in both areas and it must be only a matter of time before these disadvantages are overcome.

1.14 References

ACI COMMITTEE (440.2R-08) (2008). *Guide for the design and construction of externally bonded FRP systems for strengthening concrete structures*. Farmington Hills, MI; American Concrete Institute; 2008.

ACI COMMITTEE (440.4R.04) (2004). *Prestressing concrete structures with FRP tendons.* Farmington Hills, MI; American Concrete Institute; 2004.

ACI COMMITTEE (503.7-07) (2007). *Specification for crack repair by epoxy injection.* Farmington Hills, MI; American Concrete Institute; 2007.

ADVANCED COMPOSITES GROUP LTD (ACG), Composites House, Sinclair Close, Heanor Gate Industrial Estate, Heanor, Derbyshire, UK.

AIDOO, J., HARRIES, K. and PETROU, M. (2004). 'Fatigue behaviour of carbon fibre reinforced polymer-strengthened reinforced concrete bridge girders' *Journal of Composites for Construction*, Nov./Dec. 2004 Vol.8 Issue 6, pp 501–509.

AKLONIS, J. J. and MACKNIGHT, W. J. (1983). *Introduction to polymer viscoelasticity*, 2nd edn., Wiley, New York, pp 36–56.

ALMENARA, P. C. and THORNBURROW, P. (2004). 'A new glass fiber reinforcement for anti-corrosion composites'. *Proceedings of the ACIC 2004 Conference 'Advanced Polymer Composites for Structural Applications in Construction'*, edited by L. C. Hollaway, M. K. Chryssanthopoulos and S. S. J. Moy. Woodhead Publishing Ltd, Cambridge, UK.

ANDRÅ, P. H. and MAIER, M. (1999). 'Post strengthening of reinforced concrete structures by prestressed externally bonded carbon fibres reinforced polymer (CFRP) strips'. *Proceedings of the Conference, 'Structural Faults and Repair 1999'*, London.

APICELLA, A., MIGLIARESI, C., NICODEMO, C., NICOLAIS, L., IACCARINO., L. and ROCCOTELLI, S. (1982). 'Water sorption and mechanical properties of a glass-reinforced polyester resin' *Composites*, Vol. 13, pp 406–410.

APICELLA, A., MIGLIARESI, C., NICOLAIS, L., IACCARINO, L. and ROCCOTELLI, S. (1983). 'The water ageing of unsaturated polyester-based composites: Influence of resin chemical structure', *Composites*, Vol. 14, No. 4, pp 387–392.

ARDUINI, M. and NANNI, A. (1997). 'Behaviour of precracked RC beams strengthened with carbon FRP sheets', *Journal of Composites for Construction*, ASCE, Vol.1, No.2, pp 63–70.

ASPLUND, S. O. (1949). 'Strengthening bridge slabs with grouted reinforcement', *Journal of the American Concrete Institute*, Vol. 20. No.6. pp 397–406.

ASTM C881 (2002). *Type 1 Grade Specification for maximum filling of cracks with low viscosity resin.* Specification allows a viscosity of up to 2000 CPS.

ASTM C881 (2002). *Type IV Grade Specification for use in load-bearing applications for bonding hardened concrete to hardened concrete and other materials and as a binder for epoxy mortars and concretes.*

BALAZS G. L. and BOROSNYOI A. (2001). 'Long-term behavior of FRP'. In: *Proceedings of the International Workshop on Composites in Construction: A reality*. American Society of Civil Engineers, VA. pp 84–91.

BANK, L., LAMANNA, A. J., RAY, J. C. and VELAZQUEZ, G. I. (2002a). 'Rapid strengthening of reinforced concrete beams with mechanically fastened fiber-reinforced polymeric composite materials'. *US Army Corps of Engineers. Report Number ERDC/GSL. TR-02-4*, 93 pages.

BANK, L. C., BOROWICZ, D. T., LAMANNA, A. J., RAY, J. C. and VELAZQUEZ, G. I. (2002b). 'Rapid strengthening of full-size reinforced concrete beams with powder-actuated fastening systems and fiber-reinforced polymer (FRP) composite material' *US Army Corps of Engineers, Report Number ERDC/GSL. TR-02-12*, 110 pages.

BANK, L. C. (2004). 'Mechanically-fastened FRP (MF-FRP) – a viable alternative for strengthening RC members'. *Proceedings of FRP Composites in Construction – CICE 2004* (Ed. R. Seracino), 8–10 December 2004, Adelaide, Australia.

BANK, L. C. (2006/7). *Composites for Construction – Structural Design with FRP Materials*, John Wiley & Sons, Inc.

BARBERO, E. J. (1999). *Introduction to Composite Materials Design* (1st ed.), Taylor & Francis Inc., Philadelphia, PA.

BERSET, J. D. (1992). *Strengthening of Reinforced Concrete Structures for Shear Using Composite Materials*, M.Sc. Thesis, MIT, Cambridge, Mass., pp 105.

BEDFORD, REINFORCED PLASTICS INC., 264 Reynoldsdale Road, Bedford, PA, 15522-7401, USA. Ad Bedford PA 15522-7401 USA.

BISBY, L. A., GREEN, M. F. and KODUR, V. K. R. (2005). 'Response to fire of concrete structures that incorporate FRP', *Progress in Structural Engineering and Materials*, Vol. 7, No. 3. pp 136–149.

BLASCHKO, M. and ZILCH, K. (1999). 'Rehabilitation of concrete structures with strips glued into slits'. *Proceedings of the 12th International Conference on Composite Materials*, Paris. (CD).

BS 5400-2 (2006). *Steel, concrete and composite bridges. Specification of loads.* British Standards Institute, London, UK.

BS 8110 (1997). *Structural Use of Concrete, Part 1. Code of Practice for Design and Construction*, British Standards Institute, London, UK.

BS DD ENV 1504 PART 9 (1997). *Products and systems for the protection and repair of concrete structures.* British Standards Institute, London, UK.

CESSNA, L. C. (1971). 'Stress–time superposition for creep data for polypropylene and coupled glass reinforced polypropylene', *Polymer Engineering Science*, 11, May, pp 211–219.

CHAHROUR, ALI. and SOUDKI, K. (2005). 'Flexural response of reinforced concrete beams strengthened with end-anchored partially bonded carbon fibre-reinforced polymer strips', *J. Composites for Construction*, Vol. 9, Issue 2, pp 170–177.

CHAR, M. S., SAADATMANESH, H. and EHSANI, M. R. (1994). 'Concrete Girders Externally Prestressed with Composite Plates,' *PCI Journal*, Vol. 39, No. 3, May–June, pp 40–51.

CHEN, J.-F. and TENG, J. G. (2008). Chapter 5: 'Shear strengthening of RC beams' in *Strengthening and Rehabilitation of Civil Infrastructures using FRP Composites*, Editors L. C. Hollaway and J. G. Teng, Woodhead Publishing Ltd, Cambridge, England.

CHOWDHURY, E. U. (2005). *Performance in fire of reinforced concrete T-beams strengthened with externally bonded fibre-reinforced polymer sheets*, MSc thesis, Department of Civil Engineering, Queen's University , Kingston, ON, Canada. 273 pages.

CIVIL ENGINEERING RESEARCH FOUNDATION, EXECUTIVE REPORT 93-5011 (1993). *High performance construction materials and systems: An essential program for America and its infrastructure.*

CLARKE, J. L. (ed.) (1996). *Structural Design of Polymer Composites – EUROCOMP Design Code and Handbook*, London: E & F N Spon.

CLARKE, J. L., DARBY, A. and IBELL, T. (2004). 'Strengthening concrete structures with fibre composite materials: Updating Technical Report 55', *Proceedings of the Conference 'Advanced Polymer Composites for Structural Applications in Construction' (ACIC 2004)*. Edited L. C. Hollaway, M. K. Chryssanthopoulos and S. S. J. Moy. Woodhead Publishing Ltd, Cambridge, UK, pp 651–658.

CONCRETE SOCIETY (2000). *Design Guidance for Strengthening Concrete Structures Using Fibre Composite Materials*, TR55, 2nd ed., Camberley, UK.

CONCRETE SOCIETY (2003). *Strengthening Concrete Structures using Fibre Composite Materials: Acceptance, Inspection and Monitoring*, TR57, Camberley, UK.

CREATIVE PULTRUSIONS, 214 Industrial Lane, Alum Bank, PA – 15521, USA.

CRIPPS, A., HARRIS, B. and IBELL, I. (2002). *Fibre-reinforced polymer composites in construction*, Technical Report C564, London, CIRIA, 2002.

CSA-S806-02 (2002). *Design and Construction of Building Components with Fiber-reinforced Polymers*, CSA-S806-02, Canadian Standards Association, Toronto, Ontario, Canada.

CUSSON, D. and PAULTRE, P. (1995). 'Stress–strain model for confined high strength concrete' *Journal of Structural Engineering*, ASCE, Vol. 121, No. 3, pp 468–477.

DE LORENZIS, L. (2002). *Strengthening of RC Structures with Near Surface Mounted FRP Rods*. PhD Thesis, Universita'Degli Studi Di Lecce, Italy.

DE LORENZIS, L., MICELLI, F. and LA TEGOLA, A. (2002): 'Passive and Active Near Surface Mounted FRP Rods for Flexural Strengthening of RC Beams', *Proceedings of ICCI'02*, San Francisco, USA.

DE LORENZIS, L. and NANNI, A. A. (2002). 'A bond between near-surface mounted FRP rods and concrete in structural strengthening', *ACI Structural Journal*, Vol. 99 (2), pp 123–132.

DE LORENZIS, L. and TENG, J. G. (2007). 'Near-surface mounted reinforcement: an emerging technique for structural strengthening', *Composites Part B: Engineering*, Vol. 38, No. 2, pp 119–143.

DEMERS, C. E. (1998a), 'Fatigue strength degradation of E-glass FRP composites and carbon FRP composites' *Construction and Building Materials*, Vol. 12, No. 5, pp 311–318.

DEMERS, C. E. (1998b), 'Tension–tension axial fatigue of E-glass fiber-reinforced polymeric composites: Fatigue life diagram', *Construction and Building Materials*, Vol. 12, No. 5, pp 303–310.

DIMITRIOU, A. D. (2004). *Rehabilitation of damaged reinforced concrete beams using FRP materials*, MSc dissertation, University of Surrey, Guildford, Surrey, UK.

EGAN, J. (1998). 'Rethinking Construction: The Egan Report (1998)', published in *Construction Reports 1944–1998*, edited M. Murray and Langford, Blackwell Publishing, Oxford, UK.

EKENEL, M., RIZZO, A., MYERS, J. J. and NANNI, A. (2006). 'Flexural Fatigue Behavior of Reinforced Concrete Beams Strengthened with FRP Fabric and Precured Laminate Systems', *Journal Composites for Construction*, Vol. 10, Issue 5, pp 433–442.

EL-HACHA, R., WIGHT, R. and GREEN, M. (2003). 'Innovative system for prestressing fiber-reinforced polymer sheets' *ACI Structural Journal* Vol. 100, No. 3, pp 305–313.

EL-HACHA, R. and RIZKALLA, S. H. (2004). 'Near-surface-mounted fibre-reinforced polymer reinforcements for flexural strengthening of concrete structures.' *ACI Struct. J.*, Vol. 101(5), pp. 717–726.

EL-HACHA, R., WIGHT, R. and GREEN, M. (2001). 'Prestressed fibre-reinforced polymer laminates for strengthening structures', *Progress in Structural Engineering*, Vol. 3 Issue 2, pp 111–121.

ESHWAR, N., IBELL, T. and NANNI A. (2003). 'CFRP Strengthening of Concrete Bridges with Curved Soffits,' *Proceedings Structural Faults + Repair 2003*, London, UK. M. C. Forde (ed.), CD-ROM, 10 pp.

FARMER, N. S. (2003). 'Near Surface Mounted Carbon Fibre Reinforcement,' *Concrete*, Vol. 37, No. 1, pp 20–21.

FARMER, N. (2004). 'Near Surface Mounted Reinforcement for Strengthening – UK experience and development of best practice'. *Proceedings of the Conference Advanced Polymer Composites for Structural Applications in Construction*, (ACIC 2004), edited by L. C. Hollaway, M. K. Chryssanthopoulos and S. S. J. Moy, Woodhead Publishing Ltd. Cambridge,UK, pp 659–666.

FERRIER, E., ENNACEUR, C., BIGAUD, D. and HAMLIN, P. (2001). 'Prestressed externally bonded FRP reinforcement for RC beams', *Proceedings of the FRPRCS-5 Conference*, Vol. 1, edited by C. J. Burgoyne, Thomas Telford, London, pp 271–280.

fib (2001). *Design and Use of Externally Bonded FRP Reinforcement (FRP-EBR) for Reinforced Concrete Structures.* Bulletin 14, fib Task Group 9.3 'FRP Reinforcement for Concrete structures'. Federation Internationale du Beton, Lausanne, Switzerland.

FIBER LINE COMPOSITES A/S BARMSTEDT ALLÉ 5 DK-5500 MIDDELFART, Germany.

FIBREFORCE COMPOSITES LTD., Runcorn, Cheshire, UK.

FYFE CO. LLC, Nancy Ridge Technology Center, 6310 Nancy Ridge Drive, Suite 103, San Diego, CA 92121, USA.

FUKUYAMA, H., TUMIALAN, G. and NANNI, A. (2003). *Japanese design and construction guidelines for seismic retrofit of building structures with FRP composites*. Technical Note of National Institute for Land and Infrastructure Management. No. 104, pp 145–156.

GARDEN, H. N., HOLLAWAY, L. C. and THORNE, A. M. (1997). 'A preliminary evaluation of carbon fibre reinforced polymer plates for strengthening reinforced concrete members', *Proceedings of the Institution of Civil Engineers: Structures and Buildings*, Vol. 123, May, pp 127–142.

GARDEN, H. N., HOLLAWAY, L. C., and THORNE, A. M. (1998). 'The Strengthening and Deformation Behavior of Reinforced Concrete Beams Upgraded Using Prestressed Composite Plates', *Materials and Structures*, Vol. 31, No. 4, pp 247–258.

GARDEN, H. N. and HOLLAWAY, L. C. (1999). 'An experimental study of the failure modes of reinforced concrete beams strengthened with prestressed carbon composite plates' *Composites*, Part B, Vol. 29B, pp 411–424, 1999.

GARDEN, H. N. and HOLLAWAY, L. C. (1998). 'An experimental study of the influence of plate end anchorage of carbon fibre composite plates used to strengthen reinforced concrete beams' *Composite Structures*, Vol. 42, pp 175–188.

GHEORGHIU, C., LABOSSIÈRE, P. and PROULX, (2004a). 'Effect of a fatigue cycling on the strength of RC beams externally strengthened with CFRPs', *Proceedings of the 2nd International Conference on Advanced Polymer Composites for Structural Applications in Construction, ACIC 2004*, edited by L. C. Hollaway, M. K. Chryssanthopoulos and S. S. J. Moy, Woodhead Publishing Ltd, Cambridge, UK, pp 585–592.

GHEORGHIU, C., LABOSSIÈRE, P. and RAÎCHE, A. (2004b). Experimental fatigue and static behaviour of RC strengthened with carbon fiber-reinforced polymer', *Journal of Composites for Construction*, May/June 2004, Vol. 8, Issue 3, pp 211–218.

GHEORGHIU, C., LABOSSIÈRE, P. and PROULX, (2004c), 'Fatigue and post-fatigue quasi-static performance of RC-beams externally strengthened with CGRP', *Proceedings of the 2nd International Conference on FRP Composites in Civil Engineering – CICE 2004*, edited by R. Seracino, 8–10 December 2004, Adelaide, Australia, pp 433–438.

GRACE, N. F., SOLIMAN, A. K., ABDEL-SAYED, G. and SALEH, K. R. (1998). 'Behaviour and ductility of simple and continuous FRP reinforced beams', *Journal of Composites for Construction*, ASCE, Vol. 2, No. 4, pp 186–194.

GUYSON (1989). *Manual of Blast Media, Guyson Data Sheets*, Guyson International Ltd, Skipton, UK.

HACKMAN, I. (2007). *Nanocomposites in Civil Engineering*, PhD Thesis, University of Surrey, Guildford, UK.

HACKMAN, I. and HOLLAWAY, L. C. (2006). 'Epoxy-layered silicate nanocomposites in civil engineering', *Composites Part A*, Vol. 37, No. 8, pp 1161–1170.

HAMILTON, H. R. and DOLAN, C. W. (2000). 'Durability of FRP reinforcements for concrete', *Progress of Structural Engineering Materials*, Vol. 2, No. 2, pp 139–145.

HAQUE, A. and SHAMSUZZOHA, M. (2003). 'S2-Glass/Epoxy Polymer Nanocomposites: Manufacturing, Structures, Thermal and Mechanical Properties', *Journal of Composite Materials*, 37(20), pp 1821–1837.

HASSAN, T. (2002). *Flexural performance and bond characteristics of FRP strengthening techniques of concrete structures*. PhD Thesis, University of Manitoba, Canada.

HASSAN, T. and RIZKALLA, S. (2003). 'Investigation of Bond in Concrete Structures Strengthened with Near Surface Mounted Carbon Fiber Reinforced Polymer Strips', *Journal of Composites for Construction*, Vol. 7, No. 3, pp 248–257.

HEFFERMAN, P. J. and ERKI, M. A. (2004). 'Fatigue Behavior of Reinforced Concrete Beams Strengthened with Carbon Fiber Reinforced Plastic Laminates', *Journal of Composites for Construction*, Vol. 8, Issue 2, pp 132–140, (March/April).

HILTI CORP., FL-9494 Schaan, Principality of Liechtenstein.

HOLLAWAY, L. (1994). *Handbook of polymer composites for engineers* (1st ed.), Woodhead Publishing Limited, Cambridge, UK.

HOLLAWAY, L. C. and MAYS, G. C. (1999). 'Structural strengthening of concrete beams using unstressed composite plates' Chapter 4, of *Strengthening of Reinforced Concrete Structures using externally-bonded FRP composites in structural and civil engineering*, edited by L. C. Hollaway and M. B. Leeming, Woodhead Publishing Ltd, Cambridge, UK.

HOLLAWAY, L. C. and LEEMING, M. B. (eds) (1999). *Strengthening of Reinforced Concrete Structures using Externally-bonded FRP Composites in Structural and Civil Engineering*, Woodhead Publishing Ltd, Cambridge, UK.

HOLLAWAY, L. C. and HEAD, P. R. (2001). *Advanced Polymer Composites and Polymers in the Civil Infrastructure*, Elsevier, Oxford, UK.

HOLLAWAY, L. C., ZHANG, L., PHOTIOU, N. K., TENG, J. G. and ZHANG, S. S. (2006). 'Advances in Adhesive Joining of Carbon Fibre/Polymer Composites to Steel Members for Repair and Rehabilitation of Bridge Structures'. *Journal of Advances in Structural Engineering*, Vol. 9, No. 6, pp 101–113.

HOLLAWAY, L. C. (2007). 'Fibre-reinforced polymer composite structures and structural components: Current applications and durability issues', Chapter 10, *Durability of Composites for Civil Structural Applications*, edited by V. M. Karbhari, Woodhead Publishing, Cambridge, UK.

HOLLAWAY, L. C. (2008). Chapter 13, 'Modern developments of Advanced Polymer Composites used in Bridge Engineering' in *Manual of Bridge Engineering*, edited by G. Parke and N. Hewson, Thomas Telford, London, UK.

HOLLAWAY, L. C. and TENG, J. G. (eds) (2008). *Strengthening and Rehabilitation of Civil Infrastructures using FRP Composites*, Woodhead Publishing Ltd., Cambridge, England.

HOLLAWAY, L. C. (2009). Chapter 7.3, 'Advanced Polymer Composites' in Section 7 of the *Manual of Construction Materials*, edited by M. Forde, Thomas Telford, London, UK.

HUGHES BROTHERS, Inc. 210 N. 13th Street, Seward, NE 68464 USA.

HUTCHINSON, A. R. (1997). *Joining of Fibre-reinforced Polymer Composite Materials*, Report 46, London: Construction Industry Research and Information Association.

HUTCHINSON, A. R. and QUINN, J. (1999). 'Materials'. In *Strengthening of Reinforced Concrete Structures* (L. C. Hollaway and M. B. Leeming, eds.). Chapter 3, Woodhead Publishing, Cambridge, UK.

HUTCHINSON, A. R., BROUGHTON, J. G., FROST, S., LEE, R. J., FARMER, N., SMITH, I., LUKE, S. and CANNING, L. (2004). 'A performance-based qualification scheme for composite materials systems to be used in the civil infrastructure'. *Proceedings of the Conference Advanced Polymer Composites for Structural Applications in Construction (ACIC 2004)*, edited L. C. Hollaway, M. K. Chryssanthopoulos and S. S. J. Moy. Woodhead Publishing Ltd, Cambridge, UK, pp 667–674.

INSTITUTION OF STRUCTURAL ENGINEERS, UK (1999). *Guide to the Structural use of Adhesives*, the Institution of Structural Engineers, London, UK.

JACOBS, P. M. and JONES, F. R. (1993). 'Influence of heterogeneous crosslink density on the thermomechanical and hygrothermal properties of an unsaturated polyester resin, 2. Hygrothermal studies', *Polymer*, Vol. 34, No. 10, pp 2122–2127.

JBDPA (1999). *Seismic Retrofitting Guideline* (1999) The Japanese design and construction guidelines for seismic retrofitting of RC building materials.

KAISER, H. P. (1989). *Strengthening Reinforced Concrete with Epoxy-bonded Carbon-fibre Plastics*, Doctoral Thesis, Diss. ETH, Nr. 8918, ETH Zurich, Ch-8092 Zurich, Switzerland, (in German).

KANDOLA, B. K. and HORROCKS, A. R. (1998). 'Flame retardant composites, a review: The potential for use of intumescents'. In *Fire Retardancy of Polymers – The Use of Intumescence*, ed. R. Delobel, The Royal Society of Chemistry, London, pp 395–417.

KARBHARI, V. M., ZHAO, L., MURPHY, K. and KABALNOVA, L. (1998). 'Environmental durability of glass fiber reinforced composites – short term effects'. *Proceedings of the 1st International Conference on the Durability of Fibre Reinforced Polymer Composites for Construction (CDCC'98)*, pp 513–524.

KARBHARI, V. M. and SEIBLE, F. (2000). 'Fiber Reinforced Composites – Advanced Materials for the Renewal of Civil Infrastructure', *Applied Composite Materials*, Vol. 97, pp 95–124.

KARBHARI, V. M., CHIN, J. W., HUNSTON, D., BENMOKRANE, B., JUSKA, T., MORGAN, R., LESKO, J. J., SORATHIA, U. and REYNAUD, R. (2003). 'Durability gap analysis for fiber-reinforced polymer composites in civil infrastructure', *ASCE Journal of Composites in Construction*, Vol. 7, No. 3, pp 238–247.

KARBHARI, V. M. (ed) (2007). *Durability of Composites for Civil Structural Applications*, Woodhead Publishing Ltd, Cambridge, England.

KARBHARI, V. M., CHIN, J. W. and REYNAUD, D. (2000). 'Critical gaps in durability data FRP composites in civil infrastructure', *Proceedings 45th International SAMPE Symposium*, SAMPE, Long Beach, May 21–25.

KINLOCH, A. J. (1987). *Adhesion and Adhesives: Science and Technology*, London, Chapman and Hall.

KONTOU, E. (2005). 'Tensile behaviour of unidirectional glass fiber polymer composites', *Polymer Composites*, Vol. 26, Issue 3, pp 287–292

LAM, L. and TENG, J. G. (2002). 'Strength Models for FRP-confined Concrete', *Journal of Structural Engineering*, ASCE, Vol. 128, No. 5, pp 612–623.

LAM, L. and TENG, J. G. (2003). 'Design-oriented stress–strain model for FRP-confined concrete', *Construction and Building Materials*, Vol. 17, Nos 6 & 7, September–October, 471–489.

LAMANNA, A. J., BANK, L. C. and SCOTT, D. W. (2001a). 'Flexural strengthening of reinforced concrete beams using fasteners and fiber-reinforced polymer strips'. *ACI Structural Journal*. Vol. 98, No. 3, pp 368–376.

LAMANNA, A. J., BANK, L. C. and SCOTT, D. W. (2001b). 'Rapid flexural strengthening of RC beams using powder actuated fasteners and FRP strips'. *Proceedings of the 5th International Symposium of FRP in Reinforced Concrete Structures*, (ed C. Burgoyne), Cambridge, UK. July 16–18, pp 389–397.

LATHAM, M. (1993). *Trust and Money, Interim Report of the Joint Government/Industry Review of Procurement and Contractual Arrangements in the United Kingdom Construction Industry*, HMSO, London, UK.

LATHAM, M. (1994). *Constructing the Team, Final Report of the Joint Government/ Industry Review of Procurement and Contractual Arrangements in the United Kingdom Construction Industry*, HMSO, London, UK.

LEE, R. J., HANCOX, N. L. and HUTCHINSON, A. R. (1998). *The Assessment of Adherend Surface Quality – A Best Practice Guide*, DTI Report MTS Project 4 – Assessment of Surface Quality Prior to Adhesive Bonding.

LEEMING and DARBY (1999). 'Design and specifications for FRP plate bonding of Beams', Chapter 9 of *Strengthening of Reinforced Concrete Structures using Externally-bonded FRP Composites*, ed. L. C. Hollaway and M. B. Leeming, Woodhead Publishing, Cambridge, UK.

LIU, W., HOA, S. and PUGH, H. (2005). 'Epoxy-clay nanocomposites: Dispersion, morphology and performance', *Composites Science and Technology*, 65, pp 307–316.

LOCKETT, F. J. (1972). *Non-linear Viscoelastic Solids*, Academic Press, New York.

LOPEZ, M. M., NAAMAN, A. E., PINKERTON, L. and TILL, R. D. (2001), 'Behaviour of RC Beams Strengthened with FRP Laminates and Tested under Cyclic Loading at Low Temperatures' *Conference paper in FRP Composites in Civil Engineering*, Editor J. G. Teng, Elsevier, 12–15th December 2001, Hong Kong.

MAKSIMOV, R. D. and PLUME, É. Z. (2004). 'Creep in unidirectionally reinforced polymer composites', *Mechanics of Composite Materials*, Vol. 20, No. 2, pp 149–157.

MANDER, J. B., PRIESTLEY, M. J. N. and PARK, R. (1988). Theoretical stress–strain model for confining concrete', *Journal of Structural Engineering*, ASCE, Vol. 114, No. 8, pp 1804–1826.

MBRACE (1998). MBrace Composite Strengthening System: Engineering Design Guide-lines, Master Builders, OH. Current edition at www.mbrace.com.

MAYS, G. C. and HUTCHINSON, A. R. (1992). *Adhesives in Civil Engineering*, Cambridge University Press, Cambridge, UK.

McCRUM, N. G. (1971). *A view of the science of fibre reinforced plastics*. London, HMSO.

MEIER, U. and KAISER, H. P. (1991). *Strengthening of Structures with CFRP Laminates, Proceedings: Advanced Composite Materials in Civil Engineering Structure, ASCE Specialty Conference*, pp 224–232.

MEIER, U. (1987). *Bridge Repair with High-performance Composite Materials, Material und Technik*, Vol. 4, pp 125–128 (in German).

MEIER, U. (1995). 'Strengthening of structures using carbon fibre/epoxy composites', *Construction and Building Materials*, Vol. 9, No. 6, pp 341–351.

MILLER, C. T., CHAJES, M. J., MERTZ, D. R. and HASTINGS, N. J. (2001). 'Strengthening of a steel bridge girder using CFRP plates', *ASCE J Bridge Eng*, 6(6), pp 514–522.

MOURITZ, A. P. (2006). 'Durability of composites exposed to elevated temperature and fire', Chapter 6 of *Durability of Composites for Civil Structural Applications*, edited by V. M. Karbhari, Woodhead Publishing, Cambridge, UK.

MOURITZ, A. P. and GIBSON, A. G. (2006). *Fire Properties of Polymer Composite Materials*, Springer, Dordrecht, The Netherlands. 394 pp.

MOSALLAM, A. S. and MOSALAM, K. M. (2003). 'Strengthening of two-way concrete slabs with FRP composite laminates', *Construction and Building Materials*, 17, 43–54.

MUFTI, A., BENMOKRANE, B., BOULFIZA, M., BAKHT, B. and BREYY, P. (2005a). 'Field study on durability of GFRP Reinforcement', *International Bridge Deck Workshop*, Winnipeg, Manitoba, Canada, April 14–15, 2005.

MUFTI, A., ONOFREI, M., BENMOKRANE, B., BANTHIA, N., BOULFIZA, M., NEWHOOK, J., BAKHT, B., TADROS, G. and BRETT, P. (2005b) 'Durability of GFRP reinforced concrete in field structures', *7th International Symposium on Fiber Reinforcement for Reinforced Concrete Structures (FRPRCS-7)*, New Orleans Lousiana, USA, November 7–10, 2005.

MUFTI, A., ONOFREI, M., BENMOKRANE, B., BANTHIA, N., BOULFIZA, M., NEWHOOK, J., BAKHT, B., TADROS, G. and BRETT, P. (2005c) 'Report on the studies of GFRP durability in concrete from field demonstration structures'. *Proceedings of the Composites in Construction 2005 – 3rd International Conference*, eds., P. Hamelin, D. Bigaud, E. Ferrier, and E. Jacquelin. Lyon, France, July 11–13, 2005.

NGUYEN, D. M., CHAN, T. K. and CHEONG, H. K. (2001). 'Brittle failure and bond development length of CFRP-concrete beams'. *Journal of Composites for Construction.* ASCE, Vol. 5, No. 1, pp 12–17.

NORDIN, H. (2004). *Strengthening structures with externally prestressed tendons*, Technical Report, Luleå University of Technology, Sweden.

NSF (1993), *NSF 93-4 Engineering Brochure on Infrastructure*, US National Science Foundation, Arlington, VA.

NUTTING, P. G. (1921). ASTM 21, pp 1162–1171.

PIYONG, Y., SILVA, P. F. and NANNI, A. (2003). 'Flexural strengthening of concrete slabs by a three-stage prestressing FRP system enhanced with the presence of GFRP anchors spikes', Conference *Composites in Construction*, Cosenza, Italy, pp 239–244.

PLEVRIS, N. and TRIANTAFILLOU, T. C. (1994). 'Time-dependent behaviour of RC members strengthened with FRP laminates', *Journal of Structural Engeneering*, 120(3), pp 1016–1042.

PRIESTLEY, M. J. N. and SEIBLE, F. (1995). 'Design of retrofit measures for concrete and masonry structures', *Construction and Building Materials*, Vol. 9, No. 6, pp 365–377.

PROTA, A., NANNI, A., MANFREDI, G. and COSENZA, E. (2001). 'Design criteria for RC beam-column joints seismically upgraded with composites', *Proceedings of the International Conference on FRP Composites in Civil Engineering – CICE 2001*, ed. J.-G. Teng, Hong Kong, China, Vol. 1, pp 919–926.

QUANTRILL, R. J., HOLLAWAY, L. C. (1998). 'The flexural rehabilitation of reinforced concrete beams by the use of prestressed advanced composite plates.' *Composites Science and Technology*, Vol. 58, No. 8, pp 1259–1275.

QUANTRILL, R. J., HOLLAWAY, L. C., THORNE, A. M. (1996). 'Experimental and Analytical Investigation of FRP Strengthened Beam Response', *Mag. Conc. Res.*, 48(177), pp 331–342.

REPLARK (1999). *Replark system: Technical manual*, Mitsubishi Chemical Corporation, Sumitomo Corporation of America, New York. Current edition at www.sumitomocorp.com.

RITCHIE, P. A., THOMAS, D. A., LU, L. W. and CONNELY, G. M. (1991). 'External reinforcement of concrete beams using fiber reinforced plastics'. *ACI Structural Journal*, Vol. 67, No. 12, pp 229–233.

RIZKALLA, S., HASSAN, T. and HASSAN, N. (2003). 'Design recommendations for the use of FRP for reinforcement and strengthening of concrete structures'. *Progress in Structural Engineering and Materials*, Vol. 5, pp 16–28.

RIZZO, A. (2005). *Application in Off-system Bridges of Mechanically Fastened FRP (MF-FRP) Pre-cured Laminates*, M.Sc. Thesis, University of Missouri-Rolla, USA.

ROSTASY, F. S. and BUDELMAN, F. H. (1992). 'Strengthening of RC and PC Structures with Bonded FRP Plates', *Proceedings, Advanced Composite Materials in Bridges and Structures*, K. Neale and P. Labossière, editors, Sherbrooke, Canada, pp 253–263.

SAADATMANESH, H. and EHSANI, M. (1991a). 'RC Beams Strengthened with GFRP Plates. I: Exerimental study', *Journal of Structural Engineering*, ASCE, Vol. 117, No. 11, pp 3417–3433.

SAADATMANESH, H. and EHSANI, M. (1991b). 'RC Beams Strengthened with GFRP Plates. II: Analysis and Parametric Study', *Journal of Structural Engineering*, ASCE, Vol. 117, No. 11, pp 3434–3455.

SAADATMANESH, H., EHSANI, M. and JIN, L. (1997). 'Seismic retrofitting of rectangular bridge columns with composite straps', *Earthquake Spectra*, Vol. 13, No. 2, pp 281–304.

SCHWARTZ, M. M. (1997). *Composite materials – Properties, nondestructive testing and repair*. Vol. 1, Prentice-Hall, Upper Saddle River, NJ, USA.

SEIM, W., HÖRMAN, M., KARBHARI, V., SEIBLE, F. (2001). 'External FRP poststrengthening of scaled concrete slabs', *Journal of Composites for Construction*, ASCE, 5(2).

SHAHAWY, M. A., AROCKIASAMY, M., BEITELMAN, T. and SOWRIRAJAN, R. (1996). 'Reinforced concrete rectangular beams strengthened with CFRP laminates', *Composites: Part B*, Vol. 27B, pp 225–233.

SHANDONG QINGYUN MAOYUAN FRP PRODUCTS CO. LTD, 25 Yingbin Road, Qingyun County, Shandong Province, China 253700.

SHARIF, A., AL-SULAIMANI, G. J., BAUNBUL, I. A., BALUCH, M. H. and GHALEB, B. (1994). 'Strengthening of initially loaded reinforced concrete beams using FRP plates', *ACI Structural Journal*, Vol. 91, No. 2, pp 160–168.

SHEARD, P., CLARKE, J. L., DILL, M., HAMMERSLEY, G. and RICHARDSON, D. (1997). 'EUROCRETE – Taking account of durability for design of FRP reinforced concrete structures'. *Non-metallic (FRP) Reinforcement for Concrete Structures, Proceedings of 3rd International Symposium*, Sapporo, Vol. 2, pp 75–82.

SIKA (1997). *Sika Carbodur: Engineering Guidelines for the Use of Sika Carbodur (CFRP) Laminates for Structural Strengthening of Concrete Structures*, Sika Corporation, Lyndhurst, NJ, USA. Current edition at www.Sikaconstruction.com/

SIKA (2002). Sika Stress Head prestressed Carbodur CFRP plate system, System Information Sika, Zürick, Switzerland.

SMITH, S. T. and TENG, J. G. (2001). 'Interfacial Stresses in Plated Beams', *Engineering Structures*, Vol. 23, No. 7, July, pp 857–871.

SMITH, S. T., and TENG, J. G. (2002a), 'FRP-strengthened RC beams. I: Review of debonding strength models', *Engineering Structures*, Vol. 24, No. 4, pp 385–395.

SMITH, S. T., and TENG, J. G. (2002b), 'FRP-strengthened RC beams. II: Assessment of debonding strength models', *Engineering Structures*, Vol. 24, No. 4, pp 397–417.

SORATHIA, U. and BECK, C. (1996). 'Fire protection of glass/vinylester composites for structural applications'. *Proceedings of the 41st International SAMPE Symposium*, Anaheim, CA, March 1996. pp 687–697.

SORATHIA, U., GRACIK, T., NESS, J., DURKIN, A., WILLIAMS, F., HUNSTAD, M. and BERRY, F. (2003). 'Evaluation of intumescent coatings for shipboard fire protection', *Journal of Fire Science*, Vol. 21, pp 423–450.

S&P (1998). Clever Reinforcement Company: Schere & Partners, Brunnen, Switzerland. Current edition at www.sp-reinforcement.ch

S&P (2006). 'S&P FRP fibre reinforcement polymer'. http://www.reinforcement.ch/fibre.htm. August 2006.

STARR, T. (ed) (2000). *Pultrusion for Engineers*, Woodhead Publishing, Cambridge, UK.

STRONGWELL INC., 17, Oak Hill Drive, Basking Ridge, NF 07920, USA.

TENG, J. G. and LAM, L. (2002). 'Compressive behaviour of Carbon Fiber Reinforced Polymer-confined Concrete in Elliptical Columns'. *Journal of Structural Engineering*, Vol. 128, No. 12, pp 1535–1543.

TENG, J. G., CHEN, J. F., SMITH, S. T. and LAM, L. (2002). *FRP Strengthened RC Structures*, John Wiley and Sons Ltd, England.

TENG, J. G., SMITH, S. T. and CHEN, J. F. (2008). 'Flexural strengthening of RC beams' Chapter 4 in *Strengthening and Rehabilitation of Civil Infrastructures using FRP Composites*, editors L. C. Hollaway and J. G. Teng, Woodhead Publishing Ltd., Cambridge, England.

TONEN (1996). *Forca Towsheet Technical Manual*, Rev. 5.0, Tonen Corporation, Tokyo.

TRIANTAFILLOU, T. C. and PLEVRIS, N. (1992). 'Strengthening of RC beams with epoxy-bonded fibre-composite materials', *Materials and Structures*, Vol. 25, pp 201–211.

TRIANTAFILLOU, T. C., DESKOVIC, N., and DEURING, M. (1992). 'Strengthening of Concrete Structures with Prestressed Fiber Reinforced Plastic Sheets,' *ACI Structural Journal*, Vol. 89, No. 3, pp 235–244.

VAN GEMERT, D., AHMED, O. and BROSENS, K. (1999). 'Anchoring of externally bonded CFRP Reinforcement'. *Proceedings of the International Congress on Creating with Concrete*, Dundee, Scotland, edited by R. K. Dhir and M. J. McCarthy, Thomas Telford, London, UK, pp 81–92.

WANG, D. and WILKIE, C. A. (2006). 'Fire properties of polymer nanocomposites'. In *Fire Properties of Polymer Composite Materials*, edited by A. P. Mouritz and A. G. Gibson, Springer, AA Dordrecht, The Netherlands.

WARD, I. M. (1971). *Mechanical Properties of Solid Polymers*, Wiley, New York.

WATT, W., PHILLIPS, L. N. and JOHNSON, W. (1966). 'High strength high modulus carbon fibres', *The Engineer*, Vol. 221, pp. 815.

WIGHT, R. and ERKI, M.-A. (2003). 'Prestressed CFRP sheets for strengthening two-way slabs', *Conference Composites in Construction*, Cosenza, Italy, pp 433–438.

WIGHT, R., GREEN, M. and ERKI, M.-A. (2001). 'Prestressed FRP sheets for poststrengthening reinforced concrete beams' *Journal of Composites for Construction*, Vol. 5, No. 4, pp 214–220.

YU, P., SILVA, P. F. and NANNI, A. (2008). 'Description of a Mechanical Device for Prestressing of Carbon Fiber-reinforced Polymer Sheets – Part I', *ACI Structural Journal*, Jan/Feb 2008.

1.15 Appendix: Published design guides, codes and specifications for fibre reinforced polymer (FRP) composites in structural engineering

Currently, there is a much greater awareness of the importance of design in relation to product performance and reliability. At the end of the 20th century there were only a limited number of standard specifications and codes for practice in civil and structural engineering applications of composites, and therefore the vitally important subjects of reliability, performance and durability were difficult to address, unless through specific proven systems. There are now a number of design guidelines and design aids, particularly for the design of upgrading RC and prestressed concrete (PC) structural members, which have been written by research centres and research bodies; consequently the situation is becoming easier for the designer.

1.15.1 FRP reinforcement bars and tendons

ACI (1999), *Building Code Requirements for Structural Concrete* (ACI 318R-99), American Concrete Institute; Farmington Hills, MI, USA, pp391. (This code is recommended for RC and PC members using NSM FRP reinforcement. It is consistent with the design guides issued by ACI on the use of FRP for new construction and repair.)

ACI (2003), (440.IR-03), *Guide for the design and construction of structural concrete reinforced with FRP bars* American Concrete Institute, Farmington Hills, MI, USA.

ACI (2004), *Prestressing concrete structures with FRP Tendons*, (440.4R.04), American Concrete Institute, Farmington Hills, MI, USA.

ACI (2006), *Guide for the Design and Construction of Structural Concrete Reinforced with FRP Bars* (ACI 440. IR-06), American Concrete Institute, Farmington Hills, MI, USA.

ACI (2008), *Guide for the design and construction of externally bonded FRP systems for strengthening concrete structures* (ACI 440.2R-08), American Concrete Institute, Farmington Hills, MI, USA.

BRI (1995), *Guidelines for Structural Design of FRP Reinforced Concrete Building Structures*, Building Research Institute, Tsukuba, Japan. (See also 'Design Guidelines of FRP reinforced concrete building structures', *Journal of Composites for Construction*, Vol. 1, No. 3, pp 90–115, 1997.)

CSA (2000), *Canadian Highway Bridge Design Code*, CSA-06-00. Canadian Standards Association, Toronto, Ontario, Canada.

CSA (2002), *Design and Construction of Building Components with Fiber-reinforced Polymers*, CSA-S806-02, Canadian Standards Association, Toronto, Ontario, Canada.

ISCE (1997), *Recommendations for Design and Construction of Concrete Structures using Continuous Fiber Reinforced Materials*, Concrete Engineering Series 23, Japan Society of Civil Engineers, Tokyo, Japan.

1.15.2 FRP guidance documents for the strengthening of reinforced concrete structures

ACI 25 (1997), *Acceptance Criteria for Concrete and Reinforced and Unreinforced Masonry Strengthening using Fiber-reinforced Polymer (FRP) Composite Systems*, ICC Evaluation Service, Whittier, CA, USA.

ACI (2004), *Guide Test Methods for Fibre-reinforced Polymers (FRPs) for Reinforcing or Strengthening Concrete Structures*, ACI 440.3R-04, American Concrete Institute, Farmington Hills, MI, USA.

ACI (2004), *Prestressing Concrete Structures with FRP Tendons*, ACI 440.4R-04, American Concrete Institute, Farmington Hills, MI, USA.
ACI (2008), *Guide for the Design and Construction of Externally Bonded FRP Systems for Strengthening Concrete Structures*, ACI 440.2R-08, American Concrete Institute, Farmington Hills, MI, USA.
ACI (2008), *Seismic Strengthening of Concrete Buildings using FRP Composites*, ACI SP-258CD, American Concrete Institute, Farmington Hills, MI, USA (CD-ROM).
ASCIONE, L. and FEO, L. (2007), 'Italian Design Guidelines for the Structural Strengthening with FRP Materials', *Proceedings of the 13th International Conference on Experimantal Mechanics*, in *Experimental Analysis of Nano and Engineering Materials and Structures*, edited by E. E. Gdoutos, Alexandroupolis, Greece, July 1st – 6th 2007, pp 691–692.
FIB (2001), *Externally Bonded FRP Reinforcement for RC structures*, Task Group 9.3, Bulletin 14, International Federation for Structural Concrete, Lausanne, Switzerland.
ISCE (2001), *Recommendation for Upgrading of Concrete Structures with Use of Continuous Fiber Sheets*, Concrete Engineering Series 41, Japan Society of Civil Engineers, Tokyo.
JPDPA (1999), 'Seismic Retrofitting Guideline', *The Japanese Design and Construction Guidelines for Seismic retrofit*, The Japanese guidelines for seismic retrofitting of RC building materials.
TR 55 2nd Ed. (2000), *Design Guidance for Strengthening Concrete Structures Using Fibre Composite Materials*, The Concrete Society, Camberley, Surrey, England.
TR 57 (2003), *Strengthening Concrete Structures with Fibre Composite Materials: Acceptance, Inspection and Monitoring*, The Concrete Society, Camberley, Surrey, England.
TRB (2009), *Guide Specification for the Design of Externally Bonded FRP Systems for Repair and Strengthening of Concrete Bridge Elements*, NCHRP 10-73, Transportation Research Board. 500 Fifth St. NW, Washington, D.C. 20001.

1.15.3 Important design guides from relevant areas of the world

In recent years a significant number of design codes and specifications have been published by technical organisations in various parts of the world which provide guidance for design with FRP materials for civil engineering. The key publications are listed below (some duplication with Sections 1.15.1 and 1.15.2 is inevitable).

Britain and Europe

Structural Design of Polymer Composites, Eurocomp Design Code and handbook. Edited by John L. Clarke. (1996).
FIB Bulletin 14, Design and use of Externally Bonded FRP Reinforcement for RC Structures, Federation Internationale du Beton, (2001), Lausanne, Switzerland.

Design guidance for strengthening concrete structures using fibre composite materials, TR55, 2nd ed., Concrete Society, Camberley, UK (2000).

Strengthening Concrete Structures using Fibre Composite Materials: Acceptance, Inspection and Monitoring, TR57, Concrete Society, Camberley, UK (2003).

USA

ACI (2002), *Report on Fibre Plastic Reinforcement for Concrete Structures* 440.R-96 (Re-approved 2002).

ACI (2004), *Guide Test Methods for Fibre-Reinforced Polymers (FRP) for Reinforcing or Strengthening Concrete Structures*, 440.3R-04, American Concrete Institute, Farmington Hills, MI, USA.

ACI (2002), *Guide for the Design and Construction of Externally Bonded FRP Systems for Strengthening Concrete Structures*, 440.2R-02 American Concrete Institute, Farmington Hills, MI, USA.

Canada

AC 125 (1997), *Acceptance Criteria for Concrete and Reinforced and Unreinforced Masonry Strengthening Using Fibre Reinforced Polymer Composite Systems*, ICC Evaluation Service, Whittier, CA, USA.

CSA (2000), *Canadian Highway Bridge Design Code*, CSA-06-00, Canadian Standards Association, Toronto, Ontario, Canada.

ISIS CANADA, Design Manual No. 3, *Reinforcing Concrete Structures with Fiber Reinforced Polymers*, Canadian Network of Centers of Excellence on Intelligent Sensing for Innovative Structures, ISIS Canada Corporation, Winnipeg, Manitoba, Canada (Spring 2001).

Japan

JAPAN SOCIETY OF CIVIL ENGINEERS (JSCE) (1997), *Recommendation for Design and Construction of Concrete Structures Using Continuous Fiber Reinforced Materials*, Concrete Engineering Series 23, ed. by A. Machida, Research Committee on Continuous Fiber Reinforcing Materials, Tokyo, Japan.

JSCE (2001), *Recommendations for Upgrading of Concrete Structures with Use of Continuous Fibre Sheets.* Concrete Engineering Series 41, Japan Society of Civil Engineers, Tokyo.

2 Using fibre reinforced polymer (FRP) composites to extend the service life of corroded concrete structures

K. SOUDKI, University of Waterloo, Canada

Abstract: This chapter discusses the use of fibre reinforced polymers (FRP) to repair deterioration of concrete structures caused by corrosion of their steel reinforcement. The chapter reviews the effects of corrosion on the serviceability and strength of a corroded structure. It then discusses the effects of FRP repair on the serviceability and the strength of a corroded structure. Models for assessing the service life of a corroded structure are given. The chapter concludes with field applications for FRP repair of corroded structures.

Key words: corrosion, bond, flexure, fatigue, repair and rehabilitation, FRP reinforcement.

2.1 Introduction

Many concrete structures in adverse environments experience an unacceptable loss in serviceability or safety far earlier than anticipated due to the corrosion of their reinforcing steel, and thus need replacement, rehabilitation, or strengthening. Tuutti (1980) proposed a service life model with two distinct periods of deterioration caused by corrosion (Fig. 2.1). The first one is the initiation period, which represents the time required for CO_2 or Cl-ions to diffuse to the steel-to-concrete interface and activate corrosion. The second is the propagation period, which represents the time between corrosion initiation and corrosion cracking. If corrosion cracking can be prevented or delayed, structural strength will be maintained for a longer time in a corroding reinforced concrete member.

The use of fibre reinforced polymer (FRP) composites in civil engineering has recently emerged as an alternative to the traditional methods used for rehabilitation or reinforcement of structures. FRPs are of interest to rehabilitation engineers because of their high strength-to-weight ratio, high fatigue resistance, ease of installation and the fact that they do not corrode (ACI Committee, 2007). There are structural benefits obtained when externally bonded FRP laminates are used to repair corrosion damaged RC elements (Soudki and Sherwood, 2000; Bonacci and Maalej, 2001; Masoud

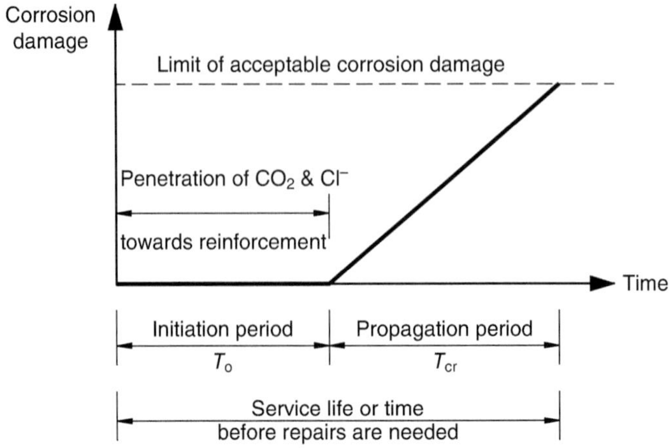

2.1 Service life model of corroded structures (Tuutti, 1980).

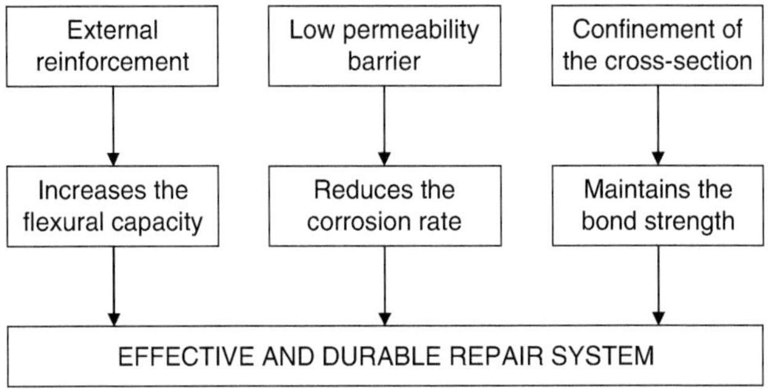

2.2 Benefits of using FRP in repairing corroded structures.

et al., 2001; Masoud and Soudki, 2006, El Maaddawy *et al.*, 2007). In addition, wrapping a corroded element with FRP laminates may reduce the corrosion activity because a FRP wrap acts as a low-permeability barrier to the further ingress of water and oxygen into concrete, which are required for corrosion reactions to continue (El Maaddawy *et al.*, 2006; Debaiky *et al.*, 2002; Soudki and Sherwood, 2000). The physical confinement may also impede the dispersion of corrosion products and thus stifle the corrosion reaction itself. Wrapping a concrete section with FRP laminates provides external confinement that resists the internal displacement caused by the expansion of the corrosion products and thus decreases corrosion and bond splitting cracks. Hence, the structural strength of corroding reinforced concrete beams is improved if they are repaired with FRP laminates. Figure 2.2 illustrates the benefits of using FRP repair for corroded RC elements.

2.2 Corrosion in reinforced concrete

The alkaline environment of the concrete provides a natural protection for the steel reinforcement against corrosion through the formation of a passive film of iron oxides. Passivation of steel can be destroyed by carbonation or by chloride attack. Once the passive film breaks down, corrosion will start, in the presence of moisture and oxygen (ACI Committee, 2002). Corrosion of steel is described as an electrochemical process involving (Fig. 2.3): (i) an anode where iron, Fe^{2+} is removed from the steel, (ii) a cathode where hydroxyl ions OH^- are produced, (iii) an electrical conductor for electrons transfer, and (iv) an electrolyte for ion transfer. The released hydroxyl ions at the cathode travel through the electrolyte to react with the ions at the anode, producing rust (Fig. 2.4a). The corrosion rate is controlled by many factors, including availability of dissolved oxygen and the concrete resistivity around the steel. Rust formation may occupy up to ten times the volume of the original steel (Broomfield, 1997). The increase in volume causes tensile stresses in the concrete that lead to longitudinal cracking and spalling of the concrete (Fig. 2.4b). For example, corrosion crack width as a function of degree of corrosion can be seen Fig. 2.5. Cracks facilitate ingress of oxygen, moisture and chlorides, and increase the rate of bar corrosion by ten times or more (Bentur *et al.*, 1997). In addition to cracking, the loss in bar section, in combination with bond loss between the steel and concrete, leads to a reduction of the capacity and an increase in deflection of the structural member.

2.2.1 Effect of corrosion on bond strength

Experimental studies reported in the literature on the effect of corrosion on the bond behaviour of reinforced concrete beams showed that, at a low

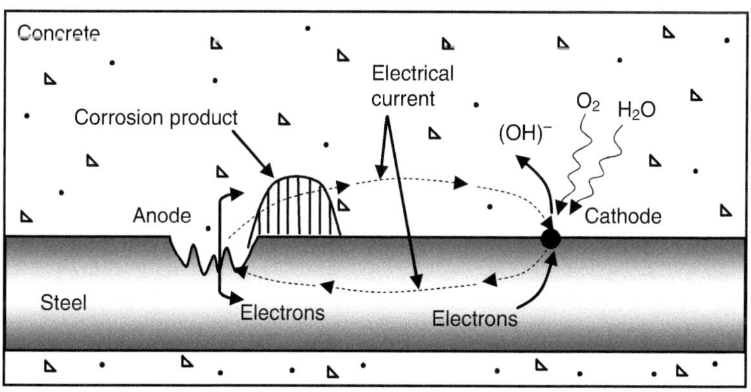

2.3 Corrosion process.

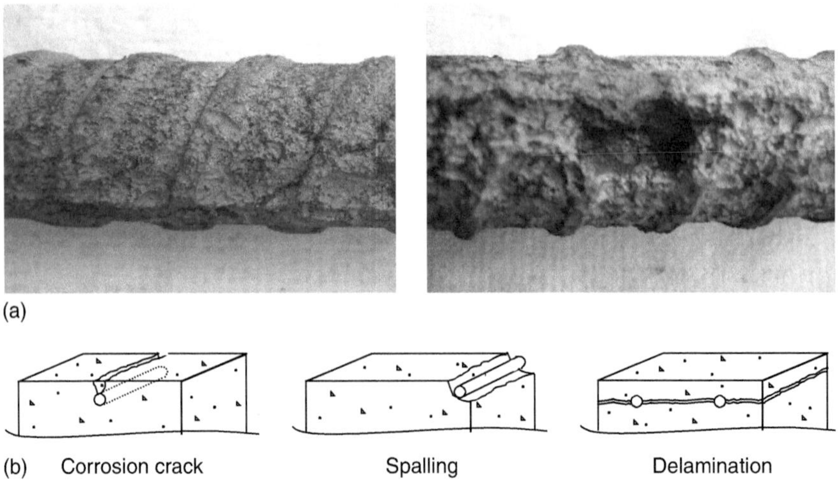

2.4 (a) Corrosion deterioration: section loss of the steel reinforcement, (b) Corrosion deterioration: concrete cracking (Bentur *et al.*, 1997).

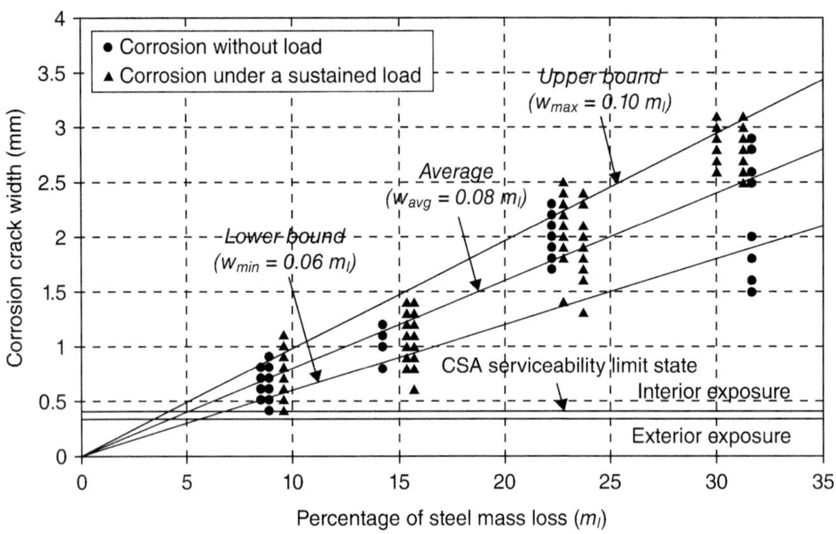

2.5 Corrosion crack width versus steel mass loss relationship (El Maaddawy, 2004).

corrosion rate (about 0.04 mA/cm^2), the bond strength initially increased, as corrosion increased, as shown in Fig. 2.6 (FIB Bulletin, 2000). This increase was attributed to the fact that the rust produced before cracking would cause a rough surface around the bar, hence increasing the friction force on the interface between the reinforcing steel and the concrete. Once the concrete was cracked due to corrosion, the bond stresses between the

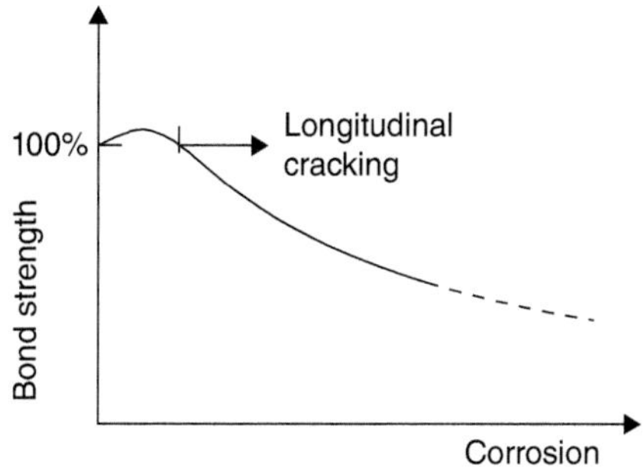

2.6 Variation of bond strength with corrosion (FIB Bulletin, 2000).

reinforcing steel and the concrete significantly decreased and the slip of the reinforcing bar relative to the concrete increased. A decrease in the bond leads to a decrease in the load carrying capacity and an increase in the deflection of the structure (ACI Committee, 2003).

2.2.2 Effect of corrosion on flexural strength

Research on the effect of corrosion on the flexural strength of concrete beams showed that corrosion leads to a decrease in the capacity of the member. The reduction in strength was found to be proportional to the reduction in the steel cross-sectional area, due to corrosion. When anchorage zones were corroded, there was bond splitting, with an increase in the deflection (Uomoto *et al.*, 1984; Lee *et al.*, 1997; Mangat and Elgarf, 1999; El Maaddawy *et al.*, 2005). A schematic of the corrosion deterioration for flexural strength is shown in Fig. 2.7.

2.2.3 Effect of corrosion on fatigue strength

Masoud *et al.* (2005) investigated the fatigue of corroded reinforcement in concrete and found that it has a significant effect on shortening the service life. Al-Hammoud *et al.* (2010) investigated the effect of corrosion on the flexural strength. They reported that the decrease in flexural fatigue was not proportional to the corrosion level; for example, a 12% mass loss reduced the fatigue strength on average by 25%, compared with that of uncorroded specimens. It was concluded that the reduction in fatigue life is proportional to the increase in the fatigue notch factor. A strain life approach using the fatigue notch factor was implemented to predict the fatigue life of the corroded beams. Rteil (2007) studied the effect of corrosion on the bond behaviour of reinforced concrete beams under repeated loading. He reported that fatigue had a more pronounced effect on the

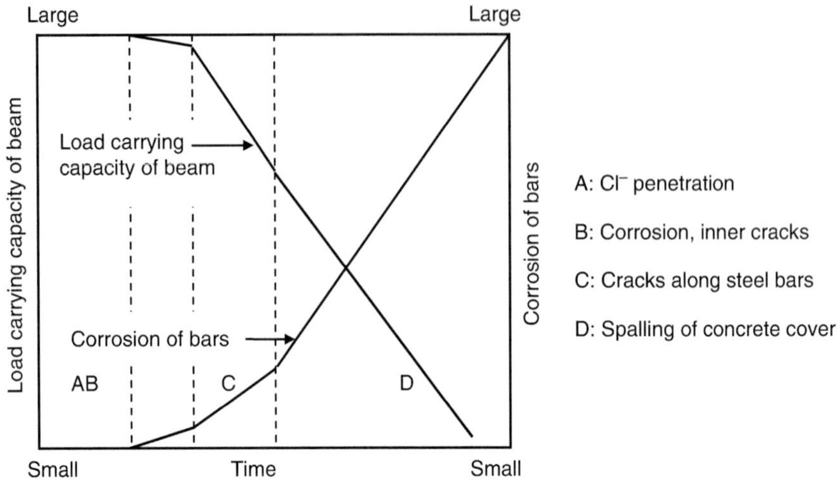

2.7 Variation of member capacity with corrosion (Uomoto *et al.*, 1984).

bond of corroded reinforcement. For example, corrosion levels of 5% and 9% decreased the fatigue bond strength on average by 19%.

2.3 Fibre reinforced polymer (FRP) repair for corrosion damage

2.3.1 Application of FRP repair

It is recommended to follow proper concrete repair procedures as per current practice before and during any FRP repair. The application of the FRP repair should be done in accordance with the procedures provided by the FRP manufacturer and repair guides. Figure 2.8 shows examples of FRP repairs on beam and column specimens in the laboratory. Prior to FRP repair, the reader should consult the relevant documents such as those published by the American Concrete Institute.

2.3.2 Effect of FRP wrapping on corrosion activity

The corrosion rate is a key element and is usually used to predict the functional service life of a corroded RC structure (Tuutti, 1980). After corrosion initiation, the corrosion rate depends mainly on the availability of oxygen and moisture at the cathode, and on the concrete resistivity (Bentur *et al.*, 1997). Fibre reinforced polymer (FRP) wraps might be beneficial in controlling corrosion by reducing the diffusion rate of oxygen and moisture into the concrete (Bonacci and Maalej, 2001; Debaiky *et al.*,

2.8 FRP repair application for corroded members.

2002; El Maaddawy *et al.*, 2006). The mechanical restraint due to the confinement effect of FRP wraps causes accumulation and densification of the corrosion products at the steel–concrete interface, which may stifle the corrosion reaction and thus retard the corrosion process (Lee *et al.*, 2000). Debaiky *et al.* (2002) reported that for concrete wrapped after 100 or 200 days of exposure, the CFRP wraps did succeed in bringing the corrosion activity from the high corrosion range to below 0.1 μA/cm^2, in the same range for concrete wrapped before exposure (Fig. 2.9). In the above study, the number of FRP layers did not affect the efficiency of the repair in reducing the corrosion rate; this indicates that the reduction in corrosion activity is due to the epoxy resin matrix providing a barrier to oxygen, not to the fibers. However, in other studies, FRP wraps were found to maintain the moisture inside the concrete, which reduces concrete resistivity and thus may increase corrosion activity (Pantazopoulou and Papoulia, 2001). Accordingly, some researchers suggest not to repair corrosion-damaged members with FRP sheets, based on the premise that

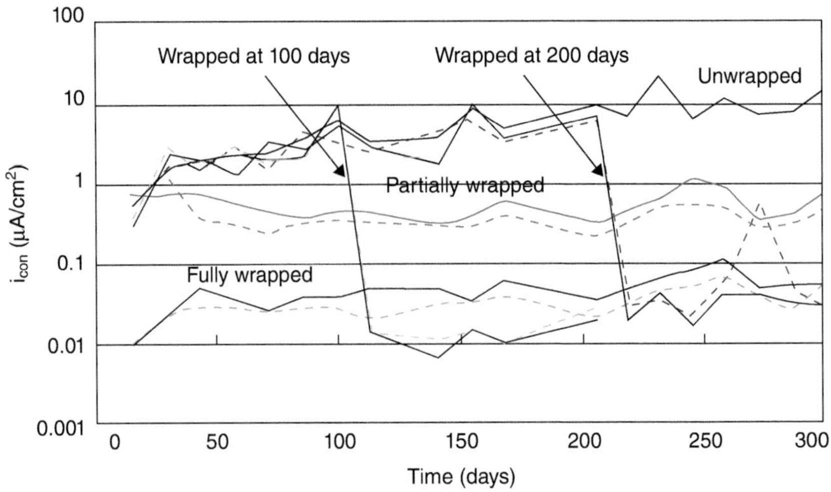

2.9 FRP repair effects on corrosion current density (Debaiky *et al.*, 2002).

FRP repair would only address the symptoms of deterioration, not the cause of the damage.

2.3.3 Effect of FRP repair on corrosion cracking

Masoud and Soudki (2006) showed that FRP repair reduced the crack opening by about 88% at the end of the corrosion process (Fig. 2.10). This implies a significant enhancement in appearance of FRP repaired and corroded beams by reducing crack opening caused by further corrosion. The rate of expansion of unconfined concrete caused by corrosion was almost twice that of FRP confined concrete. EL Maaddawy and Soudki (2005) reached similar conclusions for FRP repair with corroded beams under sustained load.

2.3.4 Effect of FRP repair on steel mass loss

Figure 2.11 shows the steel mass loss versus time relationship for FRP repaired and corroded beams. It is evident that the steel mass loss rate in the beams repaired with continuous FRP wrapping was, on average, about 32% lower than the level for the beams with intermittent FRP wrap. Beams corroded under a sustained load had connected internal microcracks and external flexural cracks, which increased the penetration of oxygen and moisture into the concrete and reduced the concrete resistivity, thus slightly increasing the steel mass loss rate to a level higher than that for the beams corroded without load.

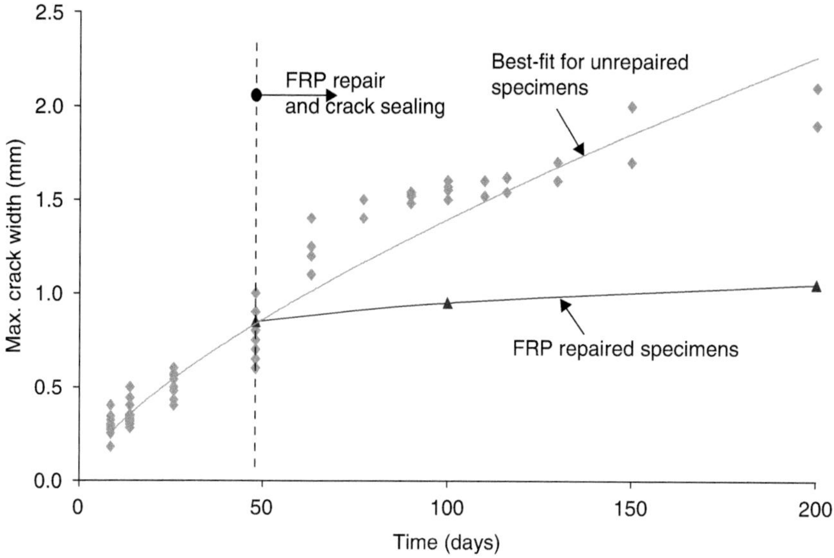

2.10 Corrosion crack width.

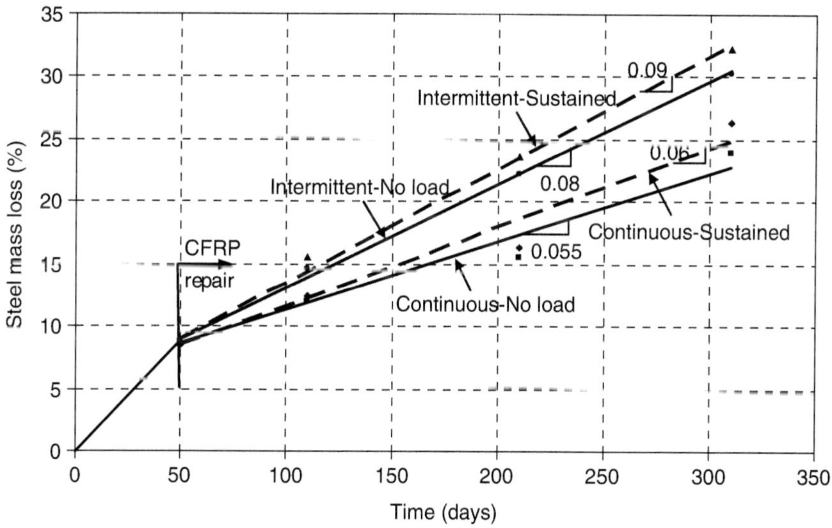

2.11 Steel mass loss versus time relationship.

2.3.5 Effect of FRP confinement on the bond

Wrapping a concrete section with FRP laminates provides external confinement that resists the internal displacement caused by the expansion of the corrosion products and thus decreases corrosion and bond splitting cracks (Soudki and Sherwood, 2003; Craig and Soudki (2005). The effect of FRP wrapping on the bond of corroded reinforcement is shown in Fig. 2.12. FRP reinforcement resists the expansion forces caused by corrosion, thus reducing crack growth and maintaining the bond interaction between the reinforcing steel and the concrete. As post-repair corrosion progressed, cracks were unable to expand due to the presence of FRP sheets. In turn, the FRP developed stresses, which increased the internal confining pressure around the reinforcing bar that counteracted the expansion stresses due to corrosion. It is important to understand the nature of failure of the FRP-confined concrete. Since no cracks are visible with the FRP wrap in place, there are no indications of failures. Even under conditions of high ultimate bond stresses, the presence of low slip initiation bond stresses indicates that failure could potentially occur prematurely by bond pullout in the case of sustained loading or creep. Therefore, caution must be used in the application of this repair method since abrupt failure of the member due to bond pullout failure could occur without warning if the repair is performed at high corrosion levels or if members were initially designed with an inadequate bond. The confining wrap may increase the bond strength but, as with all repairs, the cause of deterioration must be addressed to prevent further corrosion and deterioration.

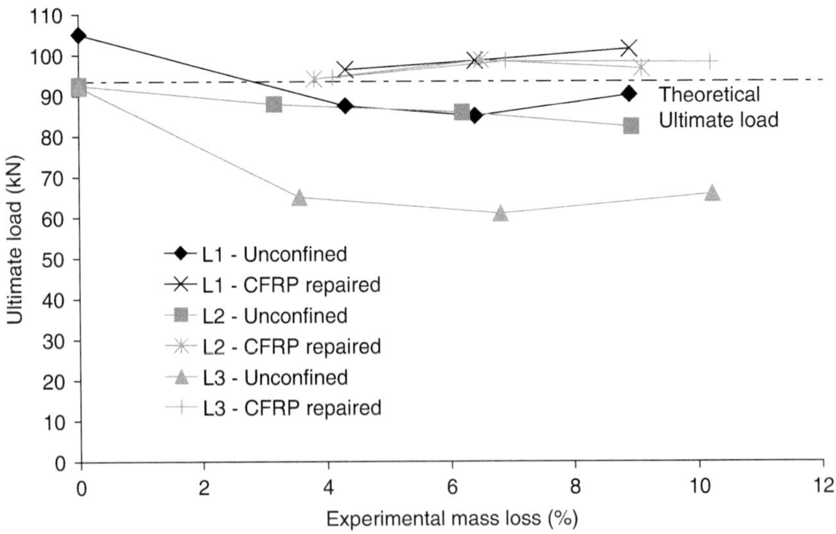

2.12 FRP repair effects on bond strength with corrosion.

2.3.6 Effect of FRP repair on static response

The effect of FRP repair on improving the flexural response of corroded concrete beams is well documented (Lee *et al.*, 1997; Bonacci and Maalej, 2001; Soudki and Sherwood, 2000; Masoud, 2002; El Maaddawy *et al.*, 2005). In general, the reduction in strength for the FRP repaired beams was proportional to the corrosion mass loss, as in the case of un-repaired beams (Section 2.2.2). The presence of sustained loads up to 60% of the yield strength during corrosion exposure had no significant effect on the behaviour of the corroded beams (see Fig. 2.13). When FRP wrapping was applied

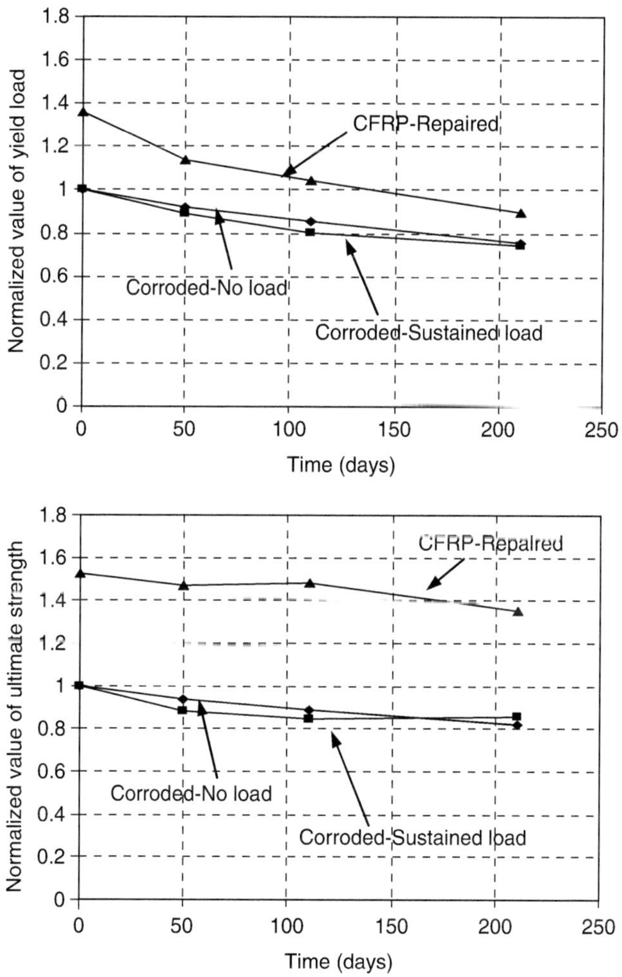

2.13 FRP repair effects on static strength.

only in the transverse direction, little or no flexural strength enhancement of the corroded beams was observed. FRP reinforcement applied in the longitudinal direction in conjunction with transverse wrapping resulted in significant increase in the load carrying capacity. When a beam was corroded before the application of the FRP laminates, it did not reach the same strength as a strengthened uncorroded beam. However, the ultimate strengths of the corroded-repaired beams were very close to that of the uncorroded-strengthened beam. The reduction of the yield load was due to the reduction of cross-sectional area of the rebar by corrosion. After yielding of the steel rebar, it is believed that the beam behaves as an arch with a tie (the longitudinal FRP).

2.3.7 Effect of FRP repair on flexural fatigue

Masoud *et al.* (2001, 2005) tested corroded beams repaired with FRP under cyclic fatigue flexural loading, and concluded that corrosion of the steel reinforcement causes a decrease of the fatigue life of a beam. Reinforcing the beam with FRP caused a reduction in the tensile stress in the steel reinforcement, which led to an increase in the fatigue life of the beam. Figure 2.14 shows the normalized fatigue life versus the average mass loss of the main reinforcing bars.

Al-Hammoud *et al.* (2010) showed that pitting of the steel reinforcement due to corrosion occurred only after about a 7% actual mass loss, which coincided with a decrease in the fatigue performance of the beam. The

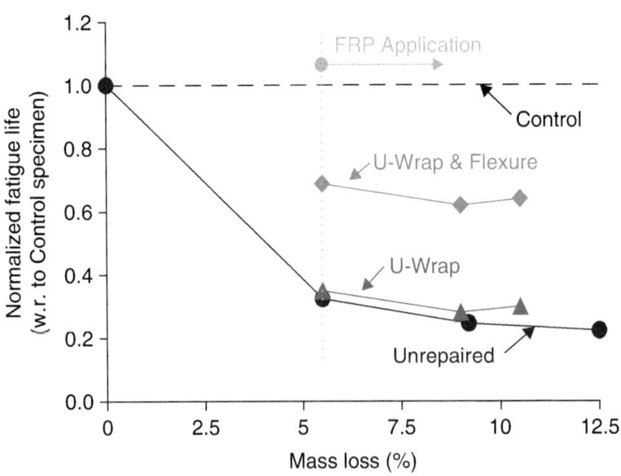

2.14 FRP repair effects on flexural fatigue life.

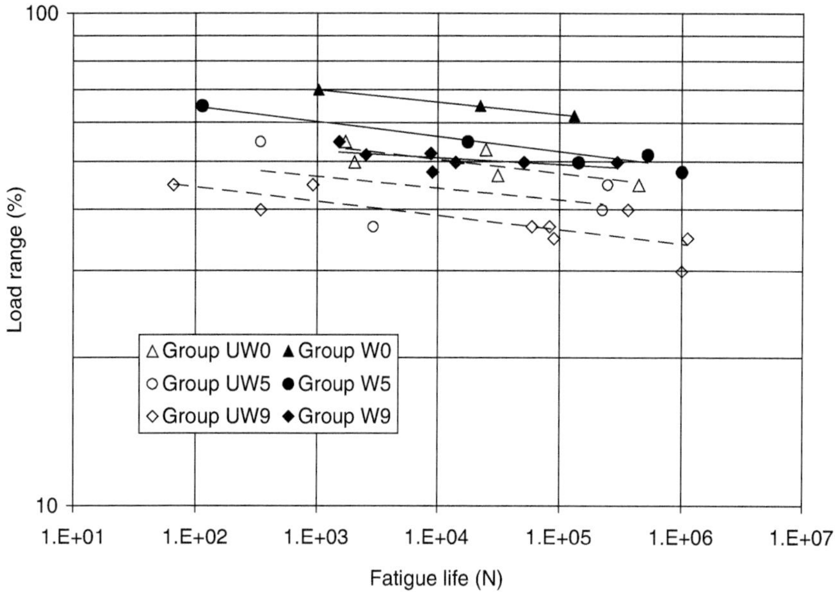

2.15 FRP repair effects on fatigue life.

controlling factor for the fatigue strength of the beams is the fatigue strength of the steel bars. FRP repair increased the fatigue capacity of the corroded beams compared to the control un-corroded, un-repaired beams. The FRP-repaired beams at a medium corrosion that were further corroded to a high corrosion level were no better in terms of fatigue performance than those repaired after a high corrosion level.

2.3.8 Effect of FRP repair on bond fatigue

Rteil (2007) reported that wrapping corroded beams with FRP sheets decreased the width and number of longitudinal cracks compared to the unwrapped beams. Corrosion caused an average reduction of 19% in the fatigue strength of the wrapped beams compared to the non-corroded wrapped beams (Fig. 2.15). However, comparing the corroded wrapped beams to the uncorroded, un-wrapped beams, the fatigue strength increased, on average, by 30%. The increase in the fatigue strength of the wrapped, un-corroded beams compared to the unwrapped, un-corroded beams was 32%. Initially, up to 8% of the wrapped beams' life, the slip increased at a decreasing rate. After that, the slip continued to increase at a slow rate until the last 10% of the beams' life when the slip rate increased exponentially.

2.4 Service life extension with fibre reinforced polymer (FRP) repair

2.4.1 Time to corrosion cracking

Prediction of time to corrosion cracking is a key element in evaluating the service life of corroding reinforced concrete (RC) structures where rehabilitation is required. El Maaddawy (2004) developed an equation to predict the time from corrosion initiation to corrosion cracking as follows:

$$T_{cr} = \left[\frac{7117.5(D+2\delta_0)(1+v+\psi)}{iE_{ef}}\right]\left[\frac{2Cf_{ct}}{D}+\frac{2\delta_0 E_{ef}}{(1+v+\psi)(D+2\delta_0)}\right] \qquad [2.1]$$

T_{cr} is the time from corrosion initiation to corrosion cracking,
C is the clear concrete cover (mm),
D is the diameter of steel reinforcing bar (mm),
E_{ef} is the effective elastic modulus of concrete = $[E_c/(1 + / \psi_{cr})]$,
E_c is the elastic modulus of concrete,
ψ_{cr} is the concrete creep coefficient,
v is the Poisson's ratio of concrete (0.18), and
δ_0 is the thickness of the porous zone, typically in the range of 10–20 μm.

2.4.2 FRP expansion due to corrosion

The lateral strain in the FRP system as a function of the corrosion mass loss could be determined using a semi-empirical equation as follows (El Maaddawy, 2004):

$$\varepsilon_f = \frac{0.04(m_{inc})^{0.4\left(\frac{w_f}{s_f}\right)}}{\left(C+\frac{d_b}{2}\right)+\left(\frac{2C}{C+d_b}\right)\left(\frac{t_f E_f}{E_{cf}}\right)}(10)^6 \qquad [2.2]$$

ε_f = the lateral strain in the CFRP laminate ($\mu\varepsilon$).
m_{inc} = the incremental steel mass loss after repair (%).
w_f = the width of the transverse CFRP laminate (mm), and
s_f = the centre-to-centre spacing between the CFRP strips (mm).
C = the concrete clear cover (mm).
d_b = the diameter of the steel reinforcing bar (mm).
t_f = the thickness of the transverse CFRP laminate (mm).
E_f = Young's modulus of the CFRP laminate (MPa), and
E_c = Young's modulus of the concrete (MPa).

2.4.3 Bond reduction due to corrosion

The bond strength of steel reinforcement in FRP-wrapped concrete, including the effects of corrosion and FRP wrapping, is accounted for by modifying the maximum bond strength, μ_{max}, expression in the bond stress-slip model for monotonic loading in the CEB Model Code (1999). Figure 2.16 shows the bond stress-slip model for corroded, and corroded FRP-repaired beams. The maximum bond strength is given as:

$$\mu_{\mathrm{max,f}} = \underbrace{(A_1 + A_2 m_1)}_{R} \underbrace{\left(0.55 + 0.24\frac{c_c}{d_b}\right)\sqrt{f_c'}}_{\mu_{\mathrm{conc}}} + \underbrace{0.191\frac{A_t}{s_s}\frac{f_{yt}}{d_b}}_{\mu_{\mathrm{st}}} + \underbrace{\frac{0.06 A_{tf} f_{ef}}{s_f d_b}\sqrt{f_c'}}_{\mu_f} \qquad [2.3]$$

R = factor for the effect of corrosion on the bond strength.
m_1 = the steel mass loss due to corrosion at time of repair (%).
A_1 and A_2 = constants for current density level taken as 1.0 and –0.030, respectively.
μ_{conc} = the contribution of concrete to the bond strength (MPa).
f_c' = the concrete compressive strength (MPa).
c_c = the smaller of the concrete clear cover and one-half the bar spacing (mm).
d_b = the steel reinforcing bar diameter (mm).
μ_{st} = the contribution of stirrups to the bond strength (MPa).

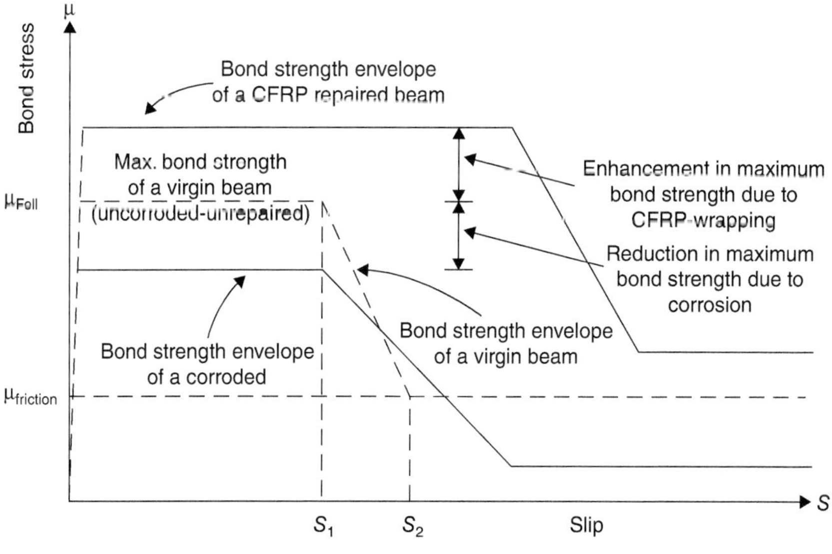

2.16 Proposed bond stress-slip model (El Maaddawy, 2004).

s_s = the centre-to-centre spacing between stirrups (mm).
A_t = the cross sectional area of one stirrup (mm).
f_{yt} = the yield strength of the stirrups (MPa).
μ_f = the contribution of FRP-wrapping to the bond strength at time of repair (MPa).
A_{tf} = the area of the transverse FRP strip (mm^2).
f_{ef} = the effective stress in the FRP laminates (MPa).
s_f = the centre-to-centre spacing between FRP strips (mm).

Rteil (2007) proposed a fatigue-life bond model, using slip-growth analysis based on fracture mechanics concepts, for FRP-repaired and corroded beams experiencing bond failure. The model is as follows:

$$N_f = \frac{s_f^{1-m/2} - s_i^{1-m/2}}{C(u_{max}P)^m(1-m/2)} \qquad [2.4]$$

s_i = slip at first fatigue cycle = $aP_{max} + b$ (a and b are variables depending on corrosion and wrapping conditions, P_{max} is the maximum applied load),
s_f = slip at final fatigue cycle = x (x depends on corrosion and wrapping conditions),
C and m are determined from the experimental load range life data and initial slip,
C varies with corrosion and wrapping conditions,
m = constant,
P is the ratio of the maximum load applied to the static capacity,
u_{max} = maximum static bond stress = $A\sqrt{f_c'}\left(\frac{c+K_c}{d_b}\right)^{2/3} \le u_1$,
A is a constant determined from experimental results,
f_c' is the concrete compressive strength,
c is the minimum clear cover to the steel bar,
K_c is a factor that takes into consideration the effects of FRP and stirrups, and
d_b is the bar diameter.

2.4.4 Strength reduction due to corrosion

The strength reduction (S_R) of a corroded beam or FRP-repaired beam at any level of steel mass loss m_l (%) can be calculated as follows (El Maaddawy *et al.*, 2005):

$$S_R = \left[1 - \frac{A_{sr}\phi_s f_y + A_f\phi_f f_{fr}}{A_{so}\phi_s f_y + A_f\phi_f f_{fr}}\right](100) \qquad [2.5]$$

$$A_{sr} = A_{so}\left(1 - \frac{m_l}{100}\right) \qquad [2.6]$$

where

S_R is the strength reduction due to corrosion

A_{sr} is the residual tensile steel reinforcement area

A_{so} is the tensile steel reinforcement area at time of repair

A_f is the amount of the longitudinal FRP reinforcement

m_l is the post-repair percentage steel mass loss

f_y is the tensile steel yield stress

f_{fr} is the CFRP rupture strength

φ_c, φ_s, and φ_f, are resistance factors of concrete, steel, and CFRP, respectively

The flexural strength of a FRP repaired and corroded beam can be determined using strain compatibility and force equilibrium. The difference from the conventional approach is that the effects of corrosion and FRP repair are explicitly incorporated in the bond stress-slip calculations (Section 2.4.3). The later is used in evaluating the elongation of the steel reinforcement and the corresponding beam deflection. The strain compatibility also relies on the bond-slip relationships. More details on this approach can be found elsewhere (El Maaddawy *et al.*, 2007).

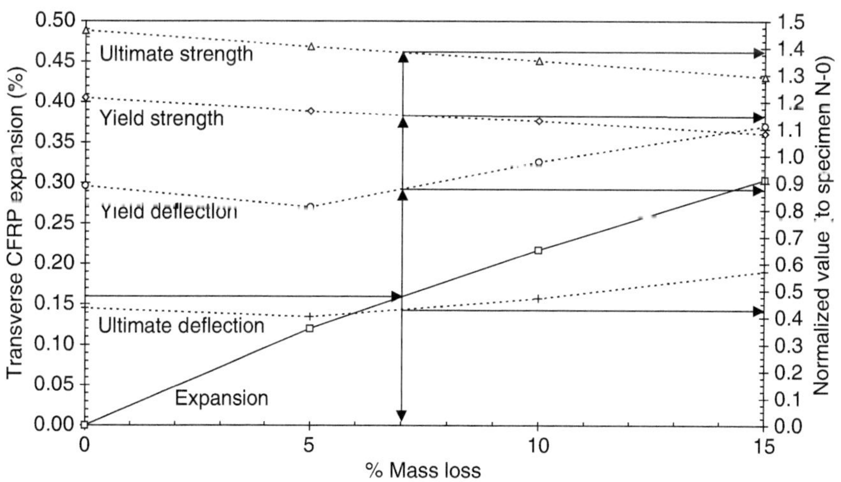

2.17 Expansion–strength/deflection–%mass loss interaction (Sherwood, 2000).

2.4.5 Interaction diagrams

Figure 2.17 shows an interaction diagram that relates the expansion of the transverse FRP laminate (external) to the degree of corrosion of the steel (internal) and the remaining strength and defection of a structural member. The strength/deflection would be normalized relative to an undamaged control beam. By knowing the transverse FRP expansion (in % strain), one can estimate the % mass loss of steel and the normalized strength and deflections of the beam. Figure 2.17 demonstrates that such design diagrams serve as a preliminary tool to estimate flexural strengths and deflections of strengthened beams undergoing active corrosion, given a known transverse FRP expansion (in %strain).

2.5 Field applications

Few field applications have implemented the use of FRP repair for corrosion damaged structures including columns, beams, piers and bridge girders. Three case studies are briefly described. The FRP repair on Highway 10 overpass columns in Quebec was among the first applications in North America and was carried by ISIS Canada-Sherbrooke in 1996. Corrosion deterioration in the 760 mm diameter column resulted from salt spray in the splash zone. The columns were repaired with carbon or glass FRP wraps and monitored using fibre optic sensors. The repair has performed very well since that time.

The Scheifele Bridge, constructed circa 1960, is on Regional Road 22, crossing the Conestogo River north of the City of Waterloo, Ontario, Canada. The bridge has a total length of 106.68 m (350 ft) and consists of four spans; the two middle spans are 30.48 m (100 ft) long and the two end spans are 22.86 m (75 ft) long. The second girder in the south-west corner of the bridge had almost one-third of its length closer to the abutment spalled, with the corroded steel reinforcement visible. CFRP repair, using SIKAWRAP 230C fabric and SIKADUR 330 epoxy, was implemented by the Regional Municipality of Waterloo to address the corrosion deterioration. Sensors were installed to monitor the condition of the girder and the repair system throughout the service life of the girder. The repair was completed in October 2005 (Fig. 2.18). Preliminary data from the monitoring sensors revealed the effectiveness of the repair system (Soudki *et al.*, 2007).

The Gandy Boulevard bridges comprise three side-by-side structures over Tampa Bay, between St. Petersburg and Tampa, FL. The bridges are approximately 2.6 miles (4.2 km) long with 275 spans supported by 254 reinforced concrete pile bents and 22 column-type piers located at the main channel crossing. Inspection showed that 77% of the 254 pile bents had

2.18 Scheifele Bridge – girder before and after repair.

been jacketed or otherwise repaired. The 20 × 20-in. (510 × 510-mm) piles were originally reinforced with eight No. 8 (25 mm) Grade 60 bars. Using an assumed uniform 20% steel loss, the amount of carbon or glass required to restore full capacity was estimated. A water-cured resin system made it possible to install an FRP wrap for the repair of piles, even in open-water conditions. The ease of installation of this system, coupled with the previously observed reduction in corrosion rate made possible by FRP wraps, suggest that this approach has tremendous merit (Mullins *et al.*, 2006).

2.6 Conclusions

Research and field demonstration projects have shown that FRP repair can mitigate corrosion damage in chloride-contaminated concrete beams, columns or piles. Models are proposed to predict the residual life of FRP repaired and corroded RC columns and beams. The field applications using FRP repair for corroded structures are limited. Therefore, until extensive field evidence becomes available that FRP repair could effectively slow corrosion rates of existing steel reinforcement, it is advisable to address the deterioration and the underlying cause of the corrosion prior to FRP wrapping. This is in line with the current recommendation of ACI 440.2R (2008) that if steel corrosion deterioration is evident, FRP reinforcement should not be applied without arresting the ongoing corrosion and repairing the deterioration of the concrete substrate.

2.7 References

ACI COMMITTEE (2002). *Corrosion of Metals in Concrete, Manual of Concrete Practice*, 222R-01 American Concrete Institute, Farmington Hills, MI, USA.

ACI COMMITTEE (2003). *Bond and Development of Straight Reinforcing Bars in Tension* (ACI 408-03). American Concrete Institute, Farmington Hills, MI, USA, 49 pp.

ACI COMMITTEE (2007). *Report on Fiber Reinforced Polymer Reinforcement for Concrete Structures*, 440R.7 American Concrete Institute, Farmington Hills, MI, USA.

ACI 440.2R-08 (2008). *Guide for the Design and Construction of Externally Bonded FRP Systems for Strengthening Concrete Structures*, (440.2R-08). American Concrete Institute, Farmington Hills, MI.

AL-HAMMOUD, R. SOUDKI, K.A., and TOPPER, T. (2010). 'Fatigue Flexural Behaviour of Corroded Reinforced Concrete Beams Repaired with CFRP Sheets,' *ASCE Journal of Composites for Construction*, in press.

BENTUR, S., DIAMOND, S., and BERKE, N.S. (1997). *Steel corrosion in concrete: Fundamental and civil engineering practice*, E & FN Spon, London.

BONACCI, J.F., and MAALEJ, M. (2001). 'Externally Bonded Fibre-reinforced Polymer for Rehabilitation of Corrosion Damaged Concrete Beams,' *ACI Structural Journal*, Vol. 97, No. 5, pp. 703–711.

BROOMFIELD, J.P. (1997). *Corrosion of Steel in Concrete: Understanding, Investigation and Repair*, E & FN Spon, London.

COMITÉ EURO-INTERNATIONAL DU BÉTON. (1999). *Durable Concrete Structures–CEB Design Guide*, Thomas Telford, London.

CRAIG, B., and SOUDKI, K.A., (2005). *Post-repair Performance of Bond Critical Corrosion Damaged Concrete Beams Repaired with CFRP*, ACI SP 230, FRPRCS-7, pp. 563–578.

DEBAIKY, A., GREEN, M., and HOPE, B., (2002). 'Carbon Fiber Reinforced Polymer Wraps for Corrosion Control and Rehabilitation of Reinforced Concrete Columns,' *ACI Materials Journal*, Vol. 99, # 2, March, pp. 129–137.

EL MAADDAWY, T. (2004). '*Performance of Corrosion-damaged Reinforced Concrete Beams Repaired with CFRP Laminates*', PhD Thesis, University of Waterloo, 283 pp.

EL MAADDAWY, T.A., and SOUDKI, K.A. (2005). 'Carbon Fiber Reinforced Polymer Repair to Extend the Service Life of Corroded Reinforced Concrete Beams,' *Journal of Composites for Construction*, ASCE, Vol. 9, No. 2, pp. 187–194.

EL MAADDAWY, T. A., SOUDKI, K. A., and TOPPER, T. (2005). 'Computer-based Mathematical Model for Performance Prediction of Corroded Beams Repaired with Fiber Reinforced Polymers,' *ASCE Journal of Composites for Construction*, Vol. 9, No. 3, pp. 227–235.

EL MAADDAWY, T. CHAHROUR, A., and SOUDKI, K. (2006). 'Effect of Fiber Reinforced Polymer Wraps on Corrosion Activity and Concrete Cracking in Chloride-contaminated Concrete Cylinders,' *ASCE Journal of Composites for Construction*, Vol. 10, No. 2, pp. 139–147.

EL MAADDAWY, T.A., SOUDKI, K.A., and TOPPER, T. (2007). 'Performance Evaluation of Carbon Fiber Reinforced Polymer-Repaired Beams Under Corrosive Environmental Conditions', *ACI Structural Journal*, Vol. 104, No. 1, pp. 3–11.

FIB BULLETIN (2000). *Bond of Reinforcement in Concrete (FIB 10)*. Fédération Internationale du Béton, Lausanne, Switzerland.

LEE, C., BONACCI, J., THOMAS, M., MAALEJ, M., KHAJEHPOUR, S., HEARN, N., PANTAZOPOUOU, S., and SHEIKH, S. (2000). 'Accelerated Corrosion and Repair of Reinforced Concrete Columns using Fibre Reinforced Polymer Sheets,' *Canadian Journal of Civil Engineering*, Vol. 27, No. 5, pp. 941–948.

LEE, H.S., TOMOSAWA, F., MASUDA, Y., and KAGE, T. (1997). 'Effect of CFRP sheets on flexural strengthening of RC beams damaged by corrosion of tension rebars.' *3rd*

Int. Symp. on Non-metallic (FRP) Reinforcement for Concrete Structures, Vol. 1, 435–442. Sage, Thousand Oaks, Calif.

MANGAT, P.S., and ELGARF, M.S. (1999). 'Flexural Strength of Concrete Beams with Corroding Reinforcement,' *ACI Structural Journal*, Vol. 96, No. 1, January, pp. 149–158.

MASOUD, S. (2002). *Behaviour of Corroded Reinforced Concrete Beams Repaired with FRP Sheets under Monotonic and Fatigue Loads*, PhD Thesis, University of Waterloo, Ontario, Canada, 417 pp.

MASOUD, S., and SOUDKI, K.A. (2006). 'Evaluation of Corrosion Activity in FRP Repaired RC Beams,' *Cement and Concrete Composites*, Volume 28, Issue 10, pp. 969–977.

MASOUD, S., SOUDKI, K., and TOPPER, T. (2001). 'CFRP-strengthened and Corroded RC Beams Under Monotonic and Fatigue Loads,' *ASCE Journal of Composites for Construction*, Vol. 5, No. 4, pp. 228–236.

MASOUD, S., SOUDKI, K., and TOPPER, T. (2005). 'Post-repair Fatigue Performance of FRP-repaired Corroded RC Beams: Experimental and Analytical Investigation,' *ASCE Journal of Composites in Construction*, Vol. 9, No. 5, September, pp. 441–449.

MULLINS, G., SEN, R., SUH, K., and WINTERS, D. (2006). 'A Demonstration of Underwater FRP Repair,' *Concrete International*, Vol. 28, No. 1, January, pp. 70–73.

PANTAZOPOULOU, S.J., and PAPOULIA, K.D. (2001). 'Modelling cover-cracking due to reinforcement corrosion in RC structures.' *Engineering Mechanics*, Vol. 127, No. 4, pp. 342–351.

RTEIL, A. (2007). *Fatigue Bond Behaviour of Corroded Reinforcement and CFRP Confined Concrete*, PhD Thesis, University of Waterloo, Canada, 136 pp.

SHERWOOD, T. (2000). *Behaviour of Corroded Reinforced Concrete Beams Strengthened with CFRP Laminates*, MASc thesis, University of Waterloo, Ontario, Canada.

SOUDKI, K.A., RTEIL, A., MCCUAIG, L., and STEPHENSON, J. (2007). 'FRP Repair of a Corroded Girder in the Scheiffle Bridge,' *3rd International Conference on Durability of Composites for Construction and Field Applications* (CDCC2007), Quebec, pp. 377–384

SOUDKI, K.A., and SHERWOOD, E.G. (2003). 'Bond Behaviour of Corroded Steel Reinforcement in Concrete Wrapped with Carbon Fibre Reinforced Polymer Sheets,' *ASCE Journal of Materials in Civil Engineering*, Vol. 15, No. 4, pp. 358–370.

SOUDKI, K.A., and SHERWOOD, T.G. (2000). 'Behaviour of Reinforced Concrete Beams Strengthened with Carbon Fibre Reinforced Polymer Laminates Subjected to Corrosion Damage,' *Canadian Journal of Civil Engineering*, Vol. 27, No. 5, pp. 1005–1010.

TUUTTI, K. (1980). *Service life of structures with regard to corrosion of embedded steel. Performance of concrete in marine environment*, ACI SP-65. Detroit, Michigan: American Concrete Institute; pp. 223–36.

UOMOTO, T., TSUJI, K., and KAKIZAWA, T. (1984). 'Deterioration Mechanism of Concrete Structures Caused by Corrosion of Reinforcing Bars,' *Transactions of the Japan Concrete Institute*, Vol. 6, pp. 163–170.

3

Areas of uncertainty in the use of fiber reinforced polymer (FRP) composites in the rehabilitation of civil engineering structures

R. A. ATADERO, Colorado State University, USA

Abstract: In order to develop reliability- and performance-based design procedures for FRP based repair, design uncertainties present in the existing structure and introduced by the FRP must be quantified. This chapter identifies the various sources of uncertainty that are present in the design of fiber reinforced polymer (FRP) based repair schemes and provides guidance in determining appropriate descriptions for these different types of uncertainty so as to facilitate the use of structural reliability methods.

Key words: structural reliability assessment, uncertainty in design variables, FRP repair.

3.1 Introduction

Uncertainty is intrinsic to all engineering endeavors; despite our best efforts it is impossible to eliminate all sources of uncertainty. Thus engineers must make allowances for the potential effects of uncertainty when creating designs. In the fields of civil and structural engineering, the consideration of uncertainty at the design stage is particularly important because civil and structural designers do not have the luxury of extensive proof testing of a design intended for mass production as is common practice in many other types of engineering (Ellingwood, 2000). Civil engineers must create a design that will, generally, be constructed just once, and that design must be based on their knowledge of structural materials and the behavior of different structural configurations. Traditionally, structural designers have compensated for many different sources of uncertainty through the use of empirical factors of safety. In the past 20 to 30 years, design of new structures has slowly moved toward design codes and procedures with a probabilistic basis. These types of procedures include Load and Resistance Factor Design (LRFD) in the United States and other partial factor formats used around the world. In contrast to the factor of safety approach, where the factors are based on previous experience and engineering judgment, the design factors used in probabilistic codes are derived, at least in part, based on explicit consideration of the likely sources of uncertainty and are meant to achieve a specified level of structural reliability. Beyond even reliability-based design, the structural engineering profession is beginning to pursue performance-based design. In the case of performance-based design, explicit performance targets for the structure to meet are defined, and many of the prescriptive features of traditional design codes are eliminated. Because the performance targets may be specified in probabilistic terms, consideration of the uncertainties present in design and evaluation of reliability is still an important aspect of performance-based design. In order to include consideration of uncertainty in the design process, or in the evaluation of service life, it is first necessary to identify the sources of uncertainty that may affect the structure.

It may appear that there would be a far lesser degree of uncertainty associated with an existing structure that has demonstrated satisfactory performance for a certain period of service. However, many of the same sources of uncertainty affecting new design are still of importance to an existing structure, and new uncertainties have been introduced during the life of the structure. The structure will have undergone years of service with unknown loading patterns in a potentially harsh climate. In the case of a structure that is undergoing repair, the new materials added to the structure provide an additional source of uncertainty. With so many sources of uncertainty, procedures for the service life prediction of all structures, and especially those undergoing repairs or changes, should be based on a probabilistic assessment of the structure's performance.

Chapter 6 of this book covers probabilistic techniques for estimating the service life of structures. However, in order to use the techniques described in Chapter 6, it is necessary to have descriptions for the many sources of uncertainty that can impact the service life. These descriptions must quantify the different sources of uncertainty in such a way that they can be readily incorporated into probabilistic computations. The objective of this chapter is to identify the various sources of uncertainty that are present in the design of fiber reinforced polymer (FRP) based repair schemes and to provide guidance in determining appropriate descriptions for these different types of uncertainties so as to facilitate the use of the probabilistic methods described in Chapter 6. The main sources of uncertainty affecting the service life of concrete structures subject to renewal with FRP are: uncertainty in the FRP itself; uncertainties in the existing structure, including those present at the time of construction as well as those developed over the previous time in service of the structure; uncertainty in the models used to describe and predict structural performance; and uncertainty in the loads acting on the structure. The remainder of this chapter will often make reference to various statistical distributions. Many books and other resources exist describing these distributions and their applications, for example Bury (1999).

3.2 Uncertainty in the fiber reinforced polymer (FRP)

The term fiber reinforced polymer or FRP can be used to refer to numerous unique and individual materials. FRPs may be composed of a variety of different polymers, reinforced with different types of fibers, and manufactured using different techniques under various conditions. Clearly, characterizing the uncertainty associated with this variety of different materials is a difficult task. In general, each unique material will have its own level of variability, which can be characterized using statistical distributions for material properties, such as ultimate strength and modulus of elasticity, and perhaps spatial or geometric quantities such as thickness or fiber orientation. In some circumstances, for designs of very high importance or that are otherwise critical in nature, it may be reasonable to conduct extensive testing on one particular FRP realization. However, for routine design and service life prediction, it is most likely necessary to narrow down the wide array of FRP possibilities into those most likely to be used in civil infrastructure applications, and to further group those materials into just a few broad categories for which generalized probability distributions can be developed.

Although there are many distinctions that can be drawn between the different types of FRP, one that is particularly relevant with respect to civil infrastructure applications, and the variability introduced to the project by the FRP, is the distinction between *prefabricated* and *field manufactured* FRP. With respect to the application of FRP to existing structures, *prefabricated FRP* refers to those composite materials where the resin and fibers are brought together and cured into FRP in a controlled factory setting, often to produce strips of composite material. Because these materials are manufactured in a more controlled (and automated) environment, they generally have high fiber volume fractions and controlled, if not elevated, temperature cures, resulting in enhanced material properties. The controlled manufacturing environments also mean that these composites generally show relatively small levels of variation. However, because these strips must be bonded to the structure in the field, there is still ample opportunity for variation in the adhesive to affect the reliability of the repair. *Field manufactured FRP* refers to those composite materials where the resin and fibers are combined in the field and the fabrication of the FRP and bonding of the FRP to the structure are combined into one step as the wetted fibers are applied directly to the structure and allowed to cure in place. Field manufactured materials typically have lower fiber volume fractions and they are cured at ambient conditions. The mechanical properties of field manufactured composites are generally less than those of prefabricated materials and, due to the loss of control in the manufacturing/application process, they are likely to show increased amounts of variation. The following sections will describe models to consider the uncertainty present in both prefabricated and field manufactured composite materials. It should be noted that the potential to introduce flaws or defects exists at all stages of the process whereby FRP is externally bonded to an existing structure, as inventoried by Kaiser and Karbhari (2003). The models presented here consider only the uncertainty in material properties and do not explicitly capture many of the uncertainties associated with application.

3.2.1 Prefabricated FRP

The uncertainty in a material, in this case prefabricated FRP, is commonly described by assigning statistical distributions to describe the variation in material properties such as strength and modulus. The Weibull distribution is a popular choice for modeling the variation in composite properties. The Composite Materials Handbook, MIL 17 (ASTM, 2002) specifies that the Weibull distribution should be used to represent composite properties, so long as a goodness-of-fit test shows that the Weibull distribution is adequate. Zureick *et al.* (2006) studied a total of 24 data sets representing strength and stiffness properties of pultruded composites manufactured from polyester or vinylester resins reinforced with E-glass fibers. They used the Anderson–Darling goodness-of-fit test to compare three possible distributions: normal, lognormal and Weibull. For strength, they found that while the Weibull distribution was rejected in three out of twelve sets, the normal and lognormal distributions could not be rejected. For stiffness, the Weibull distribution was not rejected for nine of twelve sets and the normal and lognormal were not rejected for eight of twelve sets. While these results were not conclusive with regard to which distribution was most appropriate, these authors advocated use of the Weibull distribution because it has been used commonly in the past, and because for the data they were working with, the Weibull distribution was the most conservative of the three distributions at the lower percentiles of the distribution. For variables that contribute to the resistance of a structure, the lower percentiles of a distribution are the most significant when assessing the reliability. Selection of the Weibull distribution to represent variation in composite strength has been justified in the past because it is derived based on a weakest-link theory of failure, which is typically appropriate for brittle materials (Sutherland and Soares, 1997). This fact was also noted by Zureick *et al.* (2006) as justification for their decision.

There are two types of Weibull distributions, the two-parameter model and the three-parameter model (Bury, 1999). The three-parameter model includes a location parameter which serves as a threshold for the distribution. When this location parameter is set equal to zero the resulting model is the two-parameter Weibull distribution. Alqam *et al.* (2002) compared the ability of the two distributions to model composite properties and recommended using the two-parameter model. The two-parameter model was recommended because, even though the three-parameter model may be better able to characterize the variation in properties, there were only small differences observed between the nominal design values and allowable loads predicted using the two different distributions and the two-parameter model is simpler to use. Furthermore, these authors questioned the theoretical validity of a threshold on composite properties below which no failure could occur.

When a data set of appropriate size is not available from which to determine Weibull parameters (this threshold is often set at 20 specimens), it may be necessary to relate the Weibull shape and scale parameters to estimated values for the mean and coefficient of variation (COV). The COV is defined as the standard deviation divided by the mean, and serves as a non-dimensional means for quantifying the variation of a data set. The mean value may be available from manufacturer-provided data or from material property testing of smaller sets of data. The COV may also be available or it can be estimated. In their comparison of the two-parameter and three-parameter Weibull distributions, Alqam *et al.* (2002) had an average COV of approximately 9% for tensile strength data sets and a COV of approximately 8% for tensile modulus data sets. Values of the COV used by previous researchers studying the reliability of externally bonded FRP on concrete structures are discussed in Section 3.2.3.

As mentioned previously, although prefabricated composites themselves generally exhibit reduced variation, the adhesive that is used to attach the FRP materials to the structure can be subject to more variation because it is applied and cured in the field. In this case, it may be particularly important to consider the model that is used to describe the bonding behavior, and incorporate uncertainty in both the adhesive properties and geometry (thickness of bond line) into that model. Statistics describing the degree of variation observed in field-cured adhesives are difficult to locate, but some provision for this uncertainty should be made in reliability and service life computations.

3.2.2 Field manufactured FRP

Although field manufactured FRP has the potential to display significant differences from prefabricated FRP, few researchers have made a clear distinction between field manufactured wet layup composites and prefabricated composites, and considered field manufactured materials independently. Atadero (2006) studied statistical distributions for strength, modulus and thickness for 17 data sets resulting from tensile testing of primarily carbon FRP samples cut from field manufactured panels. These 17 sets represented five different types of materials, and for most materials there were unique sets with different numbers of layers. Four different types of distributions were fitted to these sets: normal, lognormal, Weibull and Gamma. Chi-squared, Kolmogorov–Smirnov, and Anderson–Darling goodness-of-fit tests were applied to the distributions. Similar to the results for pultruded composites found by Zureick *et al.* (2006), the goodness of fit tests conducted by Atadero did not result in definitive choices to represent strength, modulus and thickness. However, the Weibull distribution was found to best represent the distribution of strength, the lognormal and Weibull distributions were suitable for modulus, and the lognormal distribution was recommended for thickness.

Atadero (2006) observed higher values for the COV for both strength and modulus than those typically observed for prefabricated composites. The COV of ultimate tensile strength was seen to range from 0.09 to 0.23, with an average value of 0.14. The COV for the modulus of elasticity ranged from 0.09 to 0.28, with an average value of 0.17. In considering the mean value of the composite strength and modulus, Atadero (2006) observed that the value found in testing often differed from values specified by the manufacturer. Several potential sources for these differences were identified including how the manufacturer reported test values, and differences in manufacture and curing between laboratory and field conditions. In order to properly use manufacturer values, it is important to know if they are reporting a mean value or a 'design value' (typically a certain percentile of test results); it is also important to know what thickness the strength and modulus are referenced to. Some sources prefer to specify strength and modulus normalized to the dry fiber thickness, while other sources specify the strength and modulus with respect to the thickness of the finished composite. Because the dry fiber thickness is always less than the composite thickness, normalized values of strength and modulus are greater than values calculated based on the actual thickness. When normalized values for strength and modulus are used either for design or reliability assessment, the thickness must be appropriate to the normalized value, otherwise it is possible to significantly overestimate the strength of a finished composite. Handling variability in thickness is a much more important consideration for wet layup composites because they are not manufactured by automated processes. Thickness variation can be handled directly by specifying non-normalized values for strength and modulus and providing a distribution for thickness. Variation in thickness can also be handled by using distributions for normalized values for strength and modulus with a deterministic value of thickness. It should be noted that, using these same data sets, Atadero and Karbhari (2009) observed changes in the mean value of strength and modulus with increases in the number of layers. These changes could not be completely explained by normalizing the strength and modulus. It was hypothesized that, as additional layers are added to a composite, there is greater opportunity for resin-rich areas to form as well as an increase in the likelihood of misaligned fibers, both resulting in slightly lower average strengths.

There are many factors that contribute to uncertainty regarding field manufactured composites. While it is not possible to fully identify and characterize these factors for an arbitrary composite, when assessing the reliability of field manufactured FRP installations, effort should be made to ensure that the models for composite properties that are being used are representative of the composite that will actually be applied to the structure. This includes possible adjustments to both the mean value and COV to consider unique features of the application and curing process, as well as the environmental conditions the FRP will experience during its lifetime.

3.2.3 Representation of FRP uncertainty in previous reliability studies

There have been a relatively small number of studies considering the reliability of civil engineering structures with externally bonded FRP. An early study was conducted by Plevris *et al.* (1995), who used a Weibull distribution to model the variation in the strain at failure for carbon FRP to evaluate the reliability of flexurally strengthened beams. The modulus of the FRP was considered deterministic. Plevris *et al.* (1995) did not specifically state whether their work considered prefabricated or field manufactured FRP; however, based on the shape parameters of the Weibull distributions they used in a parametric study considering two different types of FRP, the COVs of the materials were 0.105 and 0.075, which would likely represent prefabricated materials (or field manufactured materials with exceptional quality control). Val (2003) used a lognormal distribution with a COV of 0.15 to model the tensile strength of FRP jackets on columns; no distribution was provided for the FRP modulus. In two similar studies considering repair of reinforced (Okeil *et al.* 2002) and prestressed (El-Tawil and Okeil, 2002) concrete bridge girders, the authors used a Weibull distribution to model the failure strain of CFRP. The COV used was 0.022, which represents a very low amount of variation. These authors also did not address the variation in the FRP modulus. All of these studies considered rupture of the FRP to be the only mode of failure associated with the FRP itself, and did not consider the possibility of failure due to premature debonding of the FRP. If the FRP is anchored to the structure, this is an appropriate simplification; however, in the absence of anchorage, FRP materials are susceptible to debonding. Models for debonding of the FRP are generally sensitive to the modulus of elasticity of the composite, and therefore when debonding is a consideration, variation in the FRP modulus should also be included in the reliability analysis. Atadero and Karbhari (2008) did consider debonding of the FRP in the calibration of preliminary resistance factors for flexural repair of T-beam bridge girders. For this work, the uncertainty in the FRP was described using the data from Atadero (2006) previously discussed.

3.3 Uncertainty in the existing structure

When FRP is externally bonded to a concrete structure, it is not carrying the loads alone – it is working with the existing concrete and steel to resist the loads. Thus it is important also to consider the uncertainty present within the existing structure at the time of strengthening. Uncertainties present may include inherent variations that were present at the time of construction, as well as changes in the structure that have occurred during its previous service and which must be measured or estimated by field inspections. When predicting the service life of a repaired structure, it may also be important to consider the potential for continued deterioration as the structure continues to age.

3.3.1 Uncertainty in materials and dimensions as constructed

For reinforced concrete members, the uncertainty present at the time of construction includes variation in the material properties of the concrete and the reinforcing steel as well as variability in as-constructed dimensions. A slightly older, but very thorough study by Mirza and MacGregor (1976) provides statistical data on nearly all of the uncertainties associated with reinforced concrete members, including material properties and dimensions of several different types of members. Findings in this report are also summarized in two journal articles that may be more easily accessible (Mirza *et al.* 1979; Mirza and MacGregor, 1979). This study considers only normal weight concrete and steel reinforcing bars. More recently, Nowak and Szerszen (2003) considered the uncertainty present in reinforced concrete members for calibration of ACI 318 Building Code Requirements for Structural Concrete. Nowak and Szerszen (2003) studied ordinary, lightweight, and high-strength concrete, as well as steel rebar and prestressing steel.

All of these sources provide statistical distributions that can be used to describe various types of uncertainty as random variables. Often the distributional form, such as normal, lognormal, Weibull, etc. are provided, as well as a bias factor and COV. A bias factor is defined as the ratio of the mean value of a particular random quantity divided by the nominal value. The nominal value is the value specified for use in design, either by the designer or by the building code. For example, a designer might specify 27.5 MPa concrete, which would then be ordered from a ready mixed plant. In order to make sure that the concrete provided meets a minimum strength of 27.5, the ready mixed plant will likely provide concrete that is stronger than the minimum specified value. When using probabilistic methods to compute a reliability index or predict a service life, the most accurate assessment of material properties should be used; therefore the bias factor helps relate design values to what is actually present in the as-constructed structure. When considering the distributions used to describe uncertainty in materials, it may be appropriate to consider when the concrete structure was originally constructed, and choose distributions for material properties that are representative of that time. It should also be noted that concrete is known to continue to gain strength over long periods of time. If data is available regarding the strength of the concrete in the actual structure at the time when repair is being considered, this data may be more appropriate for use in service life estimation.

3.3.2 Uncertainty in the inspection process

When structures have spent a significant amount of time in service, they have probably undergone some degradation from their original state. For structures being repaired with externally bonded FRP, the extent of deterioration is likely to be significant, providing motivation for the repair. In order to create repair designs using FRP, it is typically necessary to calculate the deficiency in the existing structure and then determine the amount of FRP to be applied in order to overcome the deficiency. These calculations require knowledge about the existing structure which is generally obtained through inspections and testing, both non-destructive and destructive. The inspection process introduces another layer of uncertainty to the design of FRP based repairs. Data regarding statistical characterization of inspection uncertainty can be difficult to locate, and is likely to be specific to just one type of inspection. Therefore, in order to include inspection uncertainty in a reliability-based prediction of service life, it will be necessary to know what inspection methods were used to assess the existing structure.

There is some indication that the amount of uncertainty resulting from inspections may be significant. Phares *et al.* (2004) studied the variation in routine bridge inspections conducted by 49 inspectors from 25 different states. The inspections were consistent with bridge inspections conducted to satisfy US National Bridge Inspection Standards. Two in-service and five decommissioned bridges were assessed, using only visual inspection. Bridge components were rated using the FHWA Recording and Coding Guide. The amount of variation in the ratings applied to bridges by different inspectors can be expressed in terms of the COV, which ranged from approximately 0.10 to 0.20. On average, the ratings recorded by the inspectors were higher than the reference rating applied by the authors of the study, with the differences between the two ratings being statistically significant. Variation was also observed in the amount of notes recorded and pictures taken by inspectors.

While variability in visual inspections might be expected, inspections based on testing are also subject to uncertainty. Characterizing the uncertainty in an NDE technique can be complicated because, as noted by Hearn and Shim (1998), 'the significance of evaluations depends on the quality of data obtained in field tests and on the correlation of attributes measured in tests with the conditions of elements that are sought in the inspection'. Furthermore, 'NDE methods that are accurate in their measurements may still be found to be poor methods if attributes are uncertain indicators of condition of elements' (Hearn and Shim, 1998). When using reliability methods to predict the service life of structures repaired with FRP materials, it is the uncertainty in the state of the existing structure that is of importance, not that of the NDE measurement technique itself. Another consideration in the accuracy of non-destructive and destructive methods is the ability of discrete measurements to characterize conditions throughout the structure.

While data relating the uncertainty in an existing structure to various inspection techniques in a manner suitable for use in reliability assessment is largely unavailable in the literature, some initial work has shown that uncertainty in the existing structure can have clear impact on the reliability of a repair. In a preliminary study, Atadero (2008) considered uncertainty in the existing structure by varying the COV of the remaining cross-sectional area of steel. She compared the effect of uncertainty in the remaining steel area to the effect of uncertainty in the FRP applied to the structure. For the cases she considered, she found that increasing the COV of the remaining steel area reduced the reliability of the repair more than increasing the COV of the FRP strength or modulus. These results indicate that the existing structure itself may be much more important to the overall reliability than the specific materials used for repair. Thus, for accurate assessment of reliability and service life, the role of the existing structure must be considered.

3.3.3 Uncertainties associated with continued deterioration

The application of FRP to a deteriorated structure does not stop the process of degradation. When predicting the service life, it therefore may be quite important to consider the potential for continued deterioration of the existing structure. Concrete in reinforced concrete structures can be susceptible to deterioration due to a variety of mechanisms, but corrosion of the reinforcing steel is a particularly important form of deterioration affecting reinforced concrete structures. Corrosion of reinforcement weakens the structure by reducing the cross-sectional area of the steel, causing spalling and delamination of the concrete due to the expansive corrosion products, and reducing the bond between the steel and concrete (Gulikers, 2005). In fact, according to Gulikers (2005) 'there is general agreement that corrosion of the reinforcement steel is the most prevalent form of deterioration of the infrastructure, necessitating vast amounts of money for rehabilitation and repair'. While many of the deterioration mechanisms that may affect reinforced concrete are difficult to predict and model in merely a deterministic sense, corrosion of steel in concrete has been well studied, even probabilistically. For these reasons, the following discussion will focus on considering the uncertainty introduced to the existing structure by corrosion of reinforcing steel. The discussion will begin with a very brief overview of the corrosion process of steel in concrete.

Steel reinforcement within concrete is protected by a thin, passive oxide layer that develops on the surface of the steel in the presence of the highly alkaline pore solution typically found within concrete. The steel can be depassivated by a drop in pH resulting from carbonation of the cover concrete, or by chloride ingress (Bertolini *et al.*, 2004; Hunkeler, 2005). Corrosion initiated by carbonation of the cover concrete generally affects large areas of the steel surface, while corrosion initiated by chloride ions results in localized corrosion and pitting of the reinforcing bars (Bertolini *et al.*, 2004). A well-accepted model of corrosion of steel within concrete considers two stages of corrosion: initiation and propagation. During the initiation stage, the steel is not actively corroding as it has not yet been depassivated, but carbonation or chlorides are diffusing through the concrete cover. Depassivation of the steel marks the end of the initiation stage and the beginning of the propagation stage. The propagation stage lasts until the structure has reached a performance limit (Tuutti, 1982). The most important parameter for describing corrosion propagation is the corrosion rate, which is often expressed as a current density. Using Faraday's law of electrochemical equivalence, a corrosion current density of 1 $\mu A/cm^2$ (6.45 $\mu A/in^2$) can be found equivalent to a uniform penetration of 11.6 μm/yr (4.567×10^{-4} in/yr) (Val *et al.*, 2000).

A number of researchers have studied the corrosion of steel in concrete. Val and Melchers (1997) compared the effects of pitting corrosion and generalized corrosion on the reliability of a simple slab bridge, finding that general corrosion was more detrimental at early ages, but beyond approximately 50 years, pitting corrosion became a greater threat. The model used in this study considered the effect of the loss of bond between the steel and concrete for generalized corrosion as well as the variation in pit sizes for pitting corrosion. Stewart and Rosowsky (1998a) also evaluated the reliability of a RC slab bridge, but included the time-to-initiation in their study considering the exposure to chlorides through deicing salts and marine exposures. Val *et al.* (2000) considered the effect of corrosion on the probability of reaching both strength and serviceability limit states, and provided a method to update the statistical distributions of resistance variables based on data collected during a bridge inspection using Bayes' Theorem. Stewart (2004) included the effect of spatial variability in pitting corrosion, since the location of significant pitting may not necessarily coincide with the location of maximum load effect.

In most studies, the uncertainty in the amount of steel lost to corrosion is measured in terms of the corrosion current density, i_{corr}, typically expressed in $\mu A/cm^2$. Most of the studies cited herein have used a normal distribution for i_{corr}, with a mean value between 1 and 3 $\mu A/cm^2$ and a COV of 0.20 or 0.30 (Val and Melchers, 1997; Stewart and Rosowsky, 1998a; Val *et al.*, 2000). Stewart and Rosowsky (1998b) used a uniform distribution between 1 and 2 $\mu A/cm^2$. Vu and Stewart (2000) made use of a model based on properties of the concrete, such as water–cement ratio and amount of cover, to predict the rate of corrosion, and used a uniformly distributed variable with a COV of 0.2 to represent the uncertainty in this model. Another important consideration is what portion of the corrosion process should be modeled. Many of these previous studies have considered the initiation stage as well as the propagation stage. When considering new structures and perhaps optimization of scheduled maintenance, the time-to-initiation is a very important component of the service life estimation. However, for structures that are undergoing repair with FRP, it is likely that the FRP is being applied because corrosion has already been initiated, and therefore modeling the propagation stage is the most significant for predicting the future service life of the repaired structure.

3.4 Uncertainty in analytical models

Analyzing the reliability and/or predicting the service life of a structure are both analytical exercises. Although experimental testing may be used to better understand the behavior of materials or structures, there is no practical experiment that can be conducted to predict the reliability or service life of civil engineering structures. Because they are analytical operations, both reliability computation and service life prediction depend heavily on the models used to describe physical behavior and processes. The use of models introduces another source of uncertainty into the analytical computations. This is particularly true because, as stated by Ditlevsen (1982), 'model selection is guided by a balance between ability to represent reality and the pragmatic need for having such simple mathematical properties of the model that a large variety of problems can be analyzed by the model'. For some physical phenomena, it may be possible to represent reality quite closely with a very simple model; in other situations, an accurate representation may be dependent on an exceedingly complex formulation. In either case there will be differences between reality and the model, and the uncertainty that comes from using a model must be included in reliability and service life assessment. Modeling uncertainty comes from two sources (Ditlevsen, 1982). The first source is the practical necessity that not all possible variables will be considered in the model. The second source of uncertainty arises from the need to describe reality through mathematical operations.

Modeling error can be handled in reliability analysis by introducing additional random variables to describe differences between the model and reality. These random variables are introduced directly into the limit state function. Statistical properties of the random variables characterizing modeling error are often based on comparison of physical test results to model predictions. However, Ditlevsen (1982) has discussed the need to include engineering judgment in characterizing model uncertainty, particularly in the absence of test results representing a number of failure paths. Ronold and Bjerager (1992) discussed the development and use of a model uncertainty factor in the context of geotechnical engineering; the application to structural engineering would be a straightforward extension of these concepts. It must be acknowledged that, when modeling error is characterized by comparing the results of physical tests to model predictions, additional sources of uncertainty such as measurement accuracy during the physical test will also be included in the random variable representing modeling error.

Relatively little information is available regarding modeling uncertainty with respect to the application of FRP to reinforced concrete structures. In studying the reliability of concrete columns jacketed with FRP, Val (2003) did not use a single-model uncertainty factor; rather, he characterized uncertainty in the model used to predict the compressive strength of confined concrete by developing statistical distributions for two empirical constants in the model. Toutanji *et al.* (2006) conducted a study to verify models for use in the design of externally bonded FRP applied to flexural members. A database of 115 beams, tested in previous work by a number of different researchers, was developed. Although the paper does not discuss reliability, and thus does not explicitly refer to model uncertainty factors, statistics describing the relation between experimental results and a number of different design models are presented. These statistics could serve as a starting point to develop model uncertainty factors for flexural repair.

3.5 Uncertainty in load effects

The sources of uncertainty previously addressed primarily relate to the resistance portion of the reliability problem. Although modeling error is involved in the models used to describe structural loading, it is often accounted for in the distribution used to describe load uncertainty. Loading is a significant source of uncertainty affecting the reliability and service life of structures; a brief discussion of this is provided here. The probabilistic description of loading will depend heavily on the technique used to evaluate the reliability of the structure. For service life predictions, the reliability assessment will almost certainly be time dependent, and the description of load uncertainty must be compatible with the time frame of reference in the reliability analysis.

A typical reliability method used for design of new structures is a time integrated approach. For new structures, the description of resistance uncertainty of the structure is assumed to be constant over the life of the structure, and the description of loading is referenced to a specified service life using an extreme value distribution (Melchers, 1999). The extreme value distribution describes the maximum load anticipated during the service life. When a significant amount of data is available, it may be possible to directly fit a distribution to the observed maxima; more typically, a shorter record is related to a longer time interval using extreme value theory. Using extreme value theory, the probability distribution of the maximum value of a random variable can be quite simply derived following Castillo *et al.* (2005) as

$$F_{max}(x) = [F(x)]^n \qquad [3.1]$$

where $F_{max}(x)$ is the cumulative distribution function (CDF) for the maximum, $F(x)$ is the CDF for x, and n is used to relate the two in the desired fashion. For example, if $F(x)$ is the distribution of loads observed during one year, and $F_{max}(x)$ needs to describe the distribution of maximum loads during 50 years, n would be 50. As the value of n increases, the distribution of the maximum will asymptotically approach one of the extreme value distributions, depending on the type of distribution described by $F(x)$ (Bury, 1999).

Service life estimation often involves consideration of structural deterioration, and therefore the assumption that uncertainty in resistance variables is not changing with time is not acceptable. To handle structural deterioration with the time-integrated reliability approach, the distribution of maximum load during a specified time period should be compared to the distribution of minimum resistance. This technique assumes that the maximum load and minimum resistance will occur at the same time (at the end of the service life), an assumption which will typically give a conservative estimate of the reliability (Melchers, 1999). The conservatism may not be as significant if the load tends to increase in intensity or frequency over time. Because the time integrated approach is quite simple, it could be used as a preliminary check.

Other more sophisticated reliability methods are available for a more accurate assessment. One typical approach is to break the lifetime of the structure into smaller increments. Stewart and Rosowsky (1998a) used this approach to consider deterioration of reinforced concrete. In their study, the probability of failure was formulated as:

$$p_f(t_L) = 1 - \Pr[R(t_1) > S_1 \cap R(t_2) > S_2 \cap \ldots \cap R(t_n) > S_n] \qquad [3.2]$$

where $R(t_i)$ is the resistance of the structure at a specific time interval and S_i are independent load events occurring at deterministic increments during t_L, the lifetime of the structure for which the reliability is being evaluated. A common interval for consideration is one year. The distribution for loading during one year could be derived from measured data, or again extreme value analysis could be used to take the distribution of loading for a day, or week, and relate it to the distribution for a year. This formulation allows for loading that is uncertain in intensity but does not consider the random nature of the timing when loads are applied. Mori and Ellingwood (1993) describe a formulation using stochastic processes to describe loads that vary both in time and intensity.

Loads can be very structure specific. If a structure is already in service and is being repaired or strengthened with FRP, the opportunity might exist to measure loads actually acting on the structure. Even measurement over a relatively short time period can be used to develop load distributions for longer time periods using extreme value theory, or measurements can be used to update existing statistical models for load uncertainty. In the absence of experimentally measured loads, references are available to describe typical loadings for different types of structures. Statistics for building loads are described by Galambos *et al.* (1982). Bridge loading has been studied by a number of different researchers (e.g. Nowak, 1999).

3.6 Application to reliability-based design and service life estimation

Current reliability-based design techniques for new designs do not allow for the consideration of different service lives for design, nor do they explicitly consider the deterioration of the structure over time. The service life is built into the description of loading used to calibrate design factors, and the potential for degradation is accounted for through selection of the target reliability index. In service life estimation, changes in the performance of the structure over time are considered and degradation of the structure is explicitly included in the calculation. When assessing the service life of a structure in a probabilistic manner, it is likely that the performance of the structure will be measured in terms of the reliability index, and the service life will be measured as the time until the reliability index falls below a threshold value. When selecting the threshold value for the reliability index, it is important to be aware of how the reliability index should be interpreted and the many different considerations that are built into its selection. A value of the reliability index used as the target to calibrate design factors for new design may not be appropriate as a threshold value in service life estimation.

Selection of a target or threshold reliability index is complicated by the fact that most structural failures result directly from human actions, but structural reliability methods consider only statistical variation in loads and material resistances. Because human error is not included in reliability calculations, the probability of structural failure that is actually observed is much higher than the value predicted by reliability methods. (For this reason the probability of failure predicted by reliability methods is often referred to as the notional probability of failure.) This lack of direct correspondence implies that there is much more to choosing the reliability index than determining an acceptable level of failure and its corresponding index. Melchers (2001) also notes that the target or threshold value must allow for the fact that it may apply to more than just one structure and should recognize the number of approximations that may be necessary to make the code accessible to a designer.

Several different methods have been proposed for choosing an appropriate value of the reliability index. One approach is to determine an acceptable level of structural failure through comparison to risks considered acceptable for other human activities. This technique does not consider the difference between actual and notional failure probabilities, and is complicated by the need to compare risks from different sources (Ellingwood, 1994). Another approach to selection of the reliability index is to consider it as a parameter that can be optimized to provide a balance between the marginal cost of increasing safety and the marginal reduction in the risk-associated costs of failure (Madsen *et al.*, 1986). Although very rational, this method is complicated by a lack of data, as well as by the difficult issue of assigning costs to non-economic quantities. Additionally, some empirical methods for target reliability selection are discussed by Melchers (2001, p 60).

Most target reliability indices used in existing reliability-based codes for new design have been selected based on the reliability implied by previous codes (usually in the allowable stress design format) (Galambos *et al.*, 1982). This method allows experience that has produced acceptable designs in the past to have some impact on the new code. Representative structural elements are designed following the previous code and then their reliability is evaluated using the same models, distributions, and reliability methods that will be used to calibrate the new code. This process usually results in a range of different reliability indices and a good deal of subjective judgment goes into the selection of the final target reliability index. This method is attractive because, when the reliability index is used strictly as a comparative measure, many of the issues associated with modeling random variables (for example the so-called tail sensitivity problem) become less important (Melchers, 1999). The problem in applying this method to FRP strengthening is that there is not a long record of successful designs upon which to base such a reliability index.

When considering a reliability threshold for service life estimation, it is important to remember that in codes for new design, deterioration of the structure is never explicitly considered. For a new structure the target reliability that the structure must meet at the beginning of its life must be high enough such that, over the course of its service life, the structure will continue to be safe even as it deteriorates; the reliability of the structure in its deteriorated state at the end of its service life is never explicitly calculated. In service life estimation, the reliability is being continuously updated as the structure degrades. Thus the reliability index used for new design is likely to be much higher than the necessary reliability threshold for service life estimation.

There are some other considerations that should also be factored into the selection of the reliability target or threshold; one is the associated costs. For new design, a slightly higher reliability index will mean that the structure must have additional capacity, likely to require the addition of some material. For service life, if the reliability of a structure drops below the reliability threshold, the structure will have to be closed and subject to repairs. The costs associated with repair are likely to be much higher than those needed to provide modest amounts of additional material at the design stage. Another consideration is for what portion of the structure the reliability is being evaluated. Element level reliabilities are typically much lower than overall system reliabilities, due to redundancies present in most structures. Codes for new design are based on element reliabilities, largely due to the difficulties associated with calculation of system reliabilities in a calibration context. Because service life estimation is likely to apply to one specific structure, it may be more feasible to compute structure-level reliabilities in this context. The lack of ductility associated with FRP materials might also be something to consider in selection of the reliability threshold. Typically the target reliability index for ductile materials is lower than that for brittle materials. Allen (1992) summarizes adjustments made to the target reliability index for factors such as ductility and inspectability for the evaluation of bridges in Canada. In general, a target reliability index is selected considering many different factors, and persons working with the reliability index should be aware of how the selection was made. This is particularly important as the context is shifted from new design to service life estimation.

3.7 Future trends

The presence of uncertainty, and how to account for it in the design process, are issues with which engineers will always be confronted. Improved techniques for quality assurance can help reduce, but never eliminate, uncertainty. Therefore, advances in this area are likely to come in the form of greater knowledge of the types of uncertainty affecting design, and improved descriptions of uncertainty for use in service life estimation. As seen in this chapter, more information is needed regarding almost all of the variables affecting the design of FRP, renewal strategies. Adhesives used to bond prefabricated composites to reinforced concrete structures, the condition of the existing structure at the time of repair, and modeling of continued deterioration of the structure after the application of FRP, are sources of uncertainty in particular need of further study. As the external bonding of FRP reinforcements matures as a technology, more attention should be paid to levels of uncertainty present in existing techniques and models, rather than the development of new methods.

Design of structures and repair schemes to achieve a particular service life are a prime example of performance-based design. In order to facilitate the design of structures for a particular service life, descriptions of uncertainty must become more widely available and should be subject to verification by different researchers. In order to provide for comparable levels of safety across different designs and designers, it may also be necessary to standardize descriptions of uncertainty and or service life prediction techniques. Performance-based standards would thereby standardize how the performance of the structure would be evaluated, rather than standardizing how the performance must be achieved, as is the case in more traditional prescriptive standards.

3.8 Sources of further information and advice

This chapter has provided a number of references to help pursue the description of uncertainties affecting the reliability, and ultimately the service life, of reinforced concrete structures renewed with externally bonded FRP. Those references listed below that describe previous reliability assessments will be of particular value to researchers hoping to better understand the application of reliability methods to their particular areas of study. In many cases, available descriptions of uncertainty are limited in their scope and applicability or are not available in the literature at all. Researchers who have encountered uncertainty in their work with FRP, adhesives, and reinforced concrete structures are encouraged to quantify the uncertainty and share their findings through publication. Describing uncertainty in a quantified manner may require fabrication and testing of more test specimens than is necessary for a deterministic study, a fact that should be considered when the experiment is being designed. When measures of uncertainty are included in papers primarily describing other topics, researchers are also encouraged to use titles, abstracts and keywords that make it clear their publications include information about uncertainty.

3.9 References

ALLEN DE (1992), 'Canadian highway bridge evaluation: reliability index', *Canadian Journal of Civil Engineering*, 19, 987–991.

ALQAM M, BENNETT RM and ZUREICK A (2002), 'Three-parameter vs. two-parameter Weibull distribution for pultruded composite material properties', *Composite Structures*, 58, 497–503.

ASTM INTERNATIONAL (2002), *The Composite Materials Handbook MIL 17*, Volume 1, ASTM, West Conshohocken, PA, USA.

ATADERO R (2006), *Development of load and resistance factor design for FRP strengthening of reinforced concrete structures*, PhD Dissertation, University of California, San Diego, USA.

ATADERO RA (2008), 'Considering the Effect of Uncertainty in the Existing Structure on the Reliability of FRP Repair Schemes', *5th International Conference on Advanced Composite Materials in Bridges and Structures*. Winnipeg, Manitoba, Canada. Sept. 22–24.

ATADERO RA and KARBHARI VM (2008), 'Calibration of resistance factors for reliability based design of externally-bonded FRP composites', *Composites Part B: Engineering*, 39, 665–679.

ATADERO RA and KARBHARI VM (2009), 'Sources of uncertainty and design values for field manufactured FRP', *Composite Structures*, 89, 83–93.

BERTOLINI L, ELSENER B, PEDEFERRI P and POLDER R (2004). *Corrosion of Steel in Concrete*, Weinheim, Wiley-VCH.

BURY K (1999), *Statistical Distributions in Engineering*, Cambridge University Press, Cambridge, UK.

CASTILLO E, HADI AS, BALAKRISHNAN N and SARABIA JM (2005), *Extreme Value and Related Models with Applications in Engineering and Science*, Hoboken, New Jersey, USA, John Wiley & Sons, Inc., pp 154.

DITLEVSEN O (1982), 'Model uncertainty in structural reliability', *Structural Safety*, 1, 73–86.

ELLINGWOOD BR (1994), 'Probability-based codified design: Past accomplishments and future challenges', *Structural Safety*, 13, 159–176.

ELLINGWOOD BR (2000), *Load and resistance factor design (LRFD) for structures using fiber reinforced polymer (FRP) composites*, NIST GCR 00-793, US Department of Commerce.

EL-TAWIL S and OKEIL AM (2002), 'LRFD flexural provisions for prestressed concrete bridge girders strengthened with carbon fiber-reinforced polymer laminates', *ACI Structural Journal*, 99(2), 181–190.

GALAMBOS TV, ELLINGWOOD B, MACGREGOR JG and CORNELL CA (1982), 'Probability based load criteria: assessment of current design practice', *J Structural Division*, 108(ST5), 959–977.

GULIKERS J (2005), 'Numerical modeling of reinforcement corrosion in concrete', in Böhni H, *Corrosion in Reinforced Concrete Structures*, Boca Raton, FL, USA, CRC Press, pp 71–90.

HEARN G and SHIM H (1998), 'Performance of NDE methods in bridge inspection', *SPIE*, 3400, 452–463.

HUNKELER F (2005), 'Corrosion in reinforced concrete: Processes and mechanisms', in Böhni H, *Corrosion in Reinforced Concrete Structures*; Boca Raton, FL, USA: CRC Press, pp 1–45.

KAISER H and KARBHARI VM (2003), 'Identification of potential defects in the rehabilitation of concrete structures with FRP composites', *Int. J. of Materials & Product Technology*, 19(6), 498–520.

MADSEN HO, KRENK S and LIND NC (1986), *Methods of Structural Safety*, Englewood Cliffs, NJ, USA: Prentice Hall.

MELCHERS RE (1999), *Structural Reliability Analysis and Prediction*, 2nd Edition, New York: John Wiley & Sons.

MELCHERS RE (2001), 'Rational Optimization of Reliability and Safety Policies', *Reliability Engineering and System Safety*, 73, 263–268.

MIRZA SA and MACGREGOR JG (1976), *A statistical study of variables affecting the strength of reinforced normal weight concrete members*, Structural Engineering Report No. 50, University of Alberta, Department of Civil Engineering.

MIRZA SA and MACGREGOR JG (1979), 'Variations in dimensions of reinforced concrete members', *J of the Structural Division*, 105(ST4), 751–766.

MIRZA SA, HATZINIKOLAS M and MACGREGOR JG (1979), 'Statistical descriptions of strength of concrete', *J of the Structural Division*, 105(ST6), 1021–1037.

MORI Y and ELLINGWOOD BR (1993), 'Reliability-based service life assessment of aging concrete structures', *J Structural Engineering*, 119(5), 1600–1621.

NOWAK AS (1999), *Report 368: Calibration of LRFD Bridge Design Code*, Washington D.C., National Academy Press.

NOWAK AS and SZERSZEN MM (2003), 'Calibration of design code for buildings (ACI 318): Part 1 – Statistical models for resistance', *ACI Structural J*, 100(3), 377–382.

OKEIL AM, EL-TAWIL S, and SHAHAWY M (2002), 'Flexural reliability of reinforced concrete bridge girders strengthened with carbon fiber-reinforced polymer laminates', *J Bridge Engineering*, 7(5), 290–299.

PHARES BM, WASHER GA, ROLANDER DD, GRAYBEAL BA and MOORE M (2004) 'Routine highway bridge inspection condition documentation accuracy and reliability', *J Bridge Engineering*, 9(4), 403–413.

PLEVRIS N, TRIANTAFILLOU TC and VENEZIANO D (1995), 'Reliability of RC members strengthened with CFRP laminates', *J Structural Engineering*, 121(7), 1037–1044.

RONOLD KO and BJERAGER P (1992), 'Model uncertainty representation in geotechnical reliability analyses', *J Geotechnical Engineering*, 118(3), 363–376.

STEWART MG (2004), 'Spatial variability of pitting corrosion and its influence on structural fragility and reliability of RC beams in flexure', *Structural Safety*, 26, 453–470.

STEWART MG and ROSOWSKY DV (1998a), 'Time-dependent reliability of deteriorating reinforced concrete bridge decks', *Structural Safety*, 20, 91–109.

STEWART MG and ROSOWSKY DV (1998b), 'Structural safety and serviceability of concrete bridges subject to corrosion', *J Infrastructure Systems*, 4(4), 146–155.

SUTHERLAND LS and SOARES CG (1997), 'Review of probabilistic models of the strength of composite materials', *Reliability Engineering and System Safety*, 56, 183–196.

TOUTANJI H, ZHAO L and ANSELM E (2006), 'Verifications of design equations of beams externally strengthened with FRP composites', *J Composites for Construction*, 10(3), 254–264.

TUUTTI K(1982), *Corrosion of Steel in Concrete*. Stockholm: Swedish Cement and Concrete Research Institute.

VAL DV (2003), 'Reliability of fiber-reinforced polymer-confined reinforced concrete columns', *J Structural Engineering*, 129(8), 1122–1130.

VAL DV and MELCHERS RE (1997), 'Reliability of deteriorating RC slab bridges', *J Structural Engineering*, 123 (12), 1638–1644.

VAL DV, STEWART MG and MELCHERS RE (2000), 'Life-cycle performance of RC bridges: Probabilistic approach', *Computer Aided Civil and Infrastructure Engineering*, 15, 14–25.

VU KAT and STEWART MG (2000), 'Structural reliability of concrete bridges including improved chloride-induced corrosion models', *Structural Safety*, 22, 313–333.

ZUREICK A, BENNETT RM and ELLINGWOOD BR (2006), 'Statistical characterization of fiber-reinforced polymer composite material properties for structural design', *J Structural Engineering*, 132(8), 1320–1327.

4
Rehabilitation and service life estimation of bridge superstructures

L. S. LEE, University of the Pacific, USA

Abstract: The bridge superstructure acts as the primary conduit to transfer vehicles safely over a crossing. Due to harsh, changing environmental conditions and increased load demands, the safety of the structure can deteriorate over time. Fiber reinforced polymer (FRP) composites are often being used to restore the safety and extend the usable life of reinforced concrete bridge components. This chapter examines the use of FRP composites to rehabilitate flexural components, namely bridge decks and girders. A time-dependent reliability approach is presented in order to estimate the safety and remaining service life of a FRP rehabilitated bridge deck.

Key words: bridge, rehabilitation, service life, reliability, safety.

4.1 Introduction

While the public has grown to expect bridge structures to perform without interruption, the capacity of a bridge structure and its environment changes with time. Over its service life, a bridge structure is subjected to a harsh environment which includes excessive loading, fatigue, weathering, aging, and extreme events. Since interstate systems, as in the United States, were often constructed en masse, the requirements for service and repair also have been occurring en masse. Transportation officials often are required to select repairs on a bridge structure on a case-by-case basis as opposed to repairing all structural deficiencies, due to financial constraints. The increase of structural and functional deficiencies in aging bridge structures has resulted in a focus on methodologies for monitoring and maintenance of structures.

The bridge superstructure is defined as all structural elements above the bridge bearing elevation, or as the primary conduit which carries a roadway over a crossing. While the superstructure is composed of a number of elements (wearing surface, diaphragms, girders, decks, etc.), the bridge deck represents the physical extension of the roadway and often provides the first indications of distress in a bridge structure. In addition, the most common deck in use is the cast-in-place, reinforced concrete (RC) slab deck. Consequently, the use of fiber reinforced polymer (FRP) composites to repair or strengthen existing bridge decks has increased over the past decades. (Teng *et al.*, 2003; Van Den Einde *et al.*, 2003).

An increasingly popular approach to extend service life of bridge structures involves the external bonding of fiber reinforced polymer (FRP) composites to RC structures to accommodate higher design loads and mitigate deterioration-related damage. FRP composites are materials having strong, stiff fibers – continuous or discontinuous – surrounded by a polymer matrix material. The matrix of the composite serves to distribute the fibers and act as the primary means of load transfer between fibers. Common reinforcing materials include glass, aramid, and carbon fibers. Common matrices are epoxy, polyester, and vinyl ester resins.

External bonding of FRP composites for structural rehabilitation has increased because FRP composites possess advantages such as high stiffness-to-weight ratio, high strength-to-weight ratio, corrosion resistance, ease of application, and enhanced fatigue life (Karbhari and Zhao, 2000; Van Den Einde *et al.*, 2003). Literature for experimental testing and analysis of FRP applied to RC beams and slabs is extensive; here, references are provided but these are by no means exhaustive (Stallings *et al.*, 2000; Malek and Patel, 2003; Mosallam and Mosalam, 2003; Hag-Elsafi *et al.*, 2001).

Although significant evidence exists supporting the effectiveness of FRP composites, issues pertaining to durability, quality, and design of FRP composites remain. One of the primary concerns involves FRP composite durability, or the ability of FRP composites to sustain load over extended periods of time while exposed to harsh, changing environmental conditions (Karbhari *et al.*, 2003; Nanni, 2003). The lack of understanding with respect to the durability of FRP composites promotes design-related uncertainty, particularly for establishing margins of safety and remaining service life.

The uncharacterized long-term durability and lack of quality control during construction present the more significant barriers to acceptance of FRP composites as a proven technology in rehabilitation of RC structures. The uncharacterized long-term durability of FRP composites directly affects the service life estimation of the structure. Furthermore, lack of quality control standards, variations during construction due to procedure, and environment contribute to scatter in material parameters of the FRP composite. The result is uncertainty in design and unknown performance change in the rehabilitated structure, which requires a high factor of safety during rehabilitation design. The use of time-dependent, probabilistic methods to evaluate the remaining service of FRP rehabilitated structures presents a means to integrate the variability inherent in performance of an as-built structural element and the deterioration of materials over time.

Rehabilitation and service life estimation of FRP rehabilitated RC bridge superstructure, namely bridge decks and girders, are described in this chapter. First, an overview and background of rehabilitation of bridge components using FRP composites is provided. Next, a probabilistic approach to estimate the remaining service life of RC flexural members repaired or strengthened with FRP composites is described. Then, predictions of time-dependent response of FRP composites is integrated with a probabilistic approach to service life estimation. The chapter concludes with an assessment of future trends for service life estimation of FRP rehabilitated components and its application to maintenance of the bridge superstructure.

4.2 Externally bonded fiber reinforced polymer (FRP) composites

For the purposes of this chapter, the methodologies for bridge deck rehabilitation and thus for service life extension, will focus on the application of fiber reinforced polymer (FRP) composites for flexural strengthening of a deck slab. This, however, is used just as the driving example, with the methodology being developed in a manner to allow extension to other modes as well.

This section provides the background of FRP composite rehabilitation and highlights examples of flexural rehabilitation of reinforced concrete (RC) structures with FRP composites to identify the needs for performance assessment of FRP composite rehabilitation.

Externally-bonded reinforcement for the rehabilitation of concrete components initially involved the application of steel plates to the tensile face of structural components for strengthening. Flexural strengthening of existing structures with bonded steel plates was shown to be a viable option as reported by Fleming and King in 1967 (ACI, 2002). However, the use of steel plates as external reinforcement presented the following disadvantages: (i) Deterioration of the bond at the steel – concrete interface from steel corrosion; (ii) Difficulty in handling of the plates at the construction site; (iii) Increased load demand on the structure; (iv) Restrictions on length of steel plates due to weight (Meier, 2000; Karbhari and Seible, 2000; Triantafillou and Plevris, 1991). In order to develop an alternative to bonding of steel plates, the use of FRP composites for strengthening RC structures was first investigated at the Swiss Federal Laboratory for Materials Testing and Research (EMPA), where tests on RC beams strengthened with CFRP plates were conducted in 1984 (Teng *et al.*, 2003; Meier, 2000).

FRP materials are lightweight, noncorrosive, exhibit high tensile strength (ACI, 2002) and can be tailored to performance requirements via volume fraction control and/or fiber orientation in the matrix to obtain maximum efficiency (Meier, 2000; Karbhari and Seible, 2000). A rehabilitation strategy, such as flexural strengthening of a RC bridge deck with carbon FRP composites, can be performed without interference to the intended function of the structure, i.e. no interference with traffic. FRP composites are available in a variety of forms, ranging from factory-made laminates to dry fabrics that can conform to the geometry of the structure before adding resin (ACI, 2002; Teng *et al.*, 2003). FRP composite materials possess important qualities, which suggest immense potential for application to civil structures. As such, the replacement of externally-bonded steel plates with externally-bonded FRP composites has been the focus of many researchers in the recent past, to repair, strengthen, and retrofit RC structures (Bonacci and Maalej, 2000, 2001; Teng *et al.*, 2003).

Although, significant advantages are realized with FRP composites for bridge strengthening and repair, questions regarding quality and durability of FRP composites at the material level remain, as well as the effect of FRP material degradation on the long-term response of the rehabilitated component. It should be noted, however, that current codes for conventional materials do not provide an explicit method for incorporation of materials degradation and hence this in itself is a major contribution.

4.2.1 Manufacturing and application

For strengthening of civil infrastructure, composites are generally applied via wet lay-up and/or adhesive bonding (Karbhari and Seible, 2000; Meier, 2000; Teng *et al.*, 2003). The efficacy of the manufacturing/construction methods depends on the combined action of the entire system, with emphasis on the integrity of the bond and interface between the composite layer and the concrete substrate (Karbhari and Zhao, 2000). In addition, the appropriate selection of materials, based on stiffness and strength requirements, influences the overall performance and capacity of the FRP composite rehabilitated component (Karbhari and Seible, 2000).

Application of FRP composites has often occurred without prior establishment of well-documented procedures for quality control and assessment. In fact, most rehabilitation applications of FRP composites conducted to date are without a set of standardized quality assurance criteria to ensure that proper procedures and methods are adhered to during the construction process. The lack of standardized quality control procedures is of serious concern for rehabilitation applications on civil structures since (i) manufacturing processes are likely to be wet lay-up or adhesive bonding, (ii) fiber and resin are separate constituents rather than pre-impregnated materials and (iii) application procedures take place in the field, thus being susceptible to changes in the environment (Karbhari, 2000). In order to establish a set of criteria to ensure the quality of a structural rehabilitation strategy using FRP composites, an understanding of the interaction between FRP composite and the existing structure is necessary.

This section provides a summary of the procedures used in the placement of an externally-bonded FRP composite onto a concrete bridge deck.

Detailed construction guidelines can be found in NCHRP Report 514 (Mirmiran *et al.*, 2004).

4.2.2 Surface preparation

Since the efficacy of the method depends on the integrity of the bond of the FRP to the concrete substrate, it is essential that the surfaces are adequately prepared. All unsound material and surface efflorescence must be removed and the surface rebuilt, if necessary. Sandblasting is needed to expose aggregate. Details for surface preparation, including specifications, are listed in ACI 440 guidelines (2002) and NCHRP 514 construction guidelines (Mirmiran *et al.*, 2004) and these details must be followed.

4.2.3 Wet lay-up composite

For the wet lay-up process, a two-component epoxy resin is mixed, and a primer layer is first applied where the composite is to be bonded. Depending on available equipment, the fiber fabric is placed on top of the primer then impregnated by spreading resin over the fabric. If a mechanical impregnator is available, the fabric is impregnated with resin prior to application, as shown in the Fig. 4.1, then applied to the surface of the bridge deck as shown in Fig. 4.2. Compaction is performed with an aluminum roller to remove excess resin and air voids. The entire length is checked for defects during and after placement.

4.2.4 Adhesively bonded prefabricated carbon fiber reinforced polymer (CFRP) composite

For adhesive bonding, prefabricated (pultruded) CFRP composite strips are typically precut to precise dimensions for each location. A two-component

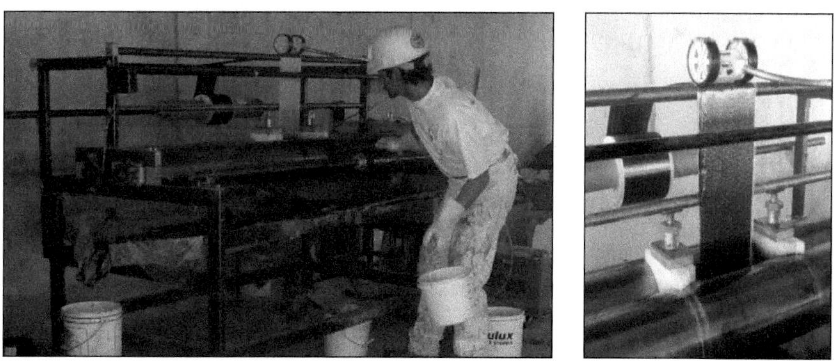

4.1 Impregnation of carbon fabric in wet lay-up process.

4.2 Application of wet lay-up CFRP composites to deck soffit.

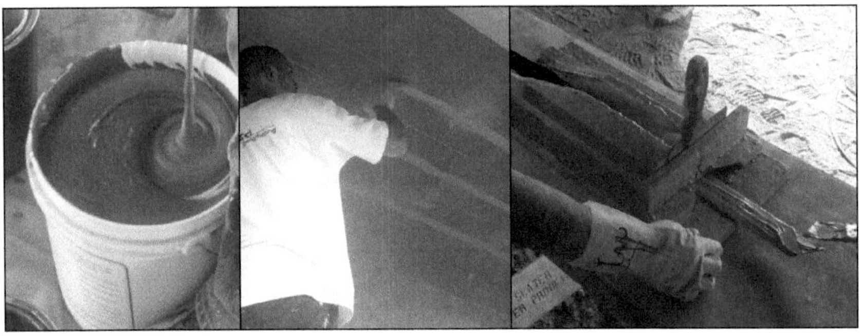

4.3 Mixing and application of epoxy adhesive.

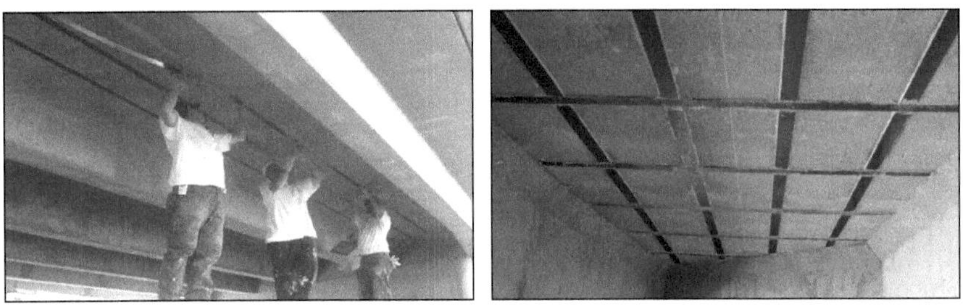

4.4 Bonding of pultruded CFRP strips to deck soffit.

epoxy adhesive is mixed using a mechanical mixer incorporating shearing action and, once mixed, the adhesive is applied to the surface of the deck slab and the prefabricated strip, as shown in Fig. 4.3. The prefabricated CFRP strips are bonded by applying pressure in a continuous manner from one end of the pultruded strip to the opposite end, in the fiber direction, as shown in Fig. 4.4. Typically, uniform pressure is applied on the CFRP strip using a polyurethane roller, and excess adhesive is removed using a squeegee edge.

4.3 Flexural strengthening

Flexural strengthening or repair of RC members with FRP composites is generally conducted by bonding composite materials on the tension side of bending members (Teng *et al.*, 2003). The design process is analogous to design of RC sections, with special considerations for FRP composite bonding (Tajlsten, 2002).

In the design or capacity evaluation of a FRP composite-strengthened RC member, the following assumptions are typically accepted in the literature (Tajlsten, 2002): (i) plane sections remain plane after bending (Bernoulli's Hypothesis); (ii) there is sufficient bond to ensure complete composite action between materials; (iii) cracked concrete retains no tensile strength; (iv) FRP composite behaves in a linear-elastic fashion up to failure.

Design of FRP composites for flexural strengthening requires consideration for seven primary failure modes, which are controlled in the design process. The failure modes are (Tajlsten, 2002; Teng *et al.*, 2003): (i) concrete crushing; (ii) yielding of tensile reinforcement; (iii) yielding of compressive reinforcement; (iv) FRP rupture; (v) shear failure; (vi) concrete cover separation; and (vii) plate and interfacial debonding.

Shear failure, though not addressed in detail here, is possible if the flexural capacity of the RC member exceeds the shear capacity (Teng *et al.*, 2003). Yielding in the steel tensile reinforcement at the ultimate limit state is assumed, while yielding of steel tensile reinforcement at the serviceability limit is not permitted for FRP rehabilitated structures (Tajlsten, 2002).

Most available literature evaluates the effectiveness of FRP composite rehabilitation based on comparisons of control specimens and specimens strengthened with FRP composites. While the application of FRP composite on the tension side of RC flexural members provides some measure of capacity increase relative to the pre-rehabilitated state of the structure, it is observed in the literature that the term 'repair' with FRP composite most often refers to restoring the intended capacity of a damaged structural component, i.e. restoring the structural member to its intended design function, whereas, 'strengthening' often refers to an increase in the load limits of the structure. In the following sections, flexural strengthening of beams, slabs, and in-field applications is summarized to show evidence of the effectiveness of FRP composites used to rehabilitate RC structures.

It is interesting to note that existing literature evaluating FRP-rehabilitated structures rarely addresses the time-dependent degradation of FRP composites when exposed to harsh, changing environments. More importantly, no analytical methods are provided to assess the change in performance of FRP-strengthened or -repaired structures that incorporates material variability and the degradation of FRP composite materials exposed to severe environmental conditions.

4.3.1 FRP repair/strengthening of beams

In this section, recent literature regarding the use of externally-bonded FRP composites to strengthen RC beam structures is provided. The intention is to provide a summary of the work as a means of showing context rather than to provide a comprehensive summary. A more detailed summary can be found in Teng *et al.* (2003) and Stallings *et al.* (2000). The intent of these summaries is to illustrate techniques to characterize the effectiveness of FRP composite rehabilitation of beam structures, estimate capacity, and incorporate durability.

Crasto *et al.* (2001) evaluated CFRP composite rehabilitation of the girders of a precast RC box-beam superstructure, and evaluated the rehabilitation one-year following application by removing an exterior strengthened beam from the bridge structure and testing in a laboratory. While the test result showed the moment capacity of the strengthened beam was approximately 30% greater than a control specimen, the removal of a bridge structure from an in-service bridge is not a feasible option for evaluating the performance of a FRP rehabilitated structure.

Reed and Peterman (2004) applied CFRP composites for flexural and shear strengthening of full-scale pre-cracked, prestressed concrete beams. Beam specimens were taken from a deteriorated bridge structure showing large shear and flexure cracks due to overloading. Beam structures were tested for control, with flexural CFRP rehabilitation, and flexural and shear CFRP composite rehabilitation. The flexural strengthening showed a moment capacity increase of 23.9% relative to control; the shear and flexure strengthened beam showed a moment capacity that was 25.6% greater than the control specimen, illustrating the effectiveness of FRP rehabilitation in a laboratory environment.

Almusallam and Al-Salloum (2001) have suggested an analysis model to predict the nominal moment capacity of RC beams strengthened with external FRP laminates. The methodology was able to establish limits on laminate thickness in order to assure tensile failure due to steel yielding and to avoid tensile failure due to laminate rupture. An experimental program was followed for strengthening 2050 mm long RC beams with glass FRP (GFRP) composites and carbon FRP (CFRP) composites using one and two layers of either GFRP composite (1.3 mm) or CFRP composite (1.0 mm), comparing glass FRP and carbon FRP materials. It was observed that theoretical predictions for load capacity were conservative and that using four layers of GFRP composite provided the same capacity as two layers of CFRP composite in this study. While the methodology provided conservative estimates of the capacity of FRP rehabilitated beams, the incorporation of material durability was not addressed.

Lau and Zhou (2001) conducted an experimental study on the performance of glass fiber composites on RC cylinders and concrete beam specimens exposed to ambient, saline water, alkaline, and acidic solutions for a six-month duration. Effects of full immersion of plain concrete and FRP-strengthened concrete beams in an acidic solution (pH 4) resulted in a 24.8% reduction in flexural strength, while specimens in other environments showed reductions in flexural strengths within 10%. In effect, Lau and Zhou conducted a durability characterization of the FRP rehabilitated beam structure; however, exposing an entire rehabilitated component neglects the intent of rehabilitation where FRP composites are bonded to an already aged or deteriorated structure.

4.3.2 FRP repair/strengthening of slabs

In this section, the literature on FRP strengthening of slabs is highlighted. Currently, flexural strengthening of RC slabs with FRP composites is designed and evaluated with the same principles used in flexural strengthening of RC beam sections. While the examples show that FRP composites on a slab structure are able to increase capacity of slabs with laboratory tests, the in-service evaluations of FRP rehabilitated slabs is not available, nor are procedures to incorporate the durability of FRP composites on the performance of a rehabilitated slab.

Seim *et al.* (2001) evaluated the response and efficiency resulting from the use of FRP strips for strengthening one-way scale slabs; specifically, load capacity with externally-bonded FRP was investigated. Slabs were strengthened with wet lay-up CFRP composites, and pultruded CFRP composite strips. Overall, the experimental tests of one-way scaled slabs using both prefabricated strips and fabric showed that, depending on material and configuration, the ultimate load capacity of the slab can be increased to 3.7 times the ultimate load capacity of as-built slabs.

Mosallam and Mosalam (2003) evaluated the strengthening of two-way RC slabs by bonding carbon and glass FRP composites. The objective of their experimental and analytical program was to confirm the effectiveness of using FRP composites in repair and rehabilitation of unreinforced and RC slabs. The results of the slab tests showed that repair of unreinforced concrete slabs with FRP composites provided five times more capacity than the unreinforced control slab. For RC slabs, repair using FRP composites doubled the capacity of damaged two-way slabs.

Seim *et al.* (2003) studied the use of externally-bonded pultruded CFRP composites to increase the load-bearing capacity of RC slabs and restore the load-bearing capacity of slabs with an accidentally cut steel rebar. Results showed that the capacity of the CFRP-strengthened slabs were twice that of the as-built slab. The objectives of the cut reinforcement specimens were to simulate damage and show the ability of the bonded CFRP composite to increase capacity in the presence of damage. The load capacity of the repaired slab was able to exceed the load capacity of as-built RC slabs. While capacity increase is shown for a damaged slab structure, the damage implemented does not reflect the deteriorated state of an in-service slab and does not correlate with rehabilitation of a deteriorated structure.

Arduini *et al.* (2004) evaluated the behavior of one-way RC slabs rehabilitated with wet lay-up CFRP composites. The objective of the research was to validate guidelines established by ACI 440.2R for flexural strengthening of one-way slabs, emphasizing modes of failure for variations in quantity of applied FRP composites and steel reinforcement. The load carrying capacity of one-way slabs was shown to increase by 122% of the control slabs, where the increase is more pronounced for slabs with low steel reinforcing ratios.

Limam *et al.* (2003) also conducted an experimental investigation on strengthening of slabs with CFRP composite strips. The strengthened RC slab was designed as a three-layer plate, using the upper bound theorem of limit analysis for a multi-layered plate to estimate the ultimate load capacity of the slab. During the experiment, the control slab failed at 48 kN while the CFRP strengthened slab failed at 120 kN, with CFRP strip debonding and steel yielding. For this failure mechanism of the strengthened slab, the limit analysis model predicts a 123 kN ultimate load. The authors propose an analytical approach to show the effectiveness of a FRP composite rehabilitation; however, it is unclear how a deteriorated slab specimen or durability of FRP composites could be incorporated into the analysis method for capacity assessment.

Oh and Sim (2004) have proposed a theoretical model to predict the punching shear capacity of RC decks with externally-bonded FRP composites, considering contributions from top, middle, and bottom layers of concrete, as well as dowel action. The authors compared the analytical model with experimental data of slabs strengthened with steel plates, carbon fiber sheets, glass fiber sheets, and GFRP laminates. A slab strengthened using CFRP composites in both directions was observed to increase the load carrying capacity of an undamaged RC slab by 15% compared with the control, the steel reinforcement yielding before punching failure.

4.3.3 FRP repair/strengthening of bridge structures in the field

The effectiveness of an FRP composite rehabilitation is typically conducted with a load test on the entire structural system, rarely addressing whether or not the design was able to meet strengthening objectives, as observed in a review of FRP composite rehabilitation of bridge structures by Meier

(2000). Assessment of load capacity alone, without consideration for field conditions and design, provides little useful information about the effectiveness of the rehabilitation and may result in inefficient FRP composite rehabilitations in the future due to degradation in FRP composites. Mere determination that capacity has increased with externally-bonded FRP, without relating to design and to material properties of the composite, fails to fully assess the quality of an FRP rehabilitation measure. A number of researchers have demonstrated increase in capacity as a result of FRP strengthening measures in both laboratory and field conditions. Bonacci and Maalej (2000) provide a comprehensive review of strengthening with FRP; here, some recent work with externally-bonded FRP is highlighted.

Hag-Elsafi *et al.* (2001) evaluated the effectiveness of FRP composites for flexural and shear strengthening of an RC T-beam bridge with the use of load tests before and after rehabilitation. The bonded FRP laminates were used by the New York State Department of Transportation in a demonstration project to repair girders of a concrete T-beam bridge to increase their flexural and shear capacities. The cost of the rehabilitation was estimated at 25% of the cost of bridge replacement. The FRP composite was designed to compensate for a 15% loss in corroded steel, using 67% of FRP ultimate strength at steel yield. From load tests, the authors observed decreases in rebar stresses, a slight increase in concrete stresses, and improvements in transverse live load distribution. The minimal changes observed in the structure were attributed to a limited load range during testing.

Stallings *et al.* (2000) have described the rehabilitation of an existing seven-simple-span, RC, T-girder bridge structure, built in 1952 on Alabama Highway 110. The bridge structure was originally designed for H15-44 design loading for girders, which corresponds to 63% of the HS20-44 design loading currently used by Alabama DOT. The objective of the repair with FRP was to mitigate the deterioration of the bridge resulting from widening of the flexural cracks due to repeated heavy traffic loading, and to increase the bending moment capacity by 20%. One span of the bridge was chosen for repair, with CFRP composite plates applied to the bottom surface of the girder and GFRP plates applied to locations on the sides of the girder to resist flexural crack opening and add stiffness to the bridge. Load tests were conducted on the structure before and after FRP application using standard trucks with a three-axle configuration and gross vehicle weight of 346 kN. Results indicated reduction in rebar stresses after FRP repair of 4 to 12%. Reductions in deflections were observed in the range of 2 to 12%, depending on the loading configuration.

Shahrooz and Boy (2004) repaired the deck of a skewed (70-degree), three-span, RC slab bridge, built in 1955. The bridge structure displayed cracks on the soffit of the deck slab and had been posted because of insufficient capacity of the slab. In order to remove the load limits of the structure, the bridge deck slab was strengthened with adhesively bonded CFRP composite strips and wide wet lay-up carbon fabric. The effectiveness of the CFRP strengthening of the bridge deck was determined before, immediately after, and one year after FRP application via load testing. From an evaluation of measured strains, the concrete strain was reduced by as much as 34% following the application of the CFRP composite. After a year of service, the concrete strains in the bridge structure did not change significantly.

4.3.4 Discussion

Literature for external bonding of FRP composites to repair or strengthen a deteriorated RC structure in flexure shows a capability to increase or maintain the load bearing capacity of structures.

For beams and slab structures, the gains in capacity of uncracked and precracked specimens with the bonding of FRP composites are well documented. However, little information is available that addresses the durability of FRP-strengthened structure. Specifically, the question remains, how long can FRP-rehabilitated structures sustain load demands in the presence of a harsh changing environment.

The evaluation of FRP rehabilitations on RC bridge structures was conducted by the use of a load test (Hag-Elsafi *et al.*, 2001; Shahrooz and Boy, 2004; Stallings *et al.*, 2000). These load tests were conducted before and after FRP application to assess changes in deflection or strain level. A reduced load deflection or reductions in measured strain levels were assumed to imply that the FRP rehabilitation was effective. The implications of varying material properties or the level of increase in performance was not specifically addressed, i.e. what is the method and standard by which an FRP rehabilitation design is deemed to achieve the intended performance level of a structure?

Durability of FRP rehabilitated bridge structures is typically addressed by conducting a load test one year after rehabilitation. The use of a load test does provide a single validation regarding the behavior and load capacity of the structure; however, a load test on the structure does not provide any indication of future performance and requires closures of the bridge structure in order to evaluate it.

Recently some researchers have attempted to apply dynamic testing to rehabilitated beam structures and bridges to evaluate the effectiveness of repairs. For instance, Cardinale and Orlando (2004) measured frequency changes before and after repairs were conducted on an RC bridge in Italy. Repairs included increasing slab thickness, installing wide flange steel beams, repairs to foundations and anchors, and the use of CFRP composites for shear strengthening of longitudinal girders. A frequency change from 2.46 to 2.7 Hz was measured after rehabilitation, and was correlated with a 20% change in flexural stiffness from a finite element model.

Bonfiglioli *et al.* (2004) used dynamic testing to measure modal parameters before and after repair of RC beam specimens. Beams were cracked by applying four cycles of loading, then strengthened in flexural. An attempt was made to correlate the static stiffness changes in the load displacement curves to changes in frequency and damping ratios from dynamic tests. Changes in frequency and damping are qualitatively correlated to stiffness increase after CFRP strengthening, with damping indicating more sensitivity to the stiffness changes than frequency measures.

While these studies indicate the potential to evaluate the effectiveness of repairs on bridges and structural components, assessments to date remain qualitative, without the ability to provide any indication for future performance of the structure. A methodology is necessary to interrogate the performance of an FRP rehabilitated structure and provide an assessment of its future performance considering deterioration and variability of FRP composites.

4.4 Service life estimation of fiber reinforced polymer (FRP) rehabilitated bridge decks

The service life of a structural component depends on the loading conditions and the ability of the materials to resist demands safely over a period of time. Variability is inherent in the loadings imposed on the structure, the material parameters used in capacity calculations, and the environment in which the structure exists.

Termination of the service life of RC bridge decks is conventionally associated with the accumulation of irreversible damage resulting from corrosion of reinforcement, freeze – thaw cycles, traffic loading, in addition to the initial damage resulting from poor design and/or construction and inadequate inspection and maintenance practices (Lounis, 2000). Corrosion of steel reinforcement is widely recognized as the most common source of strength loss in RC bridges (Enright and Frangopol, 1998). When composite materials are used to repair or strengthen bridge components, they too are subject to strength losses over time, depending on environmental factors such as temperature, moisture levels, chemical solutions, stress and time. For instance, moisture absorption into the matrix of a composite has been observed to cause plasticization and decrease the glass transition temperature (Bank *et al.*, 1995). In some cases, moisture can wick along the fiber–matrix interface, and results in debonding between the fiber and the matrix of the composite, which can lead to premature failure and loss of load transfer across fibers (Weitsman and Elahi, 2000).

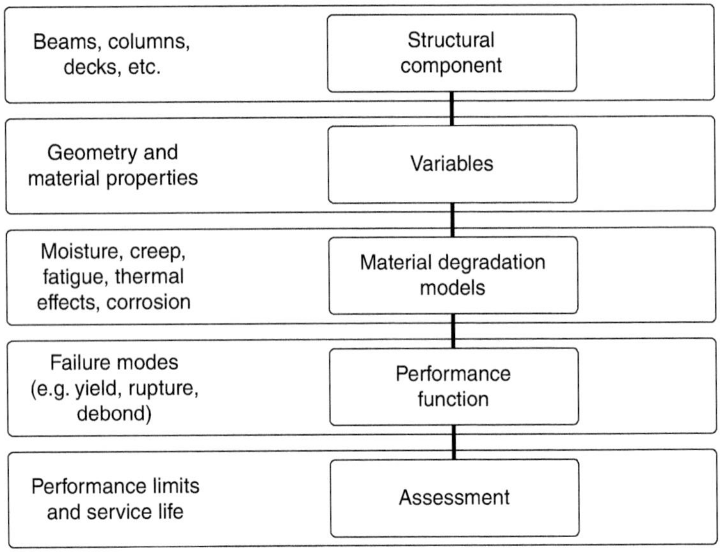

4.5 Framework for life-cycle evaluation.

Deterioration of the materials that make up the bridge superstructure, and any rehabilitation schemes, can curtail the ability of the structure to perform its intendend function. Variation in quality of construction, as well as the statistical variation inherent in material properties, leads to a level of uncertainty with regard to the safety of the structure. Design safety margins are directly affected by variation in material properties, while issues related to durability and aging of material can shorten the life of a bridge deck. A time-dependent probability based approach to service life estimation allows the integration of information related to durability of materials and the variation in material parameters that are inherent or due to manufacturing quality. Material degradation models can be integrated into life-cycle performance models, as proposed by Hastak *et al.* (2003) and Frangopol *et al.* (2003). A general framework for evaluating life-cycle performance of a component is shown in Fig. 4.5.

4.4.1 Structural reliability

A measure of performance evaluates the likelihood that a structural component is able to resist a level of demand. For example, the resistance of an RC beam section is defined by its moment capacity, while the demands are the anticipated moment-loads on the structure.

Structural reliability methods are used to compute a reliability index, β, which is based on the materials and configuration of the structure (Atadero *et al.*, 2005). A number of approximation techniques, such as mean value second moment, first-order methods, and second-order methods, are well established and available to calculate β (Melchers, 1999). The failure probability, p_f, of a structural component can be related to the reliability index, β through the standard normal distribution function, Φ.

$$p_f = \Phi(-\beta) \quad [4.1]$$

Higher values of β indicate a higher degree of structural reliability (or safety) for a given limit state equation.

Overview of the reliability problem

The structural reliability problem considers a load effect defined by S, resisted by resistance, R. These are described by probability distribution functions $f_S(\ )$ and $f_R(\ )$, respectively. In general, the resistance, R, is a function of random variables, X_i, which may represent material properties and dimensions,

$$R(X_i) \quad \text{for } i = 1, \ldots, n \text{ number of random variables.} \quad [4.2]$$

Similarly, the load, S, on a structure is a function of random variables, X_i, which may represent applied loads, material densities, and possibly geometry of the structure.

$$S(X_i) \quad \text{for } i = 1, \ldots, n \text{ number of random variables.} \quad [4.3]$$

The performance function of the reliability problem is then conveniently expressed as a function of all relevant basic variables,

$$z = \boldsymbol{g}(\boldsymbol{R}, \boldsymbol{S}) = \boldsymbol{g}(X_1, \ldots, X_n) = \boldsymbol{g}(\vec{\mathbf{X}}) \quad [4.4]$$

where $\vec{\mathbf{X}}$ is the vector of random variables with probability distributions, $f_{\vec{X}}(\ )$. With the performance function (g) or limit state function in terms of n random variables, the calculation for the probability of failure is formulated as

$$p_f = P\left[g(\vec{\mathbf{X}}) \le 0\right] = \int \ldots \int_{g(\vec{\mathbf{X}}) \le 0} f_{\vec{X}}(\vec{\mathbf{x}})\, d\vec{\mathbf{x}} \quad [4.5]$$

where $f_{\vec{X}}(\vec{\mathbf{x}})$, is the joint probability density function for the n-dimensional vector $\vec{\mathbf{X}}$ of random variables in the performance function. The region of integration $g(\vec{\mathbf{X}}) \le 0$ defines the failure domain, or 'unsafe' region, and $g(\vec{\mathbf{X}}) > 0$ is the 'safe' region. Typically, the solution for the above integral is determined via numerical simulation techniques such as Monte Carlo simulation, or by avoiding the integration problem with a transformation of the density function, $f_{\vec{X}}(\vec{\mathbf{x}})$ to a multi-normal probability density function and using its unique characteristics to approximate the probability of failure. Methods that incorporate the latter include first-order, second-moment methods, first-order reliability methods (FORM), and second-order reliability methods (SORM).

4.5 Application

In this section, the service life estimation methodology is applied to an RC bridge deck that is strengthened with FRP composites using a mean value, first-order, second-moment formulation of the reliability index where random variables are described by their respective means and variances (Melchers, 1999). Formulation of the reliability analysis for service life estimation requires the following steps:

(i) Identification of random variables with means and variances
(ii) Determination of mean μ_z and variation σ_z of the performance function, z
(iii) Evaluation of time-dependent reliability for service life assessment.

Each of the steps requires knowledge of the material composition of the structure, the load conditions and failure mechanisms of a structure. A performance function and its definition for failure are also necessary. Combining a prediction for the change in reliability as a function of time with an acceptable limit of the probability of failure effectively provides an estimate for the available service life of a structure or component. Specifically, the effect of material degradation on the performance of the structure is made available.

Time-dependent reliability, $\beta(t)$, is developed by introducing time-dependent variables within the second moment reliability equation (Sarja and Vesikari, 1996):

$$\beta(t) = \frac{\mu_z}{\sigma_Z} = \frac{\mu_R(t) - \mu_S}{\left(\sigma_R^2(t) + \sigma_S^2\right)} \qquad [4.6]$$

where μ denotes the mean value; σ^2 is the variance; R is resistance; S is demand; t is time. Specifically, time may be introduced from predictions for deterioration of material parameters such as loss in steel area under corrosion or deterioration in CFRP mechanical properties due to environmental exposure. The standard normal distribution, $\Phi(\ )$, is then used to compute the corresponding probability of failure. Utilizing the reliability index assists in evaluating the effect of variation in material properties of the applied FRP composite and existing steel reinforcement and concrete. In addition, use of the mean value first-order, second-moment reliability index provides a method to include the effects of material degradation and thus allows for determination of a time-dependent measure of reliability.

Although easily implemented, the standard second moment equation is not without its constraints. One of the primary assumptions in using the mean-value, first-order, second-moment measure of the reliability is that random variables are described only by their means and variances, and all random variables in the reliability analysis are assumed normally distributed. A second constraint is a result of the lack of invariance of the reliability index with change in the form of the performance function (Melchers, 1999). However, this constraint is resolved by using the same form of the performance function for all analyses, i.e. $Z = R - S$.

In the example bridge deck analysis, beam sections 30.48 cm (12 inches) wide were used to represent the deck slabs of a typical RC T-girder bridge in the longitudinal and transverse slab directions. Carbon FRP composites were externally bonded to the bridge deck in order to support an increased traffic load from an AAHSTO HS20 wheel load of 71.2 kN (16 kips) to a permit wheel load of 106.76 kN (24 kips), i.e. a live load increase of 50%. Geomctries of longitudinal and transverse beam sections representative of the deck slab are shown in Figure 4.6.

The flexural reinforcement consists of two No. 5 rebars in a single transverse beam width and one No. 5 rebar in a single longitudinal beam width. The area of transverse and longitudinal steel reinforcement areas per beam width are 4 cm^2 (0.62 in^2) and 2 cm^2 (0.31 in^2). In order to support the increase in live load, a wet lay-up manufactured CFRP rehabilitation design specifies 675.21 mm^2 per meter of slab (0.319 in^2/ft) in the longitudinal direction and 480.21 mm^2 per meter of slab (0.227 in^2/ft) in the transverse direction.

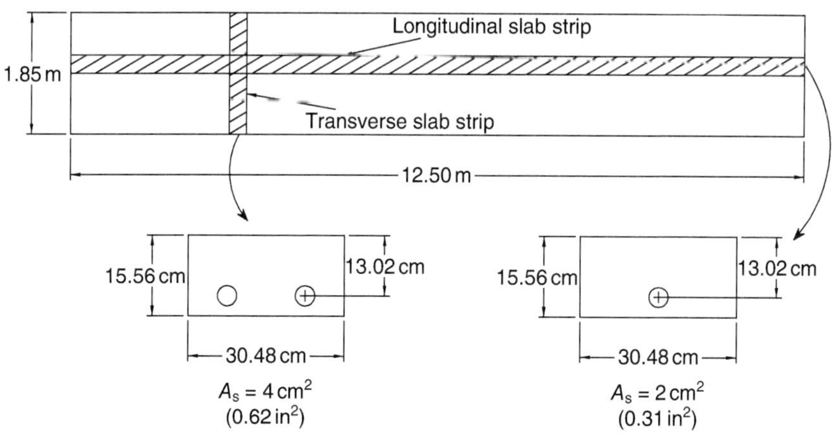

4.6 Representative beam sections for deck slab analysis.

4.5.1 Moment capacity

Means and variance of moment demands on the deck slab are assumed to remain unchanged over time. The time-dependent moment resistance and variance are provided for a beam component with externally-bonded FRP composite.

$$\mu_R(t) = \mu_R[M_R] = \mu[M_C(t)] + \mu[M_{Steel}(t)] + \mu[M_{frp}(t)] \tag{4.7}$$

$$M_C = C_c(t)\cdot\left(c_1(t) - \frac{a}{2}\right) = 0.85 f_c' \beta_1 c_1(t) b - \left(c_1(t) - \frac{\beta_1 c_1(t)}{2}\right) \tag{4.8}$$

$$M_{steel} = T_s \cdot (d - c_1(t)) = A_s(t) f_y \cdot (d - c_1(t)) \tag{4.9}$$

$$M_{frp} = T_{frp} \cdot (h - c_1(t)) = A_{frp} E_{frp}(t) \varepsilon_{frp} \cdot (h - c_1(t)) \tag{4.10}$$

$$\sigma_R^2(t) = \sigma_{M_R}^2(t) = \sigma_{M_{fc}}^2(t) + \sigma_{M_{steel}}^2(t) + \sigma_{M_{frp}}^2(t) \tag{4.11}$$

where t denotes time; $E_{frp}(t)$ is the time-dependent modulus of the FRP composite; μ denotes mean value; M_R is the moment resistance of the reinforced concrete section about the axis of zero strain, or the sum of the moment contributions from concrete, M_C, steel reinforcement, M_{steel}, and FRP composite, M_{frp}; C_c is the force acting in the concrete; f_c' is the compressive strength of concrete; β_1 is the rectangular compressive stress block factor equal to 0.85; c_1 denotes the axis of zero strain from the extreme compression fiber; b is the representative beam width; T_s, force in steel reinforcement; A_s, area of steel reinforcement; f_y, yield strength of steel; d, is the depth to steel reinforcement from the extreme compression fiber of the section; T_{frp}, force in CFRP composite; A_{frp}, area of FRP composite; E_{frp}, modulus of FRP composite; ε_{frp}, strain in FRP composite; h is the height of the section. Substitution of equations 4.7 and 4.11 into equation 4.6 yields an estimate of the reliability as a function of time.

$$\beta(t) = \frac{\mu_z}{\sigma_Z} = \frac{\mu[M_C(t)] + \mu[M_{Steel}(t)] + \mu[M_{frp}(t)] - \mu_S}{\left(\sigma_{M_{fc}}^2(t) + \sigma_{M_{Steel}}^2(t) + \sigma_{M_{frp}}^2(t) + \sigma_S^2\right)^{0.5}} \tag{4.12}$$

where variables are as described previously. Note, μ_S is the mean moment demand and σ_S^2 is the variance of the moment demand. The coefficient of variation (COV) of random variables, f_y, f_c' and E_{frp} are assumed to remain constant with respect to time, while changes in the mean moment arc a function of the time-dependent material parameters.

4.5.2 Random variables for resistance

In a reliability analysis, there are uncertainties associated with the variables that compose the resistance and demand functions. Material parameters having a direct impact on the moment capacity of a carbon FRP rehabili-

Table 4.1 Statistical descriptors for steel and concrete strengths

Variable	Design value (MPa) (ksi)	Bias factor	Mean value (MPa) (ksi)	COV (%)
f_y	413.69 (60.0)	1.1	455.05 (66.0)	10
f'_c	24.82 (3.60)	1.14	25.55 (3.71)	15

Table 4.2 CFRP composite properties

E_{frp}, Mean tensile modulus (GPa) (Msi)	Tensile modulus COV (%)	f_{frp}, Mean tensile strength (GPa) (ksi)	Strength COV (%)
78.92 (11.45)	12.09	0.920 (133.46)	10.10

tated RC deck slab are concrete compressive strength, f'_c, yield strength of steel, f_y, and CFRP composite modulus of elasticity and ultimate strength. Each parameter is described by its respective mean and coefficient of variation, COV. For mean values of steel yield strength and concrete compressive strength, bias factors are available in Barker and Puckett (1997). Statistical descriptors for steel and concrete in this analysis are shown in Table 4.1.

For the statistical descriptors of CFRP composite, the mean value and COVs of as-built composite samples manufactured during the deck rehabilitation are applied for tensile modulus and strength. Table 4.2 summarizes the means and COVs of variables used in this analysis.

4.5.3 Random variables for demand

Dead load moment is calculated by modeling a transverse segment of the slab as a simply supported beam with a continuity factor of 0.8 according to AASHTO standard bridge design specifications including a wearing surface on the top of the deck. A dead load moment bias factor of 1.05 is included for cast-in-place concrete bridges, with a coefficient of variance (COV) of 10% (Nowak, 1999). The bias factor and COV of the surface overlay are assumed to be negligible in this analysis since an overall bias factor and COV are applied to the total dead-load moment.

Consideration for live-load moment demands on the deck slab of the structure includes two load levels, i.e. HS20 and Permit wheel loads. The first is the original design load for the bridge structure, the HS20 wheel load of 71.2 kN (16 kips). The second wheel load is the Permit Truck wheel load of 106.76 kN (24 kips). A live-load bias factor of 1.2 is applied to HS20 live-load moment demands, while the COV of 18% is specified for both

Table 4.3 Summary of demand moments

Moment type	Unfactored moment (kN-m/m) (kip-ft/ft)	Live load moment reduction factor	Dynamic amplification factor	Bias factor	Mean value (kN-m/m) (kip-ft/ft)	COV (%)
Dead load	2.09 (0.47)	0.8	N/A	1.05	1.756 (0.40)	10
Live – HS20	17.98 (4.04)	0.85	1.3	1.2	23.84 (5.36)	18
Live – Permit	26.97 (6.06)	0.85	1.3	1	29.80 (6.70)	18

HS20 and Permit Truck live load moments (Nowak, 1999). In addition, the dynamic load factor of 1.3 is applied to both live load moments. A live-load moment reduction factor of 0.85 is conservatively specified for a two-lane highway bridge according to NCHRP Report 368 (Nowak, 1999). The resulting live-load moment is 7.27 kN-m per 30.48 cm of slab or 23.84 kN-m/m for the HS20 design load and 8.22 kN-m per 30.48 cm of slab or 26.97 kN-m/m with a COV of 18%. A summary of dead and live load moments and statistical descriptors is given in Table 4.3.

4.5.4 Material deterioration models

By utilizing time-dependent variables in reliability analysis, it is possible to calculate the change in reliability index as a function of time, $\beta(f)$. Specifically, time dependence may be introduced from predictions for degradation of material parameters such as potential degradation of mechanical properties of composite materials due to environmental exposure.

Prediction of the long-term response of composite materials typically uses accelerated tests, usually with higher temperatures as the forcing function in a time–temperature superposition methodology. Equations for the prediction of performance attributes (i.e. strength and modulus) using the Arrhenius method are presented in equation 4.6, which can be useful in estimating long-term response of a given limit state function (Karbhari and Abanilla, 2007).

$$P(t) = \frac{P_0}{100}[A\ln(t) + B] \quad [4.13]$$

where $P(t)$ and P_0 are performance attributes at time, t, and 0 (the unexposed condition), respectively; A is a constant that denotes degradation and B is a material constant accounting for post-cure effects ($B \geq 100$). The quantity $[A \ln(T) + B]$ represents the percentage of the performance attribute P_0 retained as a function of time.

For the purposes of the current investigation, durability of FRP composites is characterized by the use of accelerated aging procedures. Predictions for CFRP composite tensile modulus and tensile strength are modeled with an Arrhenius rate relationship for wet lay-up CFRP composites immersed in deionized water at 23°C for approximately two years (Karbhari and Abanilla, 2007). The tensile modulus and tensile strength for the CFRP composite are expressed as a function of time, t, in years.

$$E_{\text{frp}}(t) = \frac{E_{\text{frp}}}{100}[-0.4182\ln(t \cdot 365) + 100] \quad [4.14]$$

$$f_{\text{frp}}(t) = \frac{E_{\text{frp}}}{100}[-3.366\ln(t \cdot 365) + 100] \quad [4.15]$$

Corrosion-induced reduction in steel cross-sectional area is included in the service life based design procedure with the following relationship (Cheung and Kyle, 1996):

$$A_s(t) = \frac{n\pi}{4}(D_0 - C_r t)^2 \quad [4.16]$$

where $A_s(t)$ denotes the time-dependent cross-sectional area of steel reinforcement; n is the number of steel rebars in the section; D_0 is the initial diameter of a single rebar; C_r is the rate of reinforcing corrosion per year in mm/year; t is time in years. The rate of reinforcing corrosion is derived for a moderate corrosion rate of 1 μA/cm^2, which results in a loss of cross-sectional radius of approximately 0.0115 mm/year (Andrade and Alonso, 2001).

4.5.5 Results and discussion

Post rehabilitation and service life results using the reliability index are calculated with respect to the direction of the slab, i.e. the longitudinal or transverse direction of the slab. The CFRP strengthening was provided for a three-layer CFRP composite in the transverse direction and a two-layer CFRP composite in the longitudinal direction. The reliability index before strengthening with CFRP composites for all locations are shown in Table 4.4.

As reference values for the reliability index results, consider that a value of 3.5 is used in the calibration of the *AASHTO LRFD Bridge Design Specifications* for girder bridges (Nowak, 1999). The reliability index values for the non-rehabilitated deck slab show high reliabilities of 5.92 and 4.72, respectively, in the transverse slab directions when increasing the live load from an HS20 to Permit Truck wheel load. However, in the longitudinal slab directions, reliabilities of 2.07 and 0.854 for the HS20 and Permit-load

Table 4.4 Pre-rehabilitation reliability index values for deck slabs

	HS20		Permit truck	
Direction	β	P_f (%)	β	P_f (%)
Longitudinal	2.07	1.92	0.854	19.64
Transverse	5.92	1.61×10^{-7}	4.72	1.18×10^{-4}

Table 4.5 Reliability index of deck slab before and after strengthening

	Pre-strengthening		Post-strengthening	
Slab orientation	β	P_f (%)	β	P_f (%)
Longitudinal	0.854	19.6	3.59	0.0017
Transverse	4.72	1.17×10^{-4}	6.20	2.77×10^{-8}

moment demands, respectively, show a significant increase in the probability of failure when increasing the live load to a Permit-load condition.

By strengthening with CFRP composites for the live load increase of 50%, the reliability index in the transverse and longitudinal directions increased from 4.72 to 6.20 in the transverse direction and from 0.854 to 3.59 in the longitudinal direction. While the longitudinal direction initially satisfies a performance requirement of 3.5, with material degradation over time, the performance limit would be violated. However, it should be noted that the primary loading direction in a T-girder bridge deck slab is the transverse direction. Table 4.5 summarizes instantaneous reliability indices before and after strengthening of the deck slab.

One of the advantages of the reliability index evaluation is the ability to quantify the effect of FRP composite rehabilitation based solely on the variation in mechanical properties of the as-built structure and the area of composite applied to a reinforced concrete section for flexural failure. Inclusion of the degradation models can provide a graphical view of the rate of change of reliability index with respect to time. Figure 4.7 shows the time-dependent reliability results for a bridge deck undergoing moderate corrosion and a Permit truck wheel live load. Reliability curves with and without strengthening using CFRP composites are shown. At 15 years, the CFRP strengthened bridge deck reliability index decreases to 5.93 in the transverse direction and 3.24 in the longitudinal direction. Establishing a minimum acceptable performance level or the amount of acceptable risk, allows for an estimation of the remaining service life of the bridge deck. For instance, a minimum allowable reliability index of 3.0 (0.13%) would

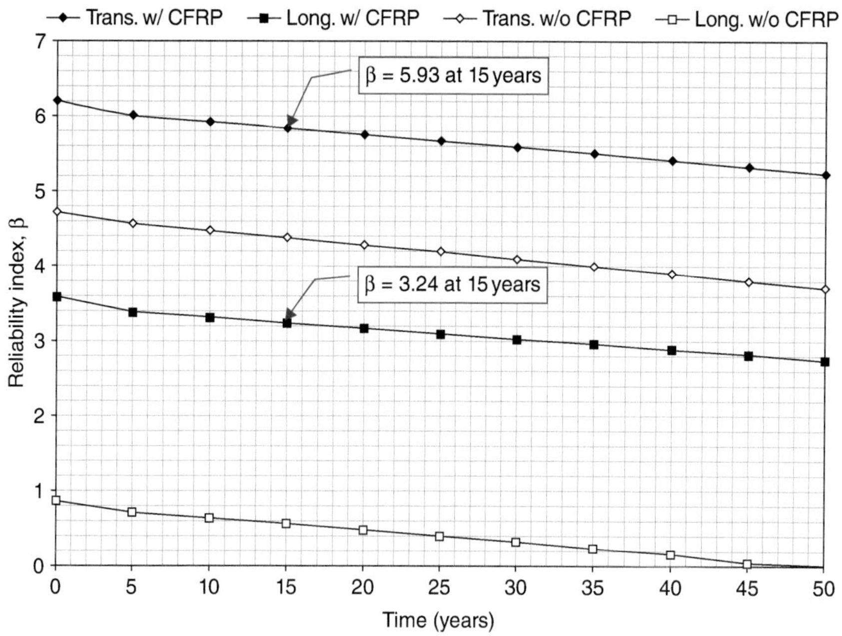

4.7 Time-dependent reliability of bridge deck subject to 50% live load increase.

be satisfied in the transverse direction even without FRP composites over a 50-year period, while in the longitudinal direction, at approximately 30 years, the minimum acceptable performance limit is violated.

Not only does an estimate of the service life of a rehabilitated component provide a quantitative measure of durability, it also provides the basis for establishing appropriate inspection intervals and evaluation of life-cycle costs, which is useful in decision making with regard to material selection, replacement, repair, or FRP rehabilitation design based on specified life extension. The estimation of the remaining service life provides a means to assess the criticality of existing structures, but can also act as a design tool to determine adequate levels of rehabilitation with respect to the reliability of the system and desired extension of service life for an existing deficient structure. Solving the reliability problem in reverse, beginning with a desired service life extension, can determine the appropriate FRP composite needed to sustain a level of performance required.

4.6 Conclusions and future trends

The use of a time-dependent probabilistic approach to service life estimation provides a means to incorporate uncertainties associated with the implementation of FRP composites to extend the service life of RC bridge structures. Uncertainties associated with material degradation, construction quality in terms of coefficient of variation, and service life can be integrated in order to characterize a component's long-term performance. Consideration of the reliability index provides a valuable method for assessment of the effectiveness of the rehabilitation, independent of the design procedure followed to determine the quantity of composite necessary to rehabilitate bridge superstructure components.

While a methodology has been presented for service life extension and rehabilitation for a bridge superstructure, future work is needed with regard to the following:

- Inclusion of additional failure modes of the service life estimation to incorporate available durability predictions for bond, flexural strength, and interlaminar shear strength of FRP composites.
- Application of durability characterization of FRP composite materials from exposure to other environments, such as immersion in saltwater solution or alkali solution to the performance of an FRP rehabilitated beam structure.
- Consideration of service life and performance for shear rehabilitation and torsional rehabilitation of bridge superstructure components with FRP composites.

4.7 References

ACI, AMERICAN CONCRETE INSTITUTE. (2002). 'Guide for the Design and Construction of Externally Bonded FRP Systems for Strengthening Concrete Structures.' *ACI 440.2R-02, Emerging Technology Series*, Reported by ACI Committee 440, October, First Printing.

ALMUSALLAM, T.H. and AL-SALLOUM, Y.A. (2001). 'Ultimate Strength Prediction for RC Beams Externally Strengthened by Composite Materials.' *Composites: Part B*, 32, 609–619.

ANDRADE, C. and ALONSO, C. (2001). 'On-site Measurements of Corrosion Rates of Reinforcements.' *Construction and Building Materials*, 15(2–3), 141–145.

ARDUINI, M., NANNI, A., and ROMAGNOLO, M. (2004). 'Performance of One-way Reinforced Concrete Slabs with Externally Bonded Fiber-reinforced Polymer Strengthening.' *ACI Structural Journal*, 101(2), 193–201.

ATADERO, R.A., LEE, L., and KARBHARI, V.M. (2005). 'Consideration of Material Variability in Reliability Analysis of FRP Strengthened Bridge Decks', *Composite Structures*, 70, pp. 430–443.

BANK, L.C., GENTRY, T.L., and BARKATT, A.R. (1995). 'Accelerated Test Methods to Determine the Long-term Behavior of FRP Composite Structures: Environmental Effects', *J. Reinforced Plastics and Composites*, 14, 559–587.

BARKER, R.M. and PUCKETT, J.A. (1997). *Design of Highway Bridges*. John Wiley & Sons, Inc., Hoboken, NJ, USA.

BONACCI, J.F. and MAALEJ, M. (2000). 'Externally Bonded FRP for Service Life Extension of RC Infrastructure.' *J. Infrastructure Systems*, 6(1), 41–51.

BONACCI, J.F. and MAALEJ, M. (2001). 'Behavioral Trends of RC Beams Strengthened with Externally Bonded FRP.' *J. Composites for Construction*, 5(2), 102–113.

BONFIGLIOLI, B., PASCALE, G., and MARTINEZ DE MINGO, S. (2004). 'Dynamic Testing of RC Beams Damaged and Repaired with Fiber Reinforced Polymer Sheets.' *J. Materials in Civil Engineering*, 16(5), 400–406.

CARDINALE, G. and ORLANDO, M. (2004). 'Structural Evaluation and Strengthening of a Reinforced Concrete Bridge.' *J. Bridge Engineering*, 9(1), 35–42.

CHEUNG, M.S. and KYLE, B.R. (1996). 'Service Life Prediction of Concrete Structures.' *Construction and Building Materials*, 10(1), 45–55.

CRASTO, A.S., KIM, R.Y., and MISTRETTA, J.P. (2001). 'The Application of Composites for the Rehabilitation of Concrete Bridge Infrastructure.' *Advances Composite Materials*, 10(2,3): 147–157.

ENRIGHT, M.P. and FRANGOPOL, D.M. (1998). 'Service life Prediction of Deteriorating Concrete Bridges.' *J. Structural Engineering*, 24(3), 309–317.

FRANGOPOL, D.M, BRUHWILER, E., FABER, M.H., and ADEY, B. (ED.) (2003). *Life-cycle Performance of Deteriorating Structures: Assessment, Design, and Management*, American Society of Civil Engineers, ASCE Publications.

HAG-ELSAFI, O., ALAMPALLI, S., and KUNIN, J. (2001). 'Application of FRP Laminates for Strengthening of a Reinforced Concrete T-beam Bridge Structure.' *Composite Structures*, 52, 453–466.

HASTAK, M., MIRMIRAN, A., and DEEPAK, R. (2003). 'A Framework for Life-cycle Costs Assessment of Composites in Construction', *J. Reinforced Plastics and Composites*, 22(15), 1409–1430.

KARBHARI, V.M. (2000). 'Determination of Materials Design Values for the Use of Fibre-reinforced Polymer Composites in Civil Infrastructure.' *Proceedings Institution Mechanical Enginess*, 214, Part L, 163–171.

KARBHARI, V.M. and ABANILLA, A. (2007). 'Design Factors, Reliability, and Durability Prediction of Wet Lay-up Carbon/Epoxy Used in External Strengthening', *Composites Part B*, 38, 10–23.

KARBHARI, V.M. and SEIBLE, F. (2000). 'Fiber Reinforced Composites – Advanced Materials for the Renewal of Civil Infrastructure.' *Applied Composite Materials*, 7, 95–124.

KARBHARI, V.M. and ZHAO, L. (2000). 'Use of Composites for 21st Century Civil Infrastructure.' *Computer Methods Applied Mechanics Engineering*, 135, 433–454.

KARBHARI, V.M., CHIN, J.W., HUNSTON, D., BENMOKRANE, B., JUSKA, T., MORGAN, R., LESKO, J.J., SORATHIA, U., and REYNAUD, D. (2003). 'Durability Gap Analysis for Fiber-Reinforced Polymer Composites in Civil Infrastructure.' *J. Composites for Construction*, 7(3), 238–247.

LAU, K. and ZHOU, L. (2001). 'Mechanical Performance of Composite-strengthened Concrete Structures.' *Composites: Part B*, 32: 21–31.

LIMAM, O., FORET, G., and EHRLACHER, A. (2003). 'RC Two-way Slabs Strengthened with CFRP Strips: Experimental Study and a Limit Analysis Approach.' *Composite Structures*, 60, 467–471.

LOUNIS, Z. (2000). 'Reliability-based Life Prediction of Aging Concrete Bridge Decks.' In *Life Prediction and Aging Management of Concrete Structures*, ed. D. Naus, RILEM Publications, France, 229–238.

MALEK, A.M. and PATEL, K. (2003). 'Flexural Strengthening of Reinforced Concrete Flange Beams with Composite Laminates.' *J. Composites for Construction*, 6(2), 97–103.

MEIER, U. (2000). 'Composite Materials in Bridge Repair.' *Applied Composite Materials*, 7, 75–94.

MELCHERS, R. (1999). *Structural Reliability Analysis and Prediction.* John Wiley & Sons, Hoboken, NJ, USA.

MIRMIRAN, A., SHAWHAWY, M., NANNI, A., and KARHBARI, V. (2004). *Bonded Repair and Retrofit of Concrete Structures Using FRP Composites – Recommended Construction Specifications and Process Control Manual.* NCHRP Report 514, Transportation Research Board.

MOSALLAM, A.S. and MOSALAM, K.M. (2003). 'Strengthening of Two-way Concrete Slabs with FRP Composite Laminates.' *Construction and Building Materials*, 17, 43–54.

NANNI, A. (2003). 'North American Design Guidelines for Concrete Reinforcement and Strengthening Using FRP: Principles, Applications, and Unresolved Issues.' *Construction and Building Materials*, 17, 439–446.

NOWAK, A.S. (1999). *NCHRP Report 368, Calibration of LRFD Bridge Design Code.* National Academy Press, Washington D.C.

OH, H. and SIM, J. (2004). 'Punching Shear Strength of Strengthened Deck Panels with Externally Bonded Plates.' *Composites: Part B*, 35: 313–321.

REED, C.E. and PETERMAN, R.J. (2004). 'Evaluation of Prestressed Concrete Girders Strengthened with Carbon Fiber Reinforced Polymer Sheets.' *J. Bridge Engineering*, 9(2), 185–192.

SARJA, A. and VESIKARI, E. (1996). *Durability Design of Concrete Structures, Report of RILEM Technical Committee 130-CSL.* E & FN Spon, London.

SEIM, W., HORMAN, M., KARBHARI, V., and SEIBLE, F. (2001). 'External FRP Poststrengthening of Scaled Concrete Slabs.' *J. Composites for Construction*, 5(2): 67–75.

SEIM, W., VASQUES, A., KARBHARI, V., and SEIBLE, F. (2003). 'Poststrengthening of Concrete Slabs: Full-scale Testing and Design Recommendations.' *J. Structural Engineering*, 129(6), 743–752.

SHAHROOZ, B.M. and BOY, S. (2004). 'Retrofit of a Three-span Slab Bridge with Fiber Reinforced Polymer Systems – Testing and Rating.' *J. Composites for Construction*, 8(3), 241–247.

STALLINGS, J.M., TEDESCO, J.W., EL-MIHILMY, M., and MCCAULEY, M. (2000). 'Field Performance of FRP Bridge Repairs.' *J. Bridge Engineering*, 5(2), 107–113.

TAJLSTEN, B. (2002). *FRP Strengthening of Existing Concrete Structures – Design Guidelines*. Lulea University Printing Office, Lulea, Sweden.

TENG, J.G., CHEN, J.F., SMITH, S.T., and LAM, L. (2003). 'Behavior and Strength of FRP Strengthened RC structures: A State-of-the-art Review.' *Structures & Buildings*, 156(1), 51–62.

TRIANTAFILLOU, T.C. and PLEVRIS, N. (1991). 'Post-strengthening of R/C Beams with Epoxy-bonded Fiber Composite Materials.' *Proceedings of the Specialty Conference on Advanced Composites Materials in Civil Engineering Structures.* Las Vegas, Nevada, January 31 thru February 1, 1991, 245–256.

VAN DEN EINDE, L., ZHAO, L., and SEIBLE, F. (2003). 'Use of FRP Composites in Civil Structural Applications.' *Construction and Building Materials*, 17, 389–403.

WEITSMAN, Y.J. and ELAHI, M. (2000). 'Effects of Fluids on the Deformation, Strength, and Durability of Polymeric Composites – An Overview', *Mechanics of Time-dependent Materials*, 4, 107–126.

Part II

Estimating the service life of civil engineering structures

5
Service life, reliability and maintenance of civil structures

D. M. FRANGOPOL and S. KIM, Lehigh University, USA

Abstract: Application of maintenance to extend the service life of civil structures is an important issue for civil structure managers. In order to establish an optimal maintenance strategy under uncertainty, it is essential to understand the concept of service life based on reliability of structures. In this chapter, concepts and applications of structural reliability and service life considering uncertainty are discussed. In addition, the general concept of optimization with conflicting objectives is introduced in order to extend the service life of structures more efficiently.

Key words: service life, reliability, maintenance, optimization, decision, uncertainty.

5.1 Introduction

Over the last century, design and construction of new civil structures were the main issues for civil infrastructure. Especially during the 1960s and 1970s, very large investments were made to construct new buildings and bridges. However, in the last decade the need for construction of new civil structures has declined. Also, a very large number of civil structures have started to show prominent signs of deterioration (Neves *et al.*, 2004). Consequently, the importance of maintenance for the deteriorating civil infrastructure has been recognized, and many engineers have made significant efforts to extend the service life of existing civil structures rather than to design new structures. Decision makers have to decide the priority of maintenance interventions on deteriorating structures (Frangopol *et al.*, 2001; Kong and Frangopol, 2003).

In this chapter, the concepts of the reliability and service life of civil structures, including time-dependent effects, are discussed. Furthermore, condition, safety and cost profiles are presented, and general concepts of optimal management using multi-criteria optimization are introduced to extend the service life of structures more efficiently.

5.2 Structural reliability and service life

5.2.1 General concepts

The main objective of structural engineering design is to create safe, economical, and serviceable structures during their specific lifetime. Due to both aleatoric uncertainty, which relates to the inherent randomness of a process, and epistemic uncertainty, which is caused by lack of information and can be reduced by additional experiments and data, there always exists a probability of structural failure (Ang and De Leon, 2005). These two types of uncertainty make the service life of a civil structure uncertain, as shown in Fig. 5.1. Therefore, these uncertainties should be treated in a rational way by using concepts and methods of probability and structural reliability theory.

In general, quantification of uncertainty may be evaluated by concepts of probability. Fig. 5.1 shows that the performance of a structure has randomness associated with some physical quantities under uncertainty. This randomness may be identified through a function of a random variable such as probability density function (PDF). The service life, which can be defined as the expected time period for which the performance of a structure is above a target level, has its own PDF.

This section introduces the concept of reliability and its application to quantify the service life of civil structures. The reliability can be defined as the probability that an item will adequately perform its specified purpose for a specified period of time under specified environmental conditions (Leemis, 1995) and, in brief, is a probabilistic measure of assurance of safe performance (Ang and Tang, 1984). In reality, the reliability problem of engineering systems can be expressed as a problem of supply and demand

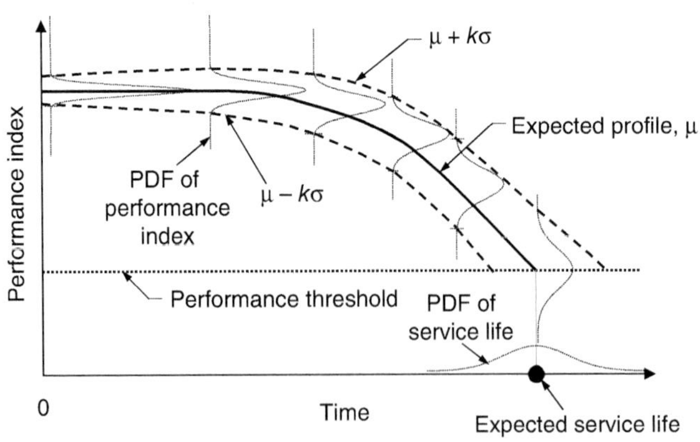

5.1 Lifetime performance of structure under uncertainty.

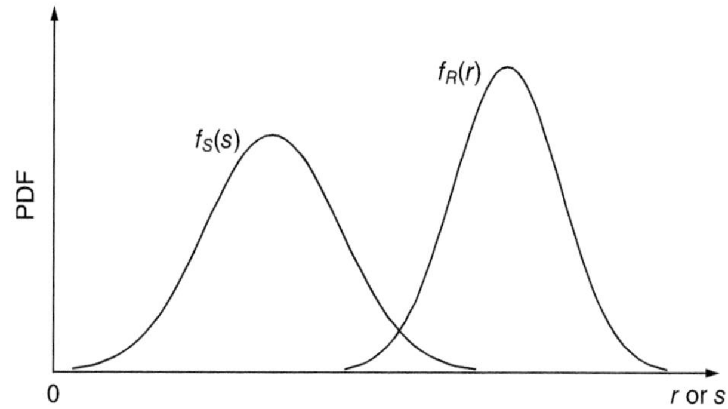

5.2 Probability density functions of *R* and *S*.

which are modeled by means of random variables. For instance, if R and S are the resistance and the load effect respectively, characterized by the PDF $f_R(r)$ and $f_S(s)$ respectively, the probability that S will not exceed R, $P(R > S)$, represents the reliability of the structural system (see Fig. 5.2). If R and S are statistically independent, the probability of failure (unreliability), $P(R < S)$, is

$$p_F = \int_0^\infty [F_R(s) f_S(s)] \mathrm{d}s \qquad [5.1]$$

where $F_R(s)$ is the cumulative distribution function (CDF) of R.

Therefore, the reliability can be formulated as:

$$p_S = 1 - \int_0^\infty [F_R(s) f_S(s)] \mathrm{d}s \qquad [5.2]$$

As a general case, if R and S are not independent, the probability of failure can be expressed in terms of joint PDF of the random variables R and S, $f_{R,S}(r, s)$, as

$$p_F = \int_0^\infty \left[\int_0^s f_{R,S}(r, s) \mathrm{d}r \right] \mathrm{d}s \qquad [5.3]$$

The corresponding probability of survival is

$$p_S = \int_0^\infty \left[\int_0^r f_{R,S}(r, s) \mathrm{d}s \right] \mathrm{d}r \qquad [5.4]$$

Safety margin

The difference between resistance and load effect can be defined as the safety margin $M = R - S$. The safety margin, M, is a random variable with PDF $f_M(m)$. As shown in Fig. 5.3, the area under the PDF upper bounded by $m = 0$ represents the probability of failure:

$$p_F = \int_{-\infty}^0 f_M(m) \mathrm{d}m \qquad [5.5]$$

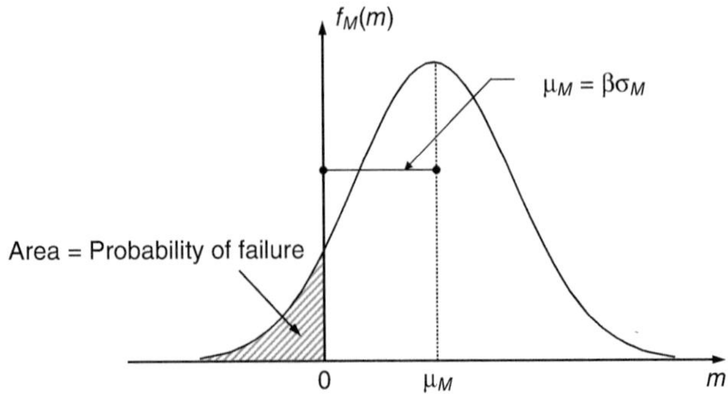

5.3 Probability density function of safety margin and the reliability index.

Reliability index

The reliability index is defined as (see Fig. 5.3)

$$\beta = \frac{\mu_M}{\sigma_M} \tag{5.6}$$

where μ_M and σ_M are the mean and standard deviation of the safety margin, respectively.

If R and S are independent, Equation 5.6 becomes

$$\beta = \frac{\mu_R - \mu_S}{\sqrt{\sigma_R^2 + \sigma_S^2}} \tag{5.7}$$

where μ_R, μ_S and σ_R, σ_S are the means and standard deviations, respectively.

Furthermore, on the assumption that the safety margin M is normally distributed, the reliability index can be expressed as follows:

$$\beta = \Phi^{-1}(p_s) = \Phi^{-1}(1 - p_F) \tag{5.8}$$

where Φ^{-1} is the inverse of the standard normal cumulative density function. The reliability index may be evaluated by using the first moment (the mean value) and the second moment (the variance). Let the reduced variables of R and S be defined by

$$R' = \frac{X - \mu_R}{\sigma_R}$$
$$S' = \frac{Y - \mu_S}{\sigma_S} \tag{5.9}$$

As shown Fig. 5.4, the minimum distance from $M = 0$ to the origin of the space of reduced variables is

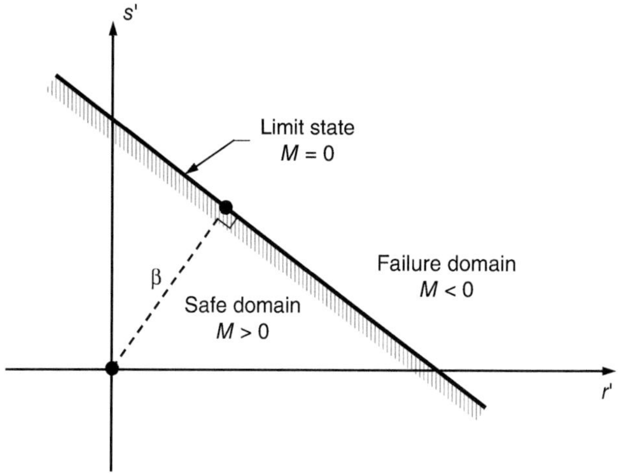

5.4 Reliability index in the space of reduced variables.

$$d = \beta = \frac{\mu_R - \mu_S}{\sqrt{\sigma_R^2 + \sigma_S^2}} \quad [5.10]$$

Performance function

The performance function (also denoted as state function) is related to the safety margin $M = R - S$. In general, the resistance and load effect consist of several variables. To generalize the problem considering these variables, the safety margin is formulated as a state function $g(\mathbf{X})$ (Ang and Tang, 1984).

$$g(\mathbf{X}) = g(X_1, X_2, \ldots, X_n) \quad [5.11]$$

where $\mathbf{X} = (X_1, X_2, \ldots, X_n)$ is a vector of design variables of the system, and the state function $g(\mathbf{X})$ determines the state of the system as

$[g(\mathbf{X}) > 0] \rightarrow$ Safe state

$[g(\mathbf{X}) < 0] \rightarrow$ Failure state

$[g(\mathbf{X}) = 0] \rightarrow$ Limit state

Considering a two-variable reduced space, the limit state, the safe domain, and the failure domain are shown graphically in Fig. 5.5.

5.2.2 System reliability

In general, structures are composed of many components. For each component, its various limit states (such as bending, shear, buckling) may need to be considered. However, reliability of the individual structural component

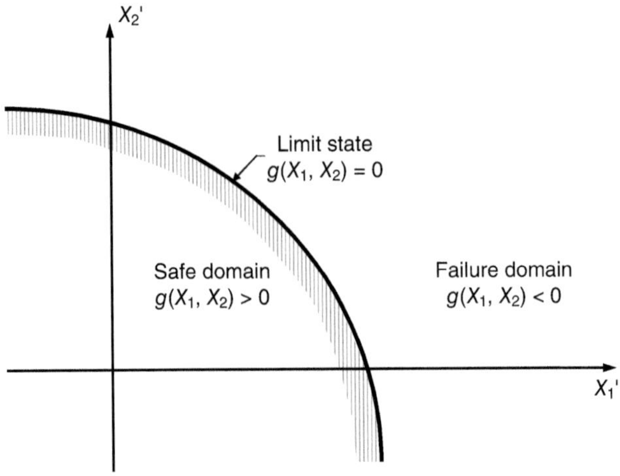

5.5 State function for limit state, safe state, and failure state in space of reduced variables X_1' and X_2'.

is not enough to guarantee the reliability of a structural system. Therefore, the problem of safety evaluation of existing structures can be correctly assessed only by considering the full structural system. In general, systems composed of multiple connected components can be classified as series systems (Fig. 5.6(a)), parallel systems (Fig. 5.6(b)), or mixed series – parallel systems (Fig. 5.6(c)).

Series systems

In a series system (see Fig. 5.6(a)), failure of any of its components constitutes the failure of the system; therefore, such a system has no redundancy and is also known as a 'weakest link' system. In other words, the reliability of the system requires that none of its components fail. The probability of failure p_F can be expressed as the probability of union of component failure events

$$p_F = p\left(\bigcup_{i=1}^{N}\{g_i(\mathbf{X}) \leq 0\}\right) \qquad [5.12]$$

The failure probability of the series system depends on the correlation among the safety margins of the components. The two extreme cases are as follows:

(a) for perfectly correlated case: $p_F = \max_{i=1}^{N} p_{Fi}$

(b) for statistically independent case: $p_F = 1 - \prod_{i=1}^{N}(1 - p_{Fi})$

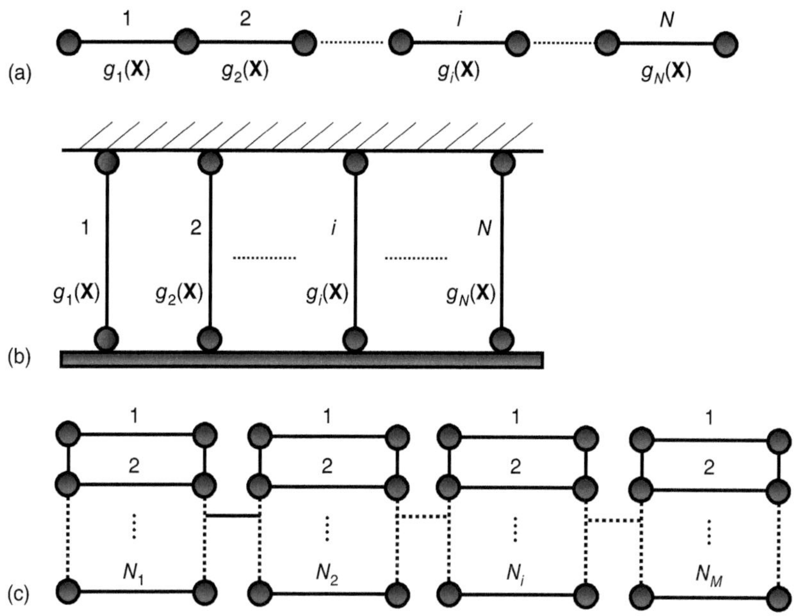

5.6 (a) Series system, (b) parallel system, and (c) combined series–parallel system.

The first-order bounds for the failure probability of a series system are (Cornell, 1967)

$$\max_{i=1}^{N} p_{Fi} \leq p_F \leq 1 - \prod_{i=1}^{N}(1 - p_{Fi}) \tag{5.13}$$

Closer bounds were developed by Ditlevsen (1979) using joint-event probabilities, which accounted for failure mode correlation:

$$p_{F1} + \sum_{i=2}^{k} \max\left[p_{Fi} - \sum_{j=1}^{i-1} p_{Fij}, 0\right] \leq p_F \leq \sum_{i=1}^{k} p_{Fi} - \sum_{i=2}^{k} \max_{j<i}\left[p_{Fij}\right] \tag{5.14}$$

where p_{Fij} is the joint probability of occurrence of the ith and jth failure modes, and k is the number of potential failure modes of a series system. Figures 5.7(a) and (b) show the safe domain, the failure domain, and the limit state of Component 1 and Component 2, respectively, and in Fig. 5.7(c), the safe domain and the failure domain are shown when these two components are linked in series.

Parallel systems

Failure of a parallel system (see Fig. 5.6(b)) requires failures of all its components. Therefore, if any one of the components survives, the system

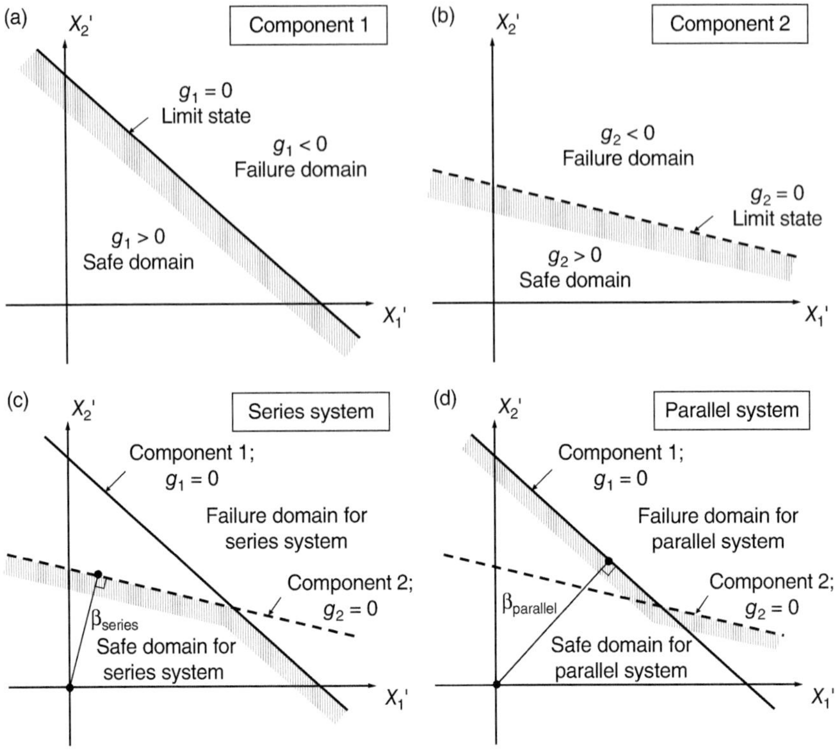

5.7 Safe and failure space for (a) Component 1; (b) Component 2; (c) series system; and (d) parallel system.

remains safe. The probability of failure of a parallel system P_F can be expressed as the probability of intersections of component failure events

$$p_F = p\left(\bigcap_{i=1}^{N}\{g_i(\mathbf{X}) \le 0\}\right) \qquad [5.15]$$

The failure of an N-component parallel system depends on the correlation among the safety margins of its components. The two extreme cases are as follows:

(a) for perfectly correlated case: $p_F = \min_{i=1}^{N} p_{Fi}$

(b) for statistically independent case: $p_F = \prod_{i=1}^{N} p_{Fi}$

The first-order bounds for the failure probability of a parallel system are (Ang and Tang, 1984)

$$\prod_{i=1}^{N} p_{\mathrm{F}i} \le p_{\mathrm{F}} \le \min_{i=1} p_{\mathrm{F}i} \qquad [5.16]$$

Practically, the first-order bounds of the failure probability of a parallel system determined by Equation 5.16 may be too wide to be useful. Therefore, an alternative approach is used as follows (Thoft-Christensen and Murotsu, 1986):

$$p_{\mathrm{F}} = \int_{\beta_1}^{\infty}\int_{\beta_2}^{\infty} \cdots \int_{\beta_N}^{\infty} \frac{1}{(2\pi)^{N/2}\sqrt{det[\rho_{\mathrm{sys}}]}} e^{-1/2\{\beta\}[\rho_{\mathrm{sys}}]^{-1}\{\beta\}^T} d\{\beta\} \qquad [5.17]$$

where $\{\beta\} = \{\beta_1, \beta_2, \ldots, \beta_N\}$, ρ_{sys} is the system correlation matrix, and N is the number of members in the system. The safe domain and the reliability index of the parallel system consisting of the two components having the safety domains shown in Figures 5.7(a) and (b) are indicated in Fig. 5.7(d). By comparing Figures 5.7(c) and 5.7(d), it can be seen that both the safety domain and the reliability index of the parallel system are larger than those of the associated series system.

Combined systems

A combined system can be modeled as a series system of parallel systems or a parallel system of series systems. Consider a series system consisting of M parallel systems, where each parallel system i has N_i components (Fig. 5.6(c)). The probability of overall system failure is given by:

$$p_{\mathrm{F}} = p\left(\bigcup_{i=1}^{M}\bigcap_{j=1}^{N_i}\{g_{ij}(\mathbf{X}) \le 0\}\right) \qquad [5.18]$$

There are several computer programs such as RELSYS (Estes and Frangopol, 1998) able to compute the probability of failure of combined systems.

Reliability importance factor

The reliability importance factor (RIF_i) of the component i of a system is defined as the impact of the ith component on the system reliability as follows (Leemis, 1995):

$$RIF_i = \frac{\partial p_{S,\mathrm{system}}}{\partial p_{S,i}} \qquad [5.19]$$

The component of an engineering system with the largest reliability importance produces the largest change in the system reliability. The associated normalized reliability importance factor (RIF_i^{norm}) of ith component of a system which consists of N individual components is (Gharaibeh *et al.*, 2002):

$$RIF_i^{\text{norm}} = \frac{RIF_i}{\sum_{j=1}^{N} RIF_j} \quad [5.20]$$

where $0 \leq RIF_i^{\text{norm}} \leq 1$. The normalized reliability importance factor can provide useful information for selecting the optimal maintenance strategy.

5.2.3 Application of structural reliability

To compute the reliability of a structural system, it is first necessary to define the system model. As an example, a four-span bridge with four girders in each span is used (see Figures 5.8(a) and (b)). It is assumed that failure of any two adjacent girders, failure of the deck, or both, result in the failure of the superstructure. A failure event of each span can be modeled by the combined model in Fig. 5.9(a) and the failure models of the four spans are connected in a series system (see Fig. 5.9(b)). The reliability analysis of a system can be extended to bridge network reliability. Evaluation of a bridge network is based on connectivity between a start point (A) and an end point (B). Such a network with six bridges is indicated in Fig. 5.10(a). The bridge network model is shown in Fig. 5.10(b).

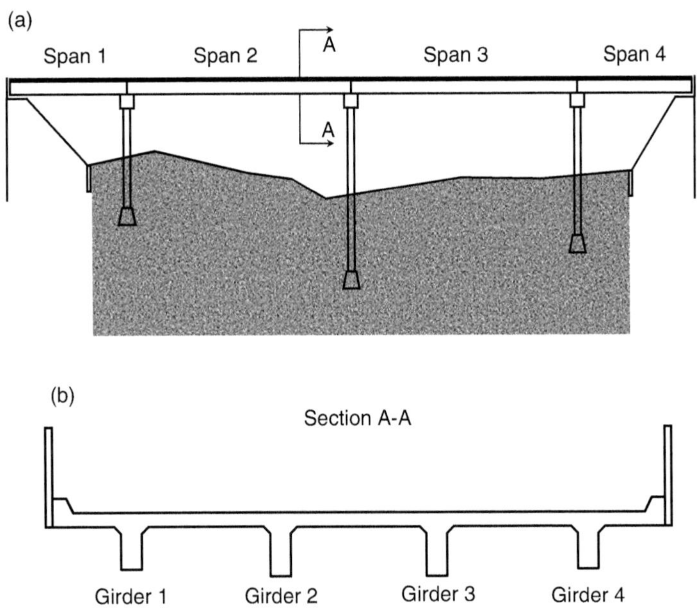

5.8 (a) Bridge elevation, and (b) cross-section.

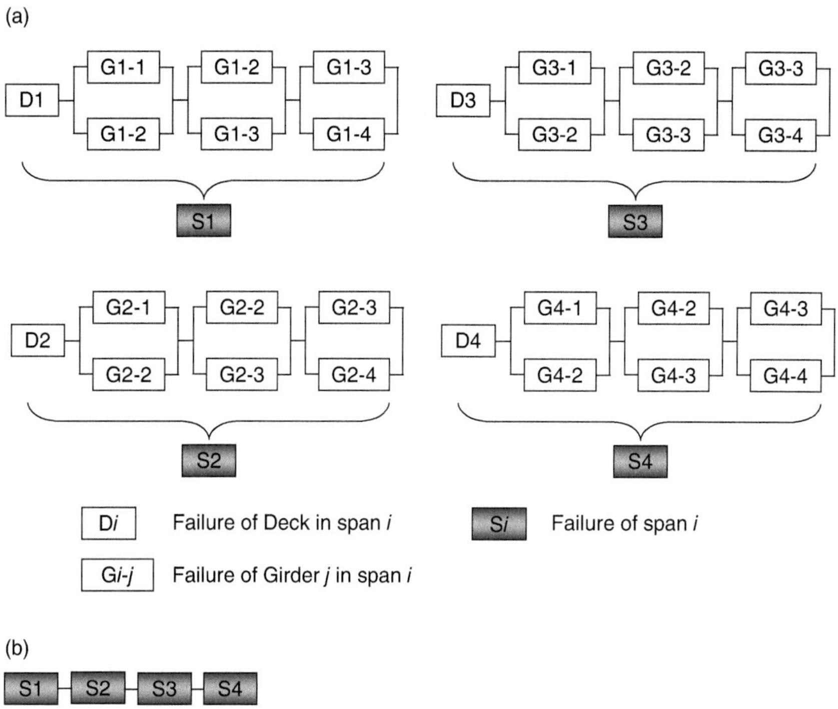

5.9 (a) Failure model of each span, and (b) entire system failure model.

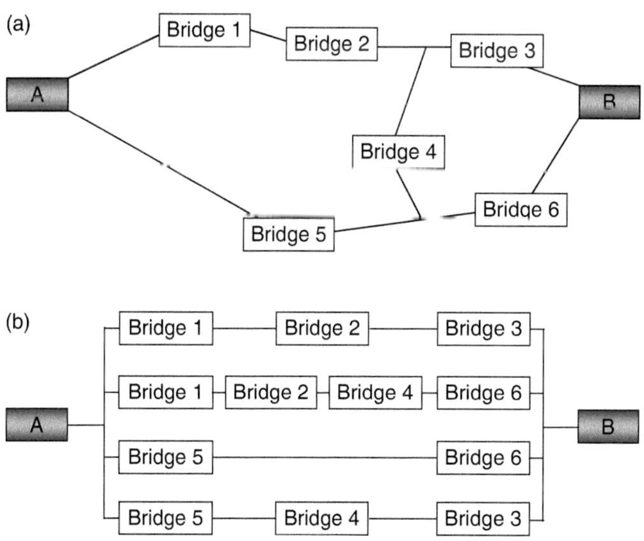

5.10 (a) Bridge network, and (b) series–parallel path model.

5.3 Time-dependent reliability and service life

5.3.1 Time-dependent effects on structures

To estimate the service life of a civil structure and to allocate the limited maintenance funds optimally for extension of its life, an accurate reliability prediction model of a deteriorating structure is necessary. To establish an accurate modeling of the structural deterioration process, it is essential to accurately model both time-dependent mechanisms of resistance and load effect. Four cases are indicated in Figures 5.11(a)–(d), as follows: (a) time-independent resistance and load effect (Fig. 5.11(a)); (b) time-dependent resistance and time-independent load effect (Fig. 5.11(b)); (c) time-independent resistance and time-dependent load effect (Fig. 5.11(c)); and (d) time-dependent resistance and load effect (Fig. 5.11(d)). The mean safety margin profiles (i.e. the difference between mean resistance and mean load effect) associated with the four cases shown in Figures 5.11(a)–(d) are indicated in Fig. 5.11(e).

Resistance

Among the factors affecting the deterioration of concrete structures, corrosion is the main factor which may produce cracking and spalling, as well as loss of bond between concrete and reinforcing steel, and loss of steel section. In general, the deterioration process of reinforced concrete due to corrosion can be described by six steps (Thoft-Christensen, 2003):(i) chloride penetration in the concrete; (ii) initiation of the corrosion of the reinforcement; (iii) evolution of corrosion of the reinforcement; (iv) initial cracking of the concrete; (v) evolution of cracks in the concrete; and (vi) spalling. Corrosion in steel structures may be a very significant performance deterioration factor because most of the components of a steel structure are exposed to the environment directly.

Fatigue in metals can be defined as the process of initiation and growth of cracks under repetitive stresses. If crack growth is allowed, failure of a steel member can occur and this process can take place at stress levels that are less than levels at which failure occurs under static loading condition. Generally, the fatigue life of a fabricated steel structure may be determined by three factors, as follows (Fisher *et al.*, 1998): (a) number of loading cycles; (b) stress range at the location of a steel member; and (c) type of detail of a steel member. All these factors can have an important effect on the service life of a steel structure.

Applied load

Maximum stress reaching yield strength and the number of stress cycles exceeding the critical number of cycles can induce failure of a structural

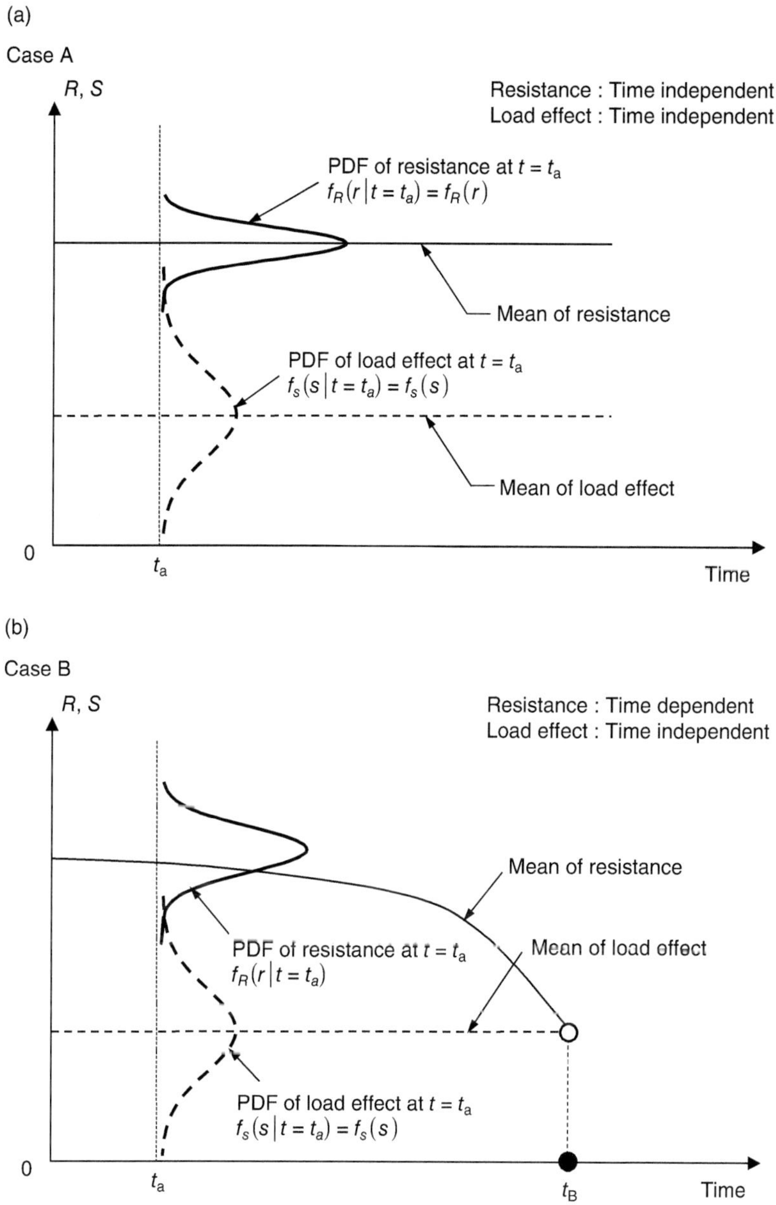

5.11 Time-independent and time-dependent *R* and *S*: (a) Time-independent; (b) time-dependent resistance and time-independent load effect; (c) time-independent resistance and time-dependent load effect; and (d) time-dependent resistance and time-dependent load effect, and (e) profiles of mean safety margin of cases (a), (b), (c), and (d).

(c)

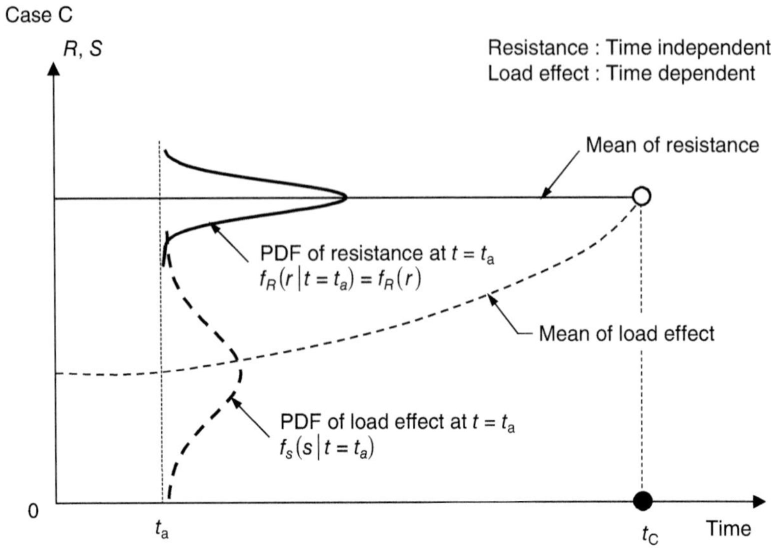

(d)

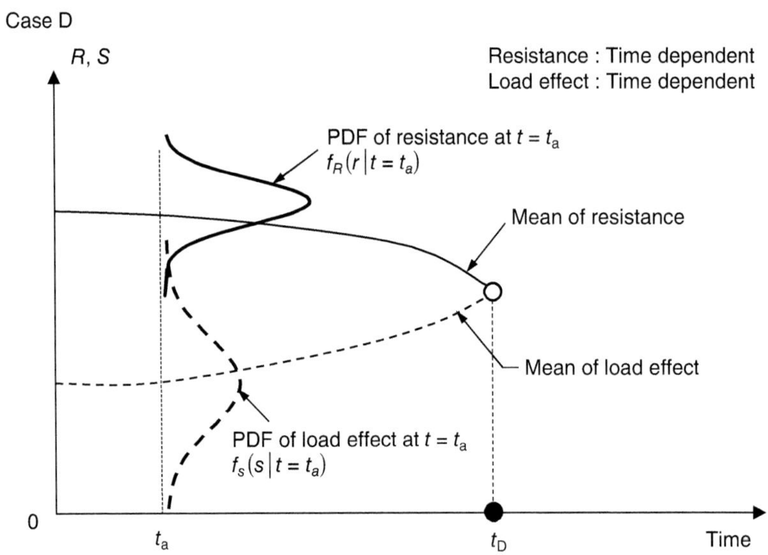

5.11 Continued

(e)

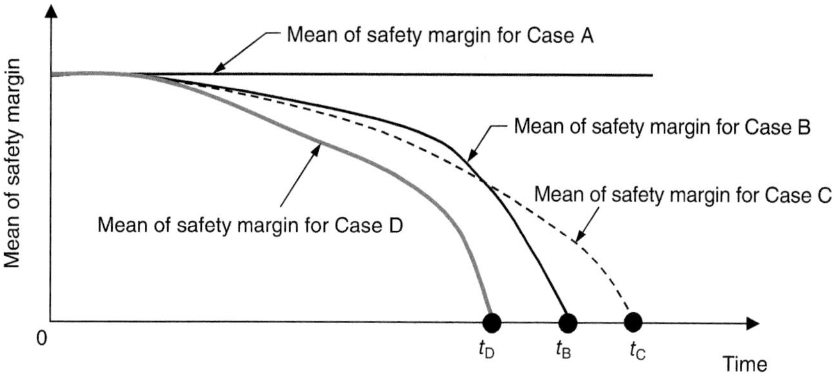

5.11 Continued

Table 5.1 Time-dependent reliability indices

Reliability index at time t (years)	
Component 1:	$\beta_1(t) = 6.0 - 0.1t$
Component 2:	$\beta_2(t) = 5.0 - 0.05t$

system. Therefore, the accuracy of the time-dependent models for prediction of the maximum stress caused by loads and number of stress cycles under live loads, and the validity of calculated reliabilities are important.

5.3.2 Analysis of reliability of deteriorating structures and service life

Performance of a civil structure decreases with time due to load increase, fatigue, and/or environmental attack such as corrosion (Ellingwood, 2005). If the stochastic models of loadings and environmental stressors are established over time accurately, the performance deterioration of structural components and of the entire system can be determined.

As an example, consider a series system and a parallel system, both consisting of two components as shown in Fig. 5.12(a). The time-dependent reliability indices of the two components are indicated in Table. 5.1. The relation between reliability index and probability of failure is determined using Equation 5.8, and the failure probabilities for the series system and parallel system are calculated by Equations 5.13 and 5.16, respectively. Figures 5.12(b) and (c) show each component reliability index and the system

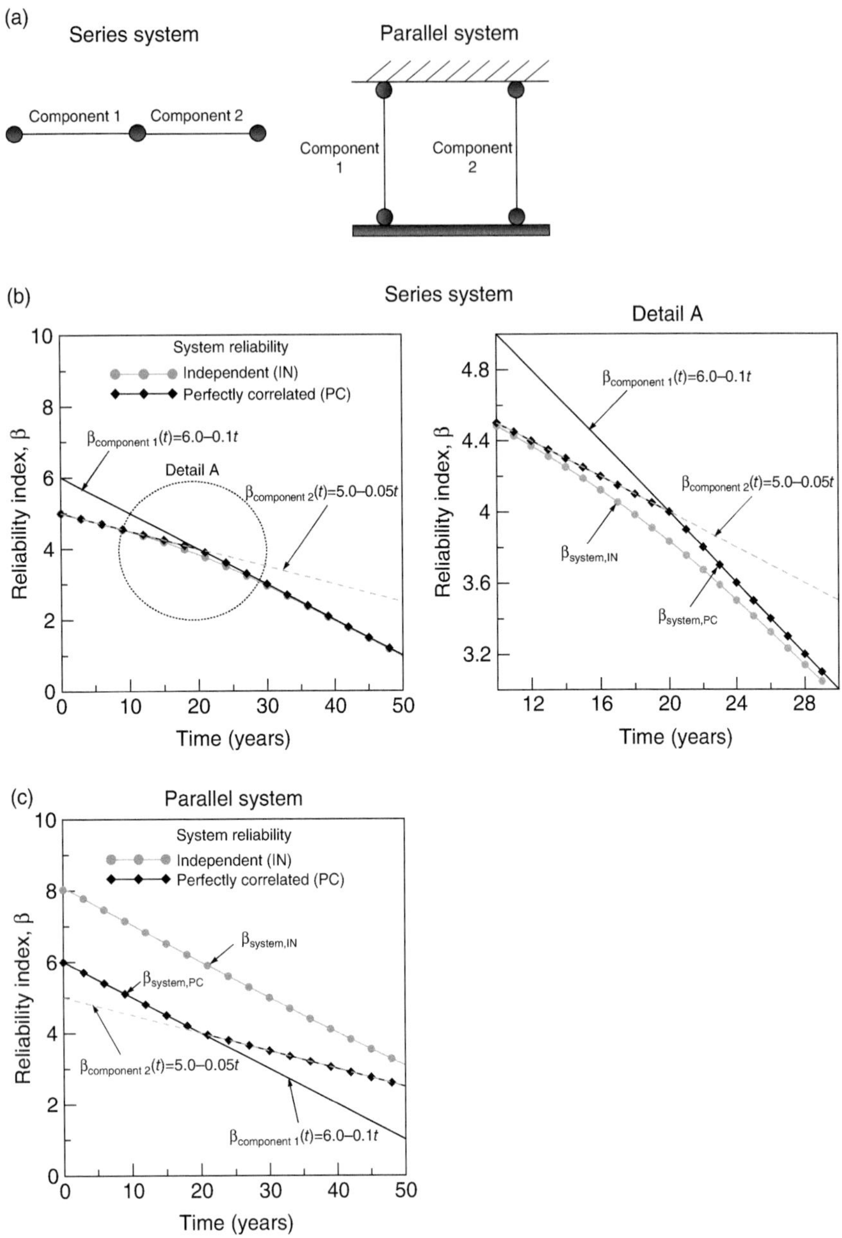

5.12 (a) Series system and parallel system for analysis of time-dependent reliability, (b) time-dependent series system reliability index, and (c) time-dependent parallel system reliability index.

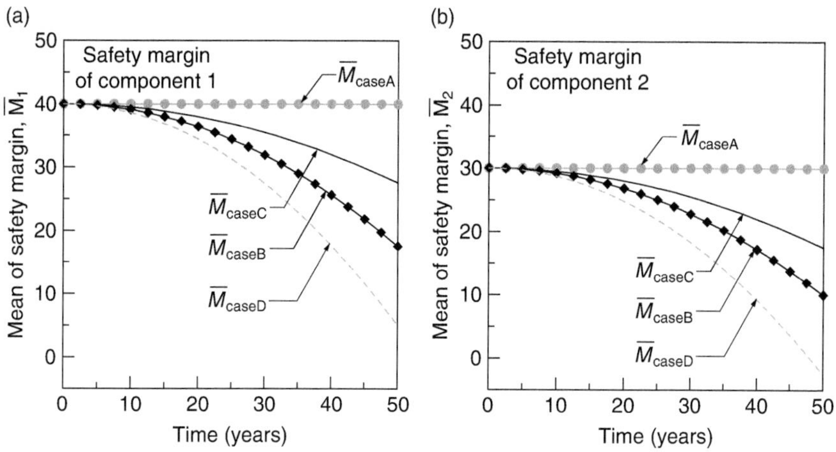

5.13 Time-dependent safety margins under the Cases A, B, C, and D in Table 5.2: (a) Component 1, and (b) Component 2.

reliability indices over time when the correlation between two members is perfect and when the safety margins of the components are statistically independent. In the perfectly correlated case, the series system reliability is equal to the smaller reliability index (i.e., the reliability index of the less safe member), and the reliability of the parallel system is equal to the larger reliability index (i.e., the reliability index of the most safe member).

As mentioned in Section 5.3.1, the deterioration process of the performance of a civil structure depends on the time-dependent resistance and load effect. Fig. 5.13 shows the change of the mean safety margin associated with each case in Table 5.2. Under three different deteriorating processes (cases B, C and D in Table 5.2), the time-dependent reliability indices are calculated by using the Monte Carlo simulation program, MONTE (Kong *et al.*, 2000). Random variables X_1 and X_3 correspond to resistance, and variables X_2 and X_4 are associated with load effect. These four random variables are considered independent. In Figures 5.14(a) to (d), the system reliability index of the series system has the lowest value, and that of the parallel system has the largest value at any time. Therefore, the time necessary to reach the target value of reliability index, $\beta_{target} = 2.0$ (i.e. the service life of the system) is the lowest in the case of the series system among all four cases. Also, the deteriorating rate associated with case D is the largest, as shown in Fig. 5.14(d).

These analyses could be utilized to determine: (a) which component has more influence on the system reliability; (b) which component needs inspection, maintenance and repair for effective improvement of system reliability; and (c) when the maintenance and repair are needed for optimal extension of service life of the system.

Table 5.2 Performance functions of components 1 and 2 with different time-dependent resistance and load effects

	Component 1	Component 2
	$g_1 = R_1(t) - S_1(t)$	$g_2 = R_2(t) - S_2(t)$
Case A		
Resistance, *R*: time independent	$R_1(t) = X_1$	$R_2(t) = X_3$
Load effect, *S*: time independent	$S_1(t) = X_2$	$S_2(t) = X_4$
Case B		
Resistance, *R*: time dependent	$R_1(t) = X_1(1\text{-}(t/100)^2)$	$R_2(t) = X_3(1\text{-}(t/100)^2)$
Load effect, *S*: time independent	$S_1(t) = X_2$	$S_2(t) = X_4$
Case C		
Resistance, *R*: time independent	$R_1(t) = X_1$	$R_2(t) = X_3$
Load effect, *S*: time dependent	$S_1(t) = X_2(1 + (t/100)^2)$	$S_2(t) = X_4(1 + (t/100)^2)$
Case D		
Resistance, *R*: time dependent	$R_1(t) = X_1(1\text{-}(t/100)^2)$	$R_2(t) = X_3(1\text{-}(t/100)^2)$
Load effect, *S*: time dependent	$S_1(t) = X_2(1 + (t/100)^2)$	$S_2(t) = X_4(1 + (t/100)^2)$
Random variables	X_1; N ($\mu_1 = 90$, $\sigma_1 = 15$) X_2; N ($\mu_2 = 50$, $\sigma_2 = 4.0$)	X_3; N ($\mu_3 = 80$, $\sigma_3 = 10$) X_4; N ($\mu_4 = 50$, $\sigma_4 = 5.0$)

Note: N = normal distribution; μ_i = mean of random variable X_i; σ_i = standard deviation of random variable X_i.

5.3.3 Maintenance

To extend the expected service life of a system, effective maintenance is necessary (Frangopol *et al.*, 1997, 2001; Moan, 2005; Frangopol and Liu, 2007). Therefore, preventive and/or essential maintenance are performed. Preventive maintenance is a time-based maintenance action, that is applied at predefined time instants to prevent the failure of the system. Preventive maintenance includes replacing small parts, patching concrete, repairing cracks, changing lubricants, and cleaning and painting exposed parts, among others. The essential maintenance is a performance-based action that it is immediately applied when some performance indicators reach predefined target values. In general, the essential maintenance follows an inspection, and includes replacing a bearing, resurfacing a deck, or modifying a girder of bridge. Preventive maintenance tends to be more frequent, less costly, and less efficient from the safety improvement viewpoint

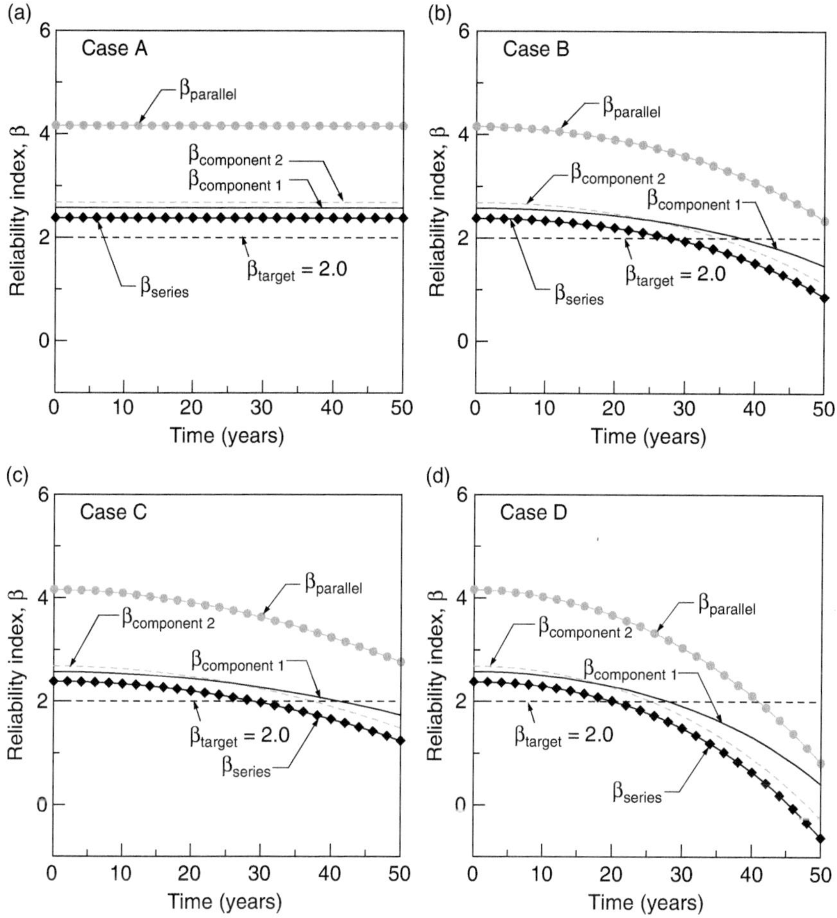

5.14 Variation of reliability index with time: (a) Case A, (b) Case B, (c) Case C, and (d) Case D.

than essential maintenance. In Figures 5.15(a)–(c), using the modified Frangopol's maintenance model (Frangopol *et al.*, 2001), a simplified maintenance scenario with two preventive maintenances and one essential maintenance is presented to show the reliability index profile affected by essential and preventive maintenance. The reliability index profile (multilinear profile) with maintenance effects in Fig. 5.15(b) is as follows: (I) structure retains the initial reliability index without any performance deterioration; (II) performance deterioration of the structure begins with the rate α until time T_{m1}; (III) first preventive maintenance is applied at time T_{m1}, performance is improved, and deterioration continues but at a lower rate α', where $\alpha > \alpha'$; (IV) the effectiveness of the first preventive maintenance ends and the rate of the deterioration becomes α; (V) at time T_{m2},

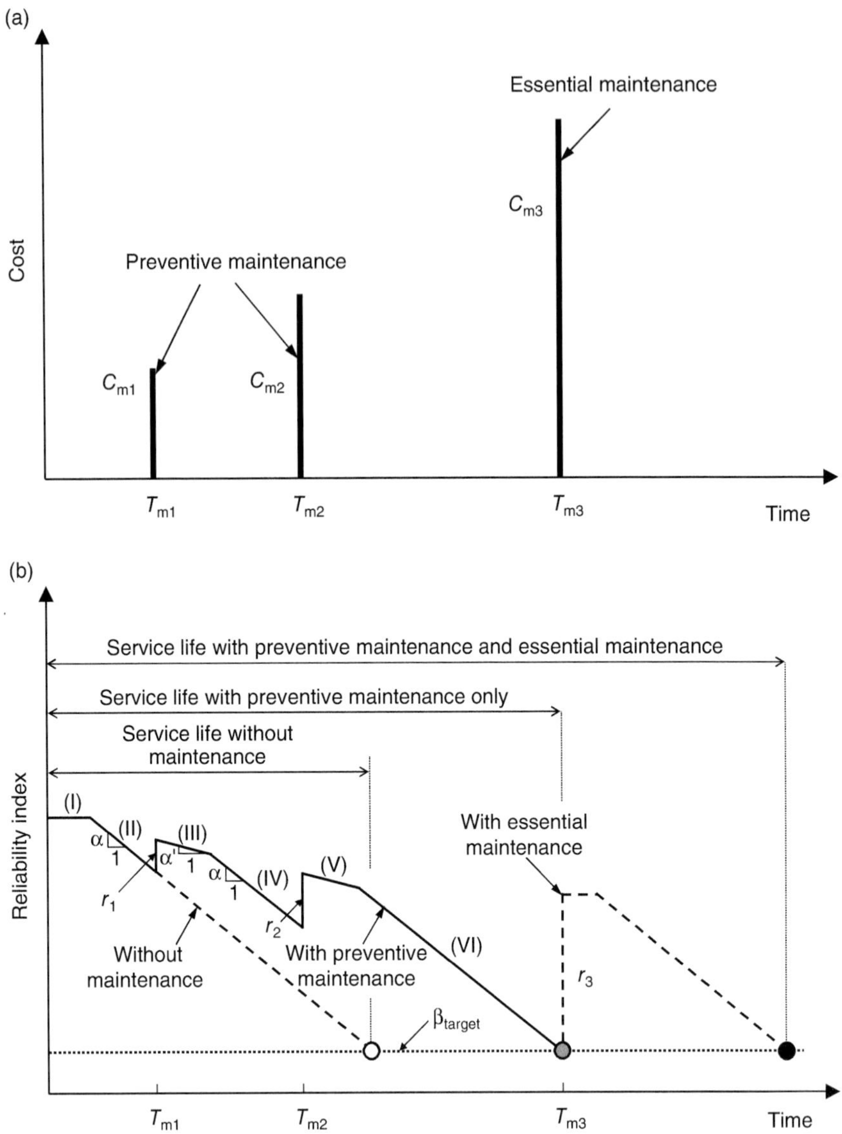

5.15 (*a*) Application time and cost of preventive and essential maintenance; (b) multi-linear reliability index profiles with and without maintenance, and (c) non-linear reliability index profiles with and without maintenance.

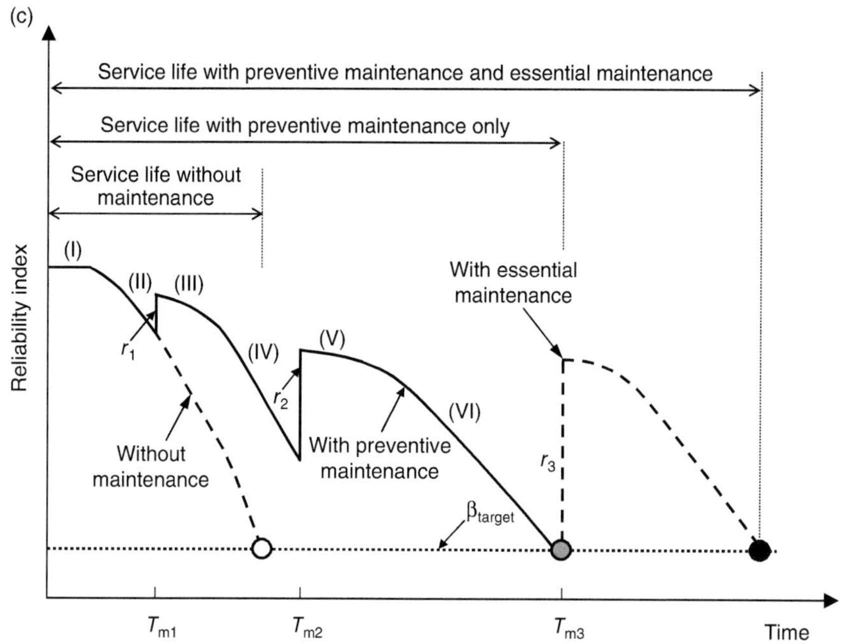

5.15 Continued

a second preventive maintenance is applied, and the steps (III) and (IV) are repeated; and (VI) when the reliability index of the structure reaches the target value, β_{target}, the essential maintenance will be applied, resulting in a substantial increase of the reliability index, r_3 (i.e., $r_3 > r_2$, $r_3 > r_1$) (see Fig. 5.15(b)). The reliability index profile associated with maintenance effects can be expressed by non-linear profiles, as shown in Fig. 5.15(c). Based on these characteristics of maintenance, effective maintenance strategy can be formulated as a multi-objective optimization problem in which the objectives are minimizing maintenance costs and maximizing the service life.

5.4 Optimum maintenance

5.4.1 Optimization of lifetime maintenance

The optimum maintenance strategy depends on many factors, including the expected service life and the associated assumptions. The reliability of an entire system provides significant information for determination of the maintenance strategy rather than the reliability of any individual component (Estes and Frangopol, 1999). To show that the maintenance strategy is based mainly on the reliability of a system, the three options indicated in

Table 5.3 Performance-based repair options

Option 1	If the reliability index of a component reaches 2.5, the component will be replaced and the reliability index of the new component will have the value of initial reliability index of the replaced component.
Option 2	If the reliability index of a system reaches 2.5, every component will be replaced and the reliability index of the system will have the value of its initial reliability index.
Option 3	If the reliability index of a system reaches 2.5 or the reliability index of a component reaches 1.0, the critical component will be replaced with a new component having the same initial reliability index as that of the replaced component. For a series system, the system reliability is at most equal to the reliability of the weakest component. Conversely, for a parallel system, the system reliability is at least equal to the reliability of the strongest component. For this reason, component will always be replaced when the reliability index of the series system reaches 2.5.

Table 5.3 are applied to two systems (series and parallel system) consisting of two perfectly correlated components with time-dependent reliability index functions, as indicated in Table 5.1.

As shown in Table 5.3, Option 1 depends on the lowest reliability index, Option 2 depends on the system reliability index, and Option 3 considers both component reliability and system reliability.

Figure 5.16(a) shows the reliability index profiles during 100 years for the series system analyzed in Section 5.3.2. It is assumed that the expected replacement cost for each component is the same, US$1000, and that the failures of the two components are perfectly correlated. The total maintenance cost, C_{m}, can be computed as follows:

$$C_{\mathrm{m}} = \sum_{i=1}^{n} \frac{C_{\mathrm{m},i}}{(1+r)^{t_i}} \qquad [5.21]$$

where n is the number of the maintenance application, $C_{\mathrm{m}i}$ is the cost of the ith maintenance, t_i is the time for the ith maintenance, and r is the discount rate of money. The annual cost discount rate is assumed $r = 2\%$.

- *Repair Option 1 for series system*: The first replacement will be conducted for the Component 1 at Year 35, the second replacement will be for the Component 2 at Year 50, and the third replacement will be for the Component 1 at Year 70. The total maintenance cost associated with the Repair Option 1 is

$$C_{\mathrm{sm1}} = \frac{1000}{(1+0.02)^{35}} + \frac{1000}{(1+0.02)^{50}} + \frac{1000}{(1+0.02)^{70}} = \text{US\$ } 1121.58 \qquad [5.22]$$

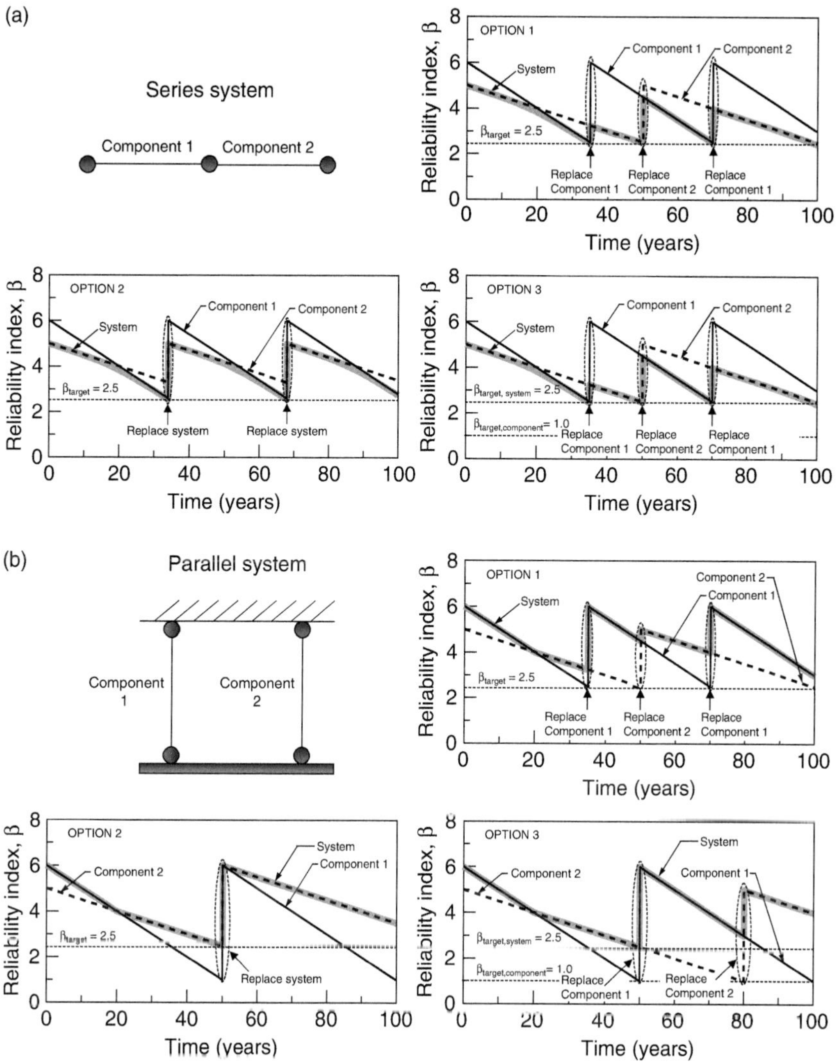

5.16 Time-dependent reliabilities under each of the three maintenance options in Table 5.3: (a) series system, and (b) parallel system.

- *Repair Option 2 for series system*: This system will be replaced entirely at Year 35 and at Year 70, when the system reliability index will reach 2.5. The total maintenance cost is

$$C_{sm2} = \frac{1000+1000}{(1+0.02)^{35}} + \frac{1000+1000}{(1+0.02)^{70}} = \text{US\$ } 1500.11 \qquad [5.23]$$

- *Repair Option 3 for series system*: Component 1 will be replaced at Year 35, Component 2 at Year 50, and Component 1 at Year 70. The total maintenance cost associated with Option 3 is as follows:

$$C_{sm3} = \frac{1000}{(1+0.02)^{35}} + \frac{1000}{(1+0.02)^{50}} + \frac{1000}{(1+0.02)^{70}} = \text{US\$ } 1121.58 \quad [5.24]$$

For the series system, Options 1 and 3 are identical. In the series system, the component with the lowest reliability is the most important and controls the system reliability. Therefore, repair Options 1 and 3 are the optimum repair strategies for this case (see Fig. 5.17).

For the parallel system made of the same two components as the associated series system, the reliability index profiles during the specified service life of 100 years are shown, for each repair option, in Fig. 5.16(b). The assumptions and discount rate for the analysis of this parallel system are the same as those used for the analysis of the associated series system.

- *Repair Option 1 for parallel system*: The first replacement will be conducted for the Component 1 at Year 35, the Component 2 will be replaced at Year 50, and the third replacement will be for the Component 1 at Year 70. The total maintenance cost is

$$C_{pm1} = \frac{1000}{(1+0.02)^{35}} + \frac{1000}{(1+0.02)^{50}} + \frac{1000}{(1+0.02)^{70}} = \text{US\$ } 1121.58 \quad [5.25]$$

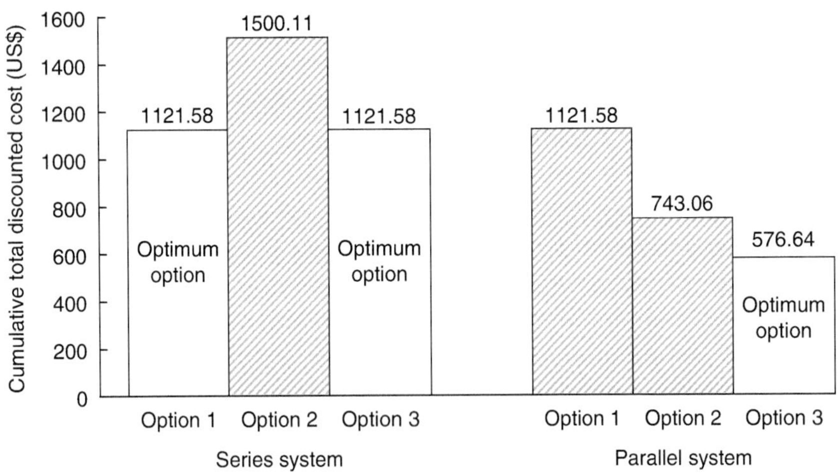

5.17 Comparison of total maintenance cost for series and parallel system under different options.

- *Repair Option 2 for parallel system*: This parallel system will be replaced at Year 50 when the system reliability index of this system will reach 2.5. The total maintenance cost is

$$C_{\text{pm2}} = \frac{1000 + 1000}{(1 + 0.02)^{50}} = \text{US\$ } 743.06 \qquad [5.26]$$

- *Repair Option 3 for parallel system*: At Year 50, the system reliability will reach $\beta_{\text{system}} = 2.5$ and the reliability of Component 1 will be 1.0. Therefore, the first replacement for Component 1 will be at Year 50, and the second for Component 2 will be at Year 80. The total maintenance cost for repair Option 3 is

$$C_{\text{pm3}} = \frac{1000}{(1 + 0.02)^{50}} + \frac{1000}{(1 + 0.02)^{80}} = \text{US\$ } 576.64 \qquad [5.27]$$

This parallel system, in which the failures of two components are perfectly correlated, has a reliability index profile that is governed by the component with the largest reliability. Therefore, repair Option 3 can be considered as the optimal maintenance strategy in this example (see Fig. 5.17).

Estes and Frangopol (1999) provided the computational platform for the time-dependent reliability analysis of existing highway bridges, and established the basis for the optimum lifetime maintenance approach under uncertainty.

5.4.2 Condition, safety and cost profiles

Condition index, C, is generally based on visual inspection. Denton (2002) classifies the condition state of reinforced concrete elements under corrosion attack in four states: (a) no chloride contamination, $C = 0$; (b) onset of corrosion, $C = 1$; (c) onset of cracking, $C = 2$; and (d) loose concrete/significant delamination, $C = 3$. A larger condition index indicates more deteriorated condition. A condition-based maintenance action may have no effect on the safety index. Safety index can be defined as indicated in Liu and Frangopol (2004). The performance indicator such as the condition index and/or safety index can be modeled as a multi-linear profile under no maintenance and under time-based (Fig. 5.18(a)) and performance-based (Fig. 5.18(b)) maintenance (Liu and Frangopol, 2004). The associated cumulative cost profiles are shown in Fig. 5.18(c). The performance profiles can also be non-linear (see Fig. 5.18(d)). However, in this section, the multi-linear profile is used.

The parameters used for the performance index profile and cost profile are provided in Table 5.4 (Denton, 2002), considering two different maintenance types for a large group of reinforced concrete bridge decks:

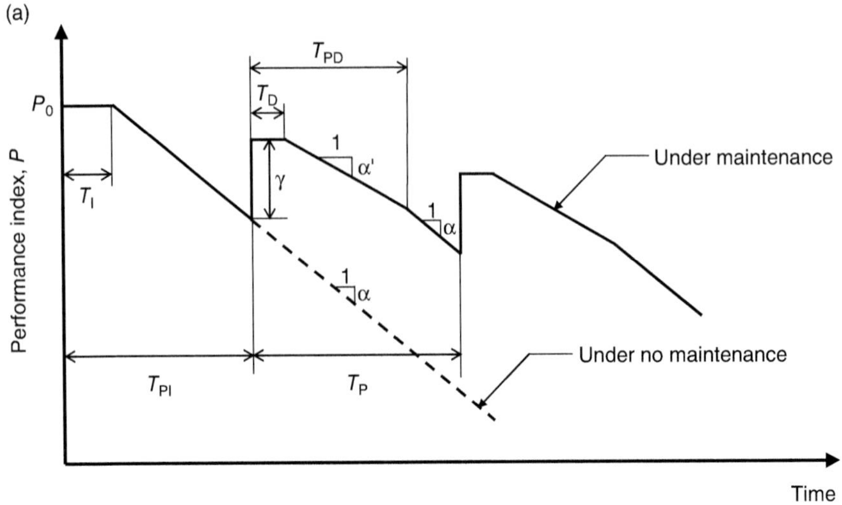

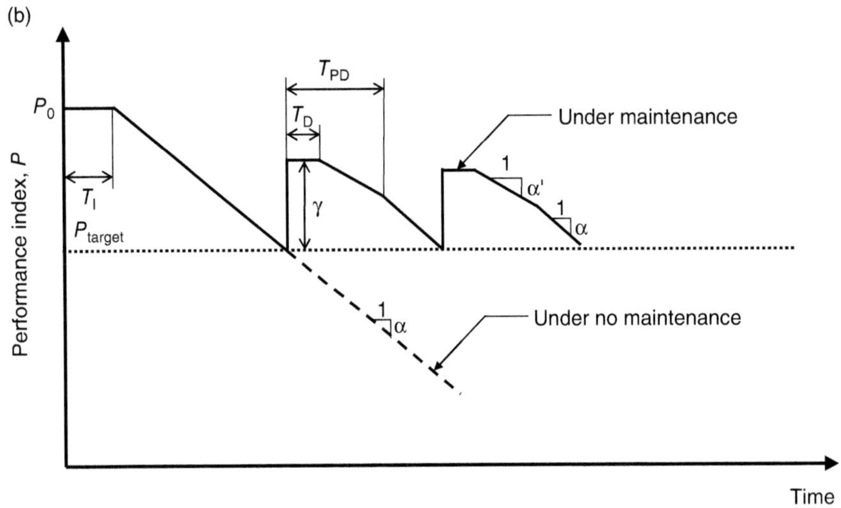

5.18 Multi-linear performance profiles; (a) time-based, and (b) performance-based maintenance interventions; (c) cumulative cost profiles, and (d) non-linear performance profile.

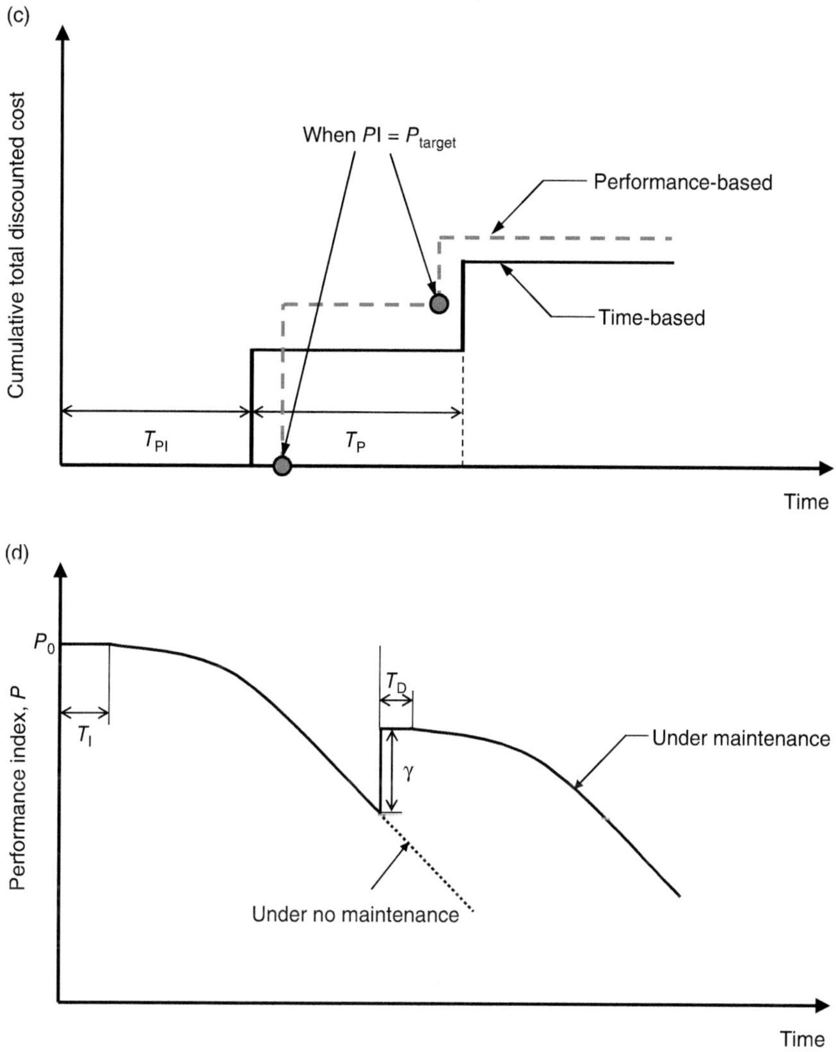

5.18 Continued

silane and rebuild. Since the silane treatment blocks chloride penetration into reinforced concrete structures, the deterioration rate of condition and safety can be reduced, but there is no delay in deterioration and no improvement. On the other hand, rebuild increases the condition and the safety to the value of the intact state, with deterioration delay, but the required cost is relatively significant. Figures 5.19(a), (b), and (c) show mean profiles of condition index, safety index, and cumulative cost, respectively. These profiles are obtained by using Monte Carlo simulation.

Table 5.4 Parameters for condition, safety, and cost profiles under silane and rebuild: All parameters are distributed triangularly – Time-based strategy (based on Denton (2002))

Parameters		Silane			Rebuild		
Time of damage initiation (years)	T_I	0			0		
Time of first application (years)	T_{PI}	Min 0.0	Mode 7.5	Max 15.0	Min 0.0	Mode 26.125	Max 50.0
Time interval of subsequent application (years)	T_P	Min 10.0	Mode 12.5	Max 15.0	0		
Duration of maintenance effect (years)	T_{PD}	Min 7.5	Mode 10.0	Max 12.5	0		
Condition index							
Initial performance index value	P_o	Min 0.00	Mode 1.75	Max 3.50	Min 0.00	Mode 1.75	Max 3.50
Deterioration rate (years^{-1})	α	Min 0.00	Mode 0.08	Max 0.16	Min 0.00	Mode 0.08	Max 0.16
Improvement	γ	0.00			Restored to condition index = 0.00		
Delay in deterioration (years)	T_D	0.00			Min 10.0	Mode 15.0	Max 30.0
Deterioration rate during effect (years^{-1})	α'	Min 0.00	Mode 0.01	Max 0.03	0		
Safety index							
Initial performance index value	P_o	Min 0.91	Mode 1.50	Max 2.50	Min 0.91	Mode 1.50	Max 2.50
Deterioration rate (years^{-1})	α	Min 0.00	Mode 0.015	Max 0.030	Min 0.00	Mode 0.015	Max 0.035
Improvement	γ	0.00			Min 1.00	Mode 1.25	Max 1.50
Delay in deterioration (years)	T_D	0.00			While condition index < 1.0		
Deterioration rate during effect (years^{-1})	α'	Min 0.00	Mode 0.007	Max 0.018	0.00		
Cost							
Cost	C	Min 0.30	Mode 39.0	Max 77.0	Min 247	Mode 7410	Max 28898

5.4.3 Multi-criteria lifetime optimization

There are many practical applications for life-cycle cost analysis where the designer may want to optimize two or more objectives simultaneously (Frangopol and Liu, 2007; Furuta *et al.*, 2006). For example, decrease of maintenance cost and increase of structural performance are two conflicting objectives. In this case, multi-criteria optimization should be applied by simultaneously minimizing a set of objective functions (Arora, 2004).

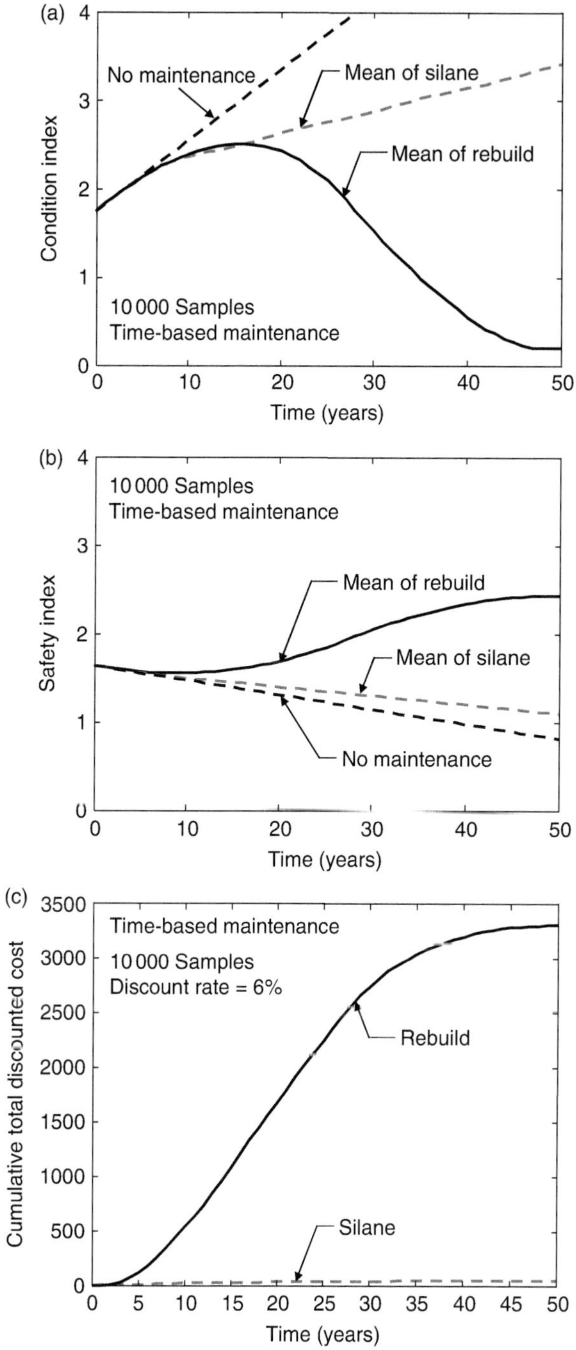

5.19 Profiles of mean values for (a) condition index, (b) safety index, and (c) cumulative total discounted cost under time-based silane and rebuild maintenances.

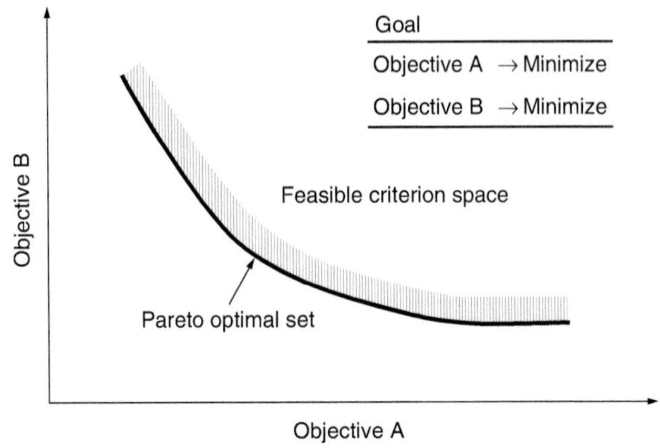

5.20 Pareto optimal sets between two objectives.

Optimum solutions which are on the boundary of the feasible criterion space are called Pareto optimal sets, as shown in Fig. 5.20. Decision makers have to compare different possible solutions and choose the best compromise.

Liu and Frangopol (2004, 2005a,b) proposed a multi-objective optimization approach with respect to condition index, safety index, and cumulative maintenance cost, and Neves *et al.* (2006a,b) considered a full probabilistic multi-objective optimization for single maintenance (silane or rebuild) and combined maintenance (silane and rebuild). The Pareto fronts between condition index and cost, and safety index and cost are qualitatively shown in Figures 5.21(a) and (b), respectively. Less cumulative maintenance cost leads to increasing the condition index, and reducing the safety index.

5.5 Conclusions

In this chapter, concepts of reliability, service life, maintenance, and optimization of structural systems are introduced. Using time-dependent reliability analysis, the service life of a structural system under uncertainty can be predicted, and optimal maintenance interventions can be made to extend its service life.

The concepts of probability and structural reliability can provide a rational tool to treat uncertainties (i.e. aleatoric and epistemic) related to structural performance quantitatively. These concepts can be extended to prediction of time-dependent reliability and service life of structural systems considering deteriorating processes. Multi-criteria optimization under uncertainty allows structure managers to actively compare different solution options and choose the one that best balances their objectives and constraints.

Goal during service life	
The largest condition index	→ Minimize
The smallest safety index	→ Maximize
The cumulative total discount cost	→ Minimize

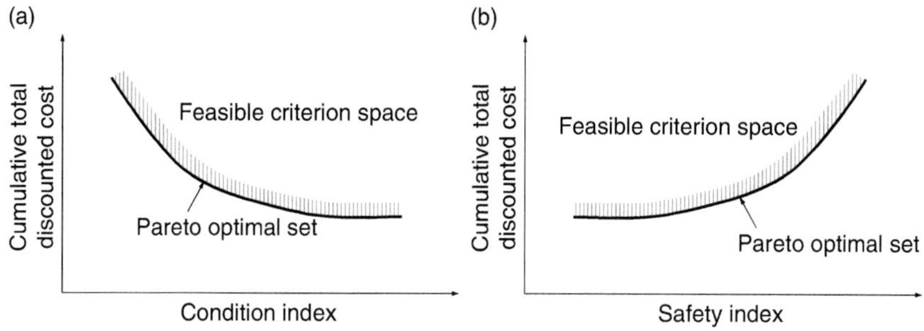

5.21 Pareto fronts between; (a) condition index and cost, and (b) safety index and cost.

5.6 Acknowledgements

The support from (a) the National Science Foundation through grants CMS-0638728 and CMS-0639428, (b) the Commonwealth of Pennsylvania, Department of Community and Economic Development, through the Pennsylvania Infrastructure Technology Alliance (PITA), (c) the U.S. Federal Highway Administration Cooperative Agreement Award DTFH61-07-H-00040, and (d) the U.S. Office of Naval Research Contract Number, N00014-08-1-0188 is gratefully acknowledged. The opinions and conclusions presented in this chapter are those of the authors and do not necessarily reflect the views of the sponsoring organizations.

5.7 Nomenclature

C_m = maintenance cost
C_{mi} = ith maintenance cost
d = minimum distance
$f_X(x)$ = probability density function
$f_{X,Y}(x,y)$ = joint probability density function
$F_X(x)$ = cumulative distribution function
$g(\mathrm{X})$ = state function
M = safety margin
$\bar{M}$ = mean of safety margin
p_F = probability of failure
p_S = probability of survival

r = discount rate
r_i = improved reliability index due to ith maintenance
R = resistance
RIF_i = reliability importance factor of ith component
RIF_i^{norm} = normalized reliability importance factor of ith component
S = load effect
T_{mi} = ith maintenance application time
X' = reduced random variable of X
α = deterioration rate without effect of maintenance
α' = deterioration rate with effect of maintenance
β = reliability index
$\beta_{\text{system,IN}}$ = system reliability index when all components are independent
$\beta_{\text{system,PC}}$ = system reliability index when all components are perfectly correlated
β_{target} = target value of reliability index
Φ^{-1} = inverse cumulative distribution function of the standard normal variable
μ = mean value
ρ_{sys} = system correlation matrix
σ = standard deviation

5.8 References

ANG, A. H.-S., and DE LEON, D. (2005), 'Modeling and analysis of uncertainties for risk-informed decisions in infrastructures engineering', *Structure & Infrastructure Engineering*, 1(1), 19–31.

ANG, A. H.-S., and TANG, W. H. (1984), *Probability Concepts in Engineering Planning and Design, Volume II*, John Wiley & Sons.

ARORA, J. S., (2004), *Introduction to Optimum Design*, Elsevier/Academic Press, 537–552.

CORNELL, C. A. (1967), 'Bounds on the reliability of structural systems', *Journal of Structural Division, ASCE*, 93(ST1), 171–200.

DENTON S., (2002), *Data estimates for different maintenance options for reinforced concrete cross heads*, Draft report for Highways Agency, U.K., Brinckerhoff Ltd.

DITLEVSEN, O. (1979), 'Narrow reliability bounds for structural systems', *Journal of Structural Mechanics, ASCE*, 7(4), 453–472.

ELLINGWOOD, B. R. (2005), 'Risk-informed condition assessment of civil infrastructure: State of practice and research issues', *Structure & Infrastructure Engineering*, 1(1), 7–18.

ESTES, A. C., and FRANGOPOL, D. M. (1998), 'RELSYS: A computer program for structural system reliability analysis', *Structure Engineering & Mechanics*, 6(8), 901–919.

ESTES, A. C., and FRANGOPOL, D. M., (1999), 'Repair optimization of highway bridges using system reliability approach', *Journal of Structural Engineering, ASCE*, 125(7), 766–775.

FISHER, J. W., KULAK, G. L., and SMITH, I. F. C. (1998), *A Fatigue Primer for Structural Engineers*, National Steel Bridge Alliance, Chicago, IL, USA.

FRANGOPOL, D. M., and LIU, M. (2007), 'Maintenance and management of civil infrastructure based on condition, safety, optimization, and life-cycle cost.' *Structure & Infrastructure Engineering*, 3(1), 29–41.

FRANGOPOL, D. M., LIN, K. Y., and ESTES, A. C. (1997), 'Life-cycle cost design of deteriorating structures', *Journal of Structural Engineering, ASCE*, 123(10), 1390–1401.

FRANGOPOL, D. M., KONG, J. S., and GHARAIBEH, E. S. (2001), 'Reliability-based life-cycle management of highway bridges', *Journal of Computing in Civil Engineering, ASCE*, 15(1), 27–34.

FURUTA, H., KAMEDA, T., NAKAHARA, K., TAKAHASHI, Y., and FRANGOPOL, D. M. (2006), 'Optimal bridge maintenance planning using improved multi-objective genetic algorithm', *Structure & Infrastructure Engineering*, 2(1), 33–41.

GHARAIBEH, E. S., FRANGOPOL, D. M., and ONOUFRIOU, T. (2002), 'Reliability-based importance assessment of structural members with applications to complex structures', *Computers and Structures*, 80(12), 1113–1131.

KONG, J. S., and FRANGOPOL, D. M. (2003), 'Evaluation of expected life-cycle maintenance cost of deteriorating structures', *Journal of Structural Engineering, ASCE*, 129(5), 682–691.

KONG, J. S., AKGUL, F., and FRANGOPOL, D. M. (2000), *User's Manual, Monte Carlo Simulation Program*, Report No. 00–1, Structural Engineering and Structural Mechanics Research Series No. CU/SR-00/1, Department of Civil, Environmental, and Architectural Engineering, University of Colorado, Boulder, November 2000.

LEEMIS, L. M. (1995), *Reliability: Probabilistic Models and Statistical Methods*, Prentice Hall.

LIU, M., and FRANGOPOL, D. M. (2004), 'Optimal bridge maintenance planning based on probabilistic performance prediction', *Engineering Structures*, 26(7), 991–1002.

LIU, M., and FRANGOPOL, D. M. (2005a), 'Bridge annual maintenance prioritization under uncertainty by multiobjective combinatorial optimization', *Computer-aided Civil and Infrastructure Engineering*, 20(5), 343–353.

LIU, M., and FRANGOPOL, D. M. (2005b), 'Multiobjective maintenance planning optimization of deteriorating bridges considering condition, safety, and life-cycle cost', *Journal of Structural Engineering, ASCE*, 131(5), 833–842.

MOAN, T. (2005), 'Reliability-based management of inspection, maintenance and repair of offshore structures', *Structure & Infrastructure Engineering*, 1(1), 33–62.

NEVES, L. A. C., FRANGOPOL, D. M., and CRUZ, P. J. S. (2004), 'Cost of life extension of deteriorating structures under reliability-based maintenance', *Computers and Structures*, 89(13–14), 1077–1089.

NEVES, L. A. C., FRANGOPOL, D. M., and CRUZ, P. J. S. (2006a), 'Probabilistic lifetime-oriented multiobjective optimization of bridge maintenance: Single maintenance type', *Journal of Structural Engineering, ASCE*, 132(6), 991–1005.

NEVES, L. A. C., FRANGOPOL, D. M., and PETCHERDCHOO, A. (2006b), 'Probabilistic lifetime-oriented multiobjective optimization of bridge maintenance: Combination of maintenance types', *Journal of Structural Engineering*, *ASCE*, 132(11), 1821–1834.

THOFT-CHRISTENSEN, P. (2003), 'Corrosion and Cracking of Reinforced Concrete', in *Life-cycle Performance of Deteriorating Structures* (edited by D.M. Frangopol, E. Brühwiler, M.H. Faber, B. Adey), ASCE, 26–36.

THOFT-CHRISTENSEN, P., and MUROTSU, Y. (1986), *Application of Structural System Reliability Theory*, Springer-Verlag.

6

Probabilistic methods for service life estimation of civil engineering structures

R.E. MELCHERS, The University of Newcastle, Australia

Abstract: This chapter describes the background to the estimation of service life for use in lifecycle cost analyses of infrastructure, including deteriorating infrastructure. Since loadings, material properties and dimensions are not known with certainty, service life estimates must be based on probabilistic analyses. The theory most useful for this, the theory of structural reliability, is outlined briefly and related to service life estimation through stochastic process theory. This requires the estimation of the probability of failure on first load application and the so-called 'out-crossing rate'. Methods to estimate these include Monte Carlo methods and moment methods. Both are revised briefly and future trends for computational methods indicated.

Key words: infrastructure, structures, probability, life, service, outcrossing, asymptotic methods.

6.1 Introduction

The worldwide investment in civil infrastructure is large and is also growing at a fast rate. While, traditionally, little attention was paid in the design process to the likely durability of the infrastructure project and its various parts, increasingly it is being recognized that infrastructure does deteriorate and wear out, and that there are very significant costs in maintenance and replacement, assuming, of course, that the infrastructure is required to remain in service. Of late it has become evident that the life-cycle costs, including those of inspection, repair, replacement and management, as well as the risk costs associated with potential disruptions, may be more important than the original cost of the asset. Ideally, the minimum lifecycle cost is of interest, perhaps subject to constraints, such as on initial total cost as well as on annual costs.

Typically, the various costs are uncertain. Also uncertain is when, precisely, future maintenance costs and replacement costs are likely to be incurred. Obviously, this depends on the rate of deterioration of materials and components of the infrastructure system. It also depends on the specified level of acceptability of performance levels, including overall system safety, safety to users and staff, performance requirements, etc. In general terms the problem may be formulated in terms of discounted expected total costs $E(C_T)$ as follows:

$$E(C_T) = E\left[C_0 + C_{QA} + C_M + \sum_i p_{fi} \cdot C_{fi}\right] \qquad [6.1]$$

where C_0 = concept, design and construction costs, C_{QA} = quality assurance and insurance costs, C_M = maintenance and on-going service costs and C_{fi} = direct and indirect costs incurred as a result of failure of the structure or system in the i-th failure mode. Also, p_{fi} is the probability of failure of the structure or system in the i-th failure mode during the time $(0, t_L)$. Here t_L may be defined as the 'service life'. Assuming that none of the failure modes is acceptable, interest lies in the minimum time likely to elapse before the occurrence of the first of any of the failure modes. Let the time to the occurrence of this event be termed t_1; and for the system to be satisfactory it is required that $t_1 \geq t_L$. Evaluation of t_1 is a key component in the evaluation of Equation 6.1.

To allow t_1 to be estimated, three components are required to be known or defined. The first is the demand or loading $Q(t)$ on the system or component being considered. The second is the capacity or resistance $R(t)$ of that system or component. The third is the service or safety requirement, expressed here as a limit state function $G(\mathbf{X})$, where the vector $\mathbf{X}$ embraces all the variables necessary to define the parameters involved in the service or safety requirement, including $Q(t)$ and $R(t)$. Note that $G(\mathbf{X})$ need not be confined simply to ultimate loading and ultimate capacity as has been the traditional application in structural reliability theory applied to design code calibration (Melchers, 1999). It may encompass any design requirement that can be formulated mathematically in terms of demand(s) and capacity(s). This includes matters such as limits on fatigue and corrosion damage, deflection limits, creep design requirements, etc. In addition, for other than extremely simple structural systems, there will be multiple members or components each with their own limit state functions. Thus, in general, there will be a set of limit state functions, denoted $\mathbf{G}(\mathbf{X})$.

Typically $Q(t)$ is a random process, one realization of which is shown in Fig. 6.1 (It is sometimes known as a 'fast' process). In the real problem there are infinitely many such realizations possible, although the random nature of $Q(t)$ means that the realization that occurs is known only probabilistically. Also typically, $R(t)$ is a random variable that changes with time (usually slowly) as a result of fatigue or corrosion or similar deterioration processes. (Compared to $Q(t)$, its variation is much slower, is often positive semi-definite and it is sometimes known as a 'slow' process). It is represented, schematically, in Fig. 6.1. Note that the variability of $R(t)$ is shown as increasing with time. This is the usual situation for deteriorating systems.

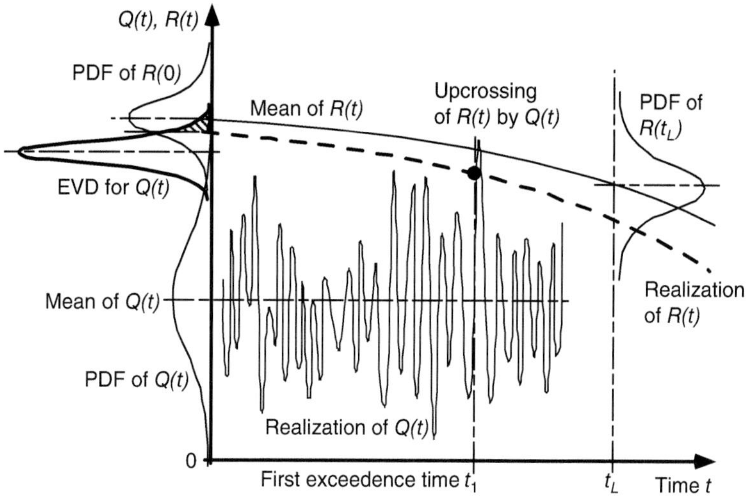

6.1 Realization of a stochastic process $Q(t)$ having an instantaneous probability distribution function (PDF) and mean as shown, and an extreme value distribution (EVD) for the maxima, together with a realization of the random variable resistance $R(t)$ having changing PDFs as a function of time, as shown. A typical up-crossing event is shown at time t_1 within the service life t_L.

In the simplest case of just one load process and one resistance, survival can be stated as the requirement that $Q(t)$ must not exceed $R(t)$ at any time $0 < t < t_L$ where t_L is as defined above. Using the conventional notation, the limit state function G(**X**) becomes:

$$G[\mathbf{X}(t)] = 0 = R(t) - Q(t) \ \forall (0 < t < t_L) \qquad [6.2]$$

where $G[\mathbf{X}(t)] < 0 \ \forall \ (0 < t < t_L)$ denotes failure and $G[\mathbf{X}(t)] \geq 0 \ \forall \ (0 < t < t_L)$ survival.

Figure 6.1 shows both a realization of the load process $Q(t)$ and a realization of the deteriorating resistance $R(t)$, together with a nominal up-crossing event at time $t_u = t_i$. The essential matter of interest may now be restated as $P(t_1 < t_L)$, that is, the probability that the first up-crossing event occurs before the service life has been reached. Usually it is desired that this probability should be as low as reasonably achievable consistent with other design requirements.

In general, either or both of $Q(t)$ and $R(t)$ may be multi-component, denoted by random vector (processes) $\mathbf{Q}(t)$ and $\mathbf{R}(t)$ respectively. In this case, the up-crossing notion just introduced generalizes to the 'out-crossing' notion, meaning the event when the vector process $\mathbf{Q}(t)$ 'crosses-out' of the safe domain specified at time t by the limit state function(s) and $\mathbf{R}(t)$. This

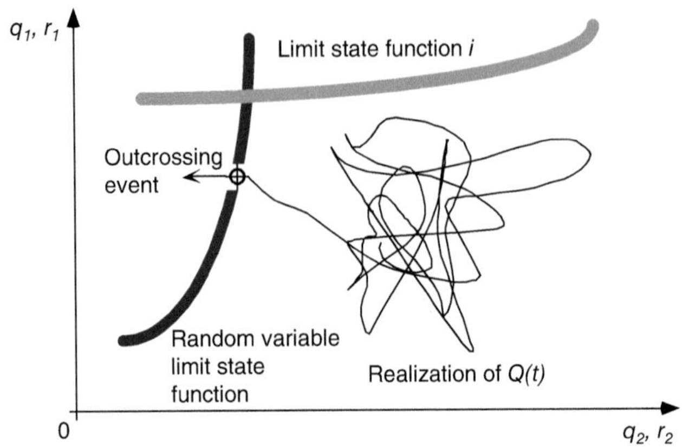

6.2 Schematic example of an out-crossing event for a realization of the vector load process $Q(t)$ showing also the multiple limit state functions. The time axis is perpendicular to the q_1,q_2 plane.

may be illustrated easily for a system in which $\mathbf{Q}(t)$ and $\mathbf{R}(t)$ are each of just two components, when viewed along the time axis (Fig. 6.2).

6.2 Direct numerical evaluation

Rather than estimate the probability $P(t_u > t_L)$ directly, it is more convenient to consider the probability of structural system failure within a given service period (or design life) $(0, t_L)$.

$$P(t_l > t_L) = P\{G[\mathbf{X}](t) > 0\} \; \forall (0 < t < t_L) \qquad [6.3]$$

This means that for the service life t_L to be reached, it is necessary that the loads do not exceed the resistance (see Eqn 6.2) or, more generally, that the limit state function is not violated, at any time during $(0, t_L)$.

One simple way of estimating Eqn 6.3 is to use Monte Carlo methods, providing the system of interest is not too large. The process is to generate values of $Q(t)$ and $R(t)$ for various t and check the limit state function. Since Q is a random process (either stationary or non-stationary) and R is a random variable that depends on time, their values also will be random, and can be generated using standard Monte Carlo techniques assuming their probability density functions $f_{Q(t)}(\;)$ and $f_{R(t)}(\;)$ are known. In principle, the process is that a value for t is selected randomly from a uniform distribution and for that random values of $Q(t)$ and $R(t)$ are generated from $f_{Q(t)}(\;)$ and $f_{R(t)}(\;)$ respectively. Note that each is a function of time to allow for possible load process non-stationarity and for resistance deterioration. Substitution of these random values into the limit state function will then show whether the structure survives. The proportion of such survivals compared with the total number of trials will produce, in the limit, an estimate for $P(t_1 > t_L)$. The procedure is readily extended to situations with more random variables (loads, resistances, etc.) forming the vector **X** and to situations where the resistances deteriorate with time. However the computational time for this is extremely large (e.g. Moarefzadeh and Melchers, 1996). Another approach is to generate the random processes numerically and to estimate the probability of the vector process leaving a deterministic domain (Hasofer *et al.*, 1987; Shinozuka, 1987).

The main drawback for these direct approaches is the sheer computational effort, particularly if the probability of not achieving the service life $P(t_u < t_L)$ is low. For example, if this probability is one in a milllion, at least 10 million samples would be required to provide some level of confidence in the outcome of the Monte Carlo analysis. The situation may be improved somewhat by invoking variance reduction techniques such as importance sampling (Ditlevsen *et al.*, 1987) or by using conditional expectation (Mori and Ellingwood, 1993).

6.3 Asymptotic solution

A different approach to estimating the probability $P(t_u < t_L)$ is again to consider the probability of structural system failure within a given service period (or design life) $(0, t_L)$ but now recognizing that this probability can be considered as the sum of probability that the structure will fail when it is first loaded, denoted as $p_f(0)$, and the probability that it will fail subsequently, given that it has not failed earlier (that is, within the time period $(0, t_L)$). This may be expressed as:

$$p_f(0, t_L) \approx p_f(0) + [1 - p_f(0)].\,[1 - e^{-\nu T}] \qquad [6.4]$$

where ν is the so-called 'out-crossing rate' (see Fig. 6.2). Equation 6.4 is based on the assumption that the probability of failure during the lifetime is very small and that therefore failure events are 'rare', and thus sufficiently far apart in time to be considered independent (this is also known as the asymptotic assumption). The practical result is that this sequence of assumptions allows the occurrence of such events to be represented by a Poisson distribution, shown by the second [ ] term. It also means that Eqn 6.4 is approximate and actually an upper bound to the correct result (Leadbetter *et al.*, 1983). It is progressively a more accurate upper bound as the probability of failure becomes smaller, or, equivalently, as the limit state functions are further removed from the loads (see Fig. 6.2). A further limitation on Eqn 6.4 is that it assumes that the process involved is 'narrow-banded'. This means that its frequency range is confined to a limited range rather than

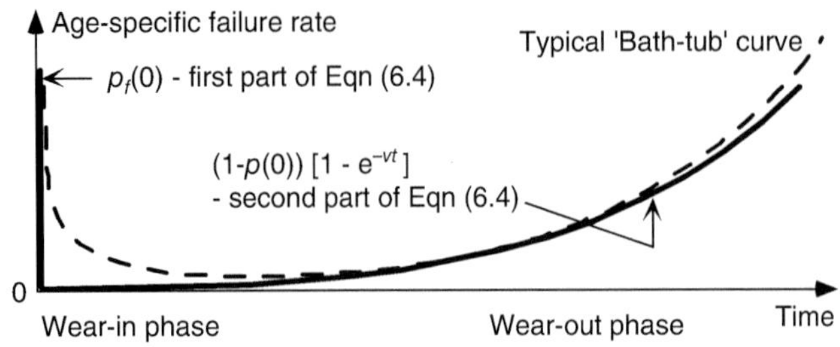

6.3 Nominal representation of the two components of Eqn 6.4 and comparison to the conventional 'bathtub' curve using the age-specific failure rate of mechanical components.

over a very wide range. For most processes of interest in structural engineering this is not a significant restriction.

In general, it may be expected that a structure will slowly deteriorate with time or with loading cycles. Thus, the out-crossing rate will be a function of time $v = v(t)$ and will be expected to increase with t. This means that the time dependence both of $\mathbf{Q}(t)$ and $\mathbf{R}(t)$ is required to be known. In turn, this means that both will be 'non-stationary' functions in that their means as well as their variance and various higher moments will not be constant with time. This is an important differentiation from much of the theory for stationary processes available in the structural dynamics literature.

Equation 6.4 may be sketched in the general case as shown in Fig. 6.3. Evidently, the first term $p_f(0)$ represents the spike at $t = 0$ and the increase in probability thereafter is the result of the second [ ] term in Eqn 6.4. It increases only slightly at first. A generally similar type of behaviour has long been recognized in mechanical systems, commonly represented schematically by the 'bath-tub' curve. It has, early on, a high probability of failure, representing the 'wear-in' phase, and a gradual increase later, representing the 'wear-out' phase. It can be argued, then, that Eqn 6.4 is an idealization since it assumes that $p_f(0)$ occurs precisely at $t = 0$, whereas in practice it may take some time for the complete full loading to be applied.

The evaluation of Eqn 6.4 requires evaluation or estimation of two parameters: (i) the probability of failure when first loaded $p_f(0)$ and (ii) the out-crossing rate ω. Both may be estimated using conventional techniques in structural reliability analysis, as will now be reviewed briefly.

6.4 Estimation of initial failure probability

For the initial failure probability, interest lies in the description of the load process as a random variable at $t = 0$. Let this be denoted simply $Q(t)|_{t=0}$ with known probability density function $f_Q(\ )$. This implies that all the moments also are known. Similarly, the resistance will be $R(t)|_{t=0}$ with known probability density function $f_R(\ )$.

Consider now the probability that the resistance $R(t)|_{t=0} = R(0)$ is less than some particular value of the load, say $x(t = 0) = x(0)$. This is given by $P(R(0) > x(0)) = F_R(x)|_{t=0} =$ where $F_R(\ )$ is the cumulative distribution function defined, as usual, as

$$F_R(r) = \text{Prob}(R < r) = \int_{-\infty}^{r} f_R(x)dx \qquad [6.5]$$

To obtain the total probability involved, all contributions for all possible values of x are required. However, each must also be weighted by the probability that the load $Q(0)$ corresponds to that x value. This can be stated (mathematically somewhat loosely) as $x(0) = f_Q(x)|_{t=0}$. As a result we have that the probability is given by

$$p_f(0) = \int_{-\infty}^{+\infty} F_R(x) \cdot f_Q(x)dx \qquad [6.6]$$

where it is understood that the evaluation is at $t = 0$. This integral is also known as a 'convolution' integral. Substituting Eqn 6.5 into Eqn 6.6 produces the more general form

$$p_f(0) = \iint_{D_f} f_Q(x) \cdot f_R(x)dx \qquad [6.7]$$

where D_f represents the 'failure domain', that is, the region over which failure occurs as defined by the limit state function(s). This expression can be generalized further by using the notion, introduced already in Eqn 6.2, that the random vector $\mathbf{X}$ collects all the random variables in the reliability problem. Moreover, where there are multiple limit state functions, these can be represented by their union $D_f = \bigcup_i [G_i(\mathbf{X}) < 0]$, where as before, $G_i(\mathbf{X}) < 0$ denotes failure in the ith limit state function. Here, D_f represent the 'failure domain'. Finally, Eqn 6.7 may be generalized to

$$p_f(0) = \int \dots \int_{D_f} f_X(\mathbf{x})d\mathbf{x} \qquad [6.8]$$

Some analytical solutions are available for Eqn 6.6. Also, Equations 6.6, 6.7 or 6.8 may be solved using numerical integration but this is usually not feasible for dimensions of $\mathbf{X}$ greater than about 5. In general, however, Equations 6.6, 6.7 or 6.8 can be solved only through approximate methods. One option is to use Monte Carlo methods. These approximate the integration process by using random numbers (e.g. see above). This usually incurs heavy computational demands. Another option is to use the first order second moment (FOSM) methods and their refinements. These approximate the integrand, thereby usually obviating the integration process. In the case of FOSM, the calculations are very simple but the approximation often severe, including that all random variables are represented only by their means and variances and requiring that the limit state function, if not linear, can be approximated by a linear function (Melchers, 1999). The refinements can remove these restrictions. However, this can add considerably to the computational requirements, in the limit approaching those of the refined Monte Carlo methods (Ditlevsen and Madsen, 1996).

6.5 Estimation of out-crossing rate

The out-crossing rate at any time t may be estimated by assuming that the random (load) processes continue indefinitely and have a 'stationary' statistical nature (e.g. in the simplest case, their means and variances do not change with time). This means that long-term average values can be estimated, including the outcrossing rate. It is convenient and conventional to do so in the space of the vector (load) processes $\boldsymbol{x}$:

$$\nu = \int_{\text{safe domain}} E\left(\dot{X}_n \middle| \mathbf{X} = \mathbf{x}\right)^{+} f_{\mathbf{X}}(\mathbf{x})\, d\mathbf{x} \qquad [6.9]$$

where $(\ )^{+}$ denotes the positive component of the vector process $\mathbf{X}$, that is, the component that moves out of the safe domain (the other component moves back in and therefore is of no interest). For mathematical completeness the term $E(\dot{X}_n|\mathbf{X} = \mathbf{x}) = \dot{x}_n = \mathbf{n}(t).\dot{\mathbf{x}}(t) > 0$ represents the outward normal component of the vector process at the domain boundary. The term $f_{\mathbf{x}}(\mathbf{X})$ in Eqn 6.9 represents the probability that the process has reached the boundary (since otherwise it is not possible for it to cross-out from the safe domain). When the structural capacity, defined through the limit state functions and hence by the boundaries of the safe domain, reduces with time, Eqn 6.9 can be considered conditional on time, and hence the out-crossing rate becomes time-dependent $\nu = \nu(t)$. One way of thinking about this is to consider the time dependence as a series of steps during each of which the process is 'stationary', allowing Eqn 6.9 to be applied. More generally, the concept of egodicity may be invoked, particularly when estimates of $\nu = \nu(t)$ could be made from observations. One theoretical difficulty is that dependency would be expected, in general, between the out-crossing rate at time t_1 and time $t_2 > t_1$. Reasons for this include, for example, the fact that the same resistance components are involved at each time point and that essentially the same deterioration process is involved. In most practical situations, however, it will be necessary (and mostly sufficiently accurate) to estimate the out-crossing rate at different time points, ignoring possible dependence.

Estimation of the out-crossing rate using Eqn 6.9 is not necessarily straight-forward. Except for some very specialized domains and particular vector processes, **X**, there are no analytical solutions available (e.g. Hasofer, 1974; Veneziano *et al.*, 1977) and generally reliance must be placed on numerical procedures. FOSM theory has been applied for continuous Gaussian and Poisson pulse processes (Breitung and Rackwitz, 1982; Grigoriu, 1984; Hohenbichler and Rackwitz, 1986; Rackwitz, 1993). Monte Carlo techniques also have been used for continuous Gaussian processes and for Poisson pulse processes, including the use of importance sampling techniques to reduce the computational burden (Melchers, 1992, 1995).

6.6 Generalization of out-crossing methods

As noted, the use of Eqn 6.4 is predicated on the out-crossings being 'rare' events. This is adequate for considerations of structural safety, for which the probability of failure usually is very low, and thus failure events, if they occur, can be considered to be independent events, rendering the Poisson approximation in Eqn 6.4 an upper bound. However, violations of serviceability criteria are much more frequent. This may be visualized in Fig. 6.2 as the limit state functions 'closing-in' on the load process. It is obvious that, as the limit state functions come closer to the mean of the load process, the potential for out-crossings to occur increases and thus, over time, the rate of out-crossing for continuous processes becomes greater and hence the rate of structural failure increases. It also means that the independence assumption, and hence the Poisson assumption become increasingly less appropriate. The upper bound provided by Eqn 6.4 also becomes less accurate.

Moreover, it is possible for some groups of out-crossings to occur (close) together in time (usually termed 'clumps') (Fig. 6.4). Evidently the out-

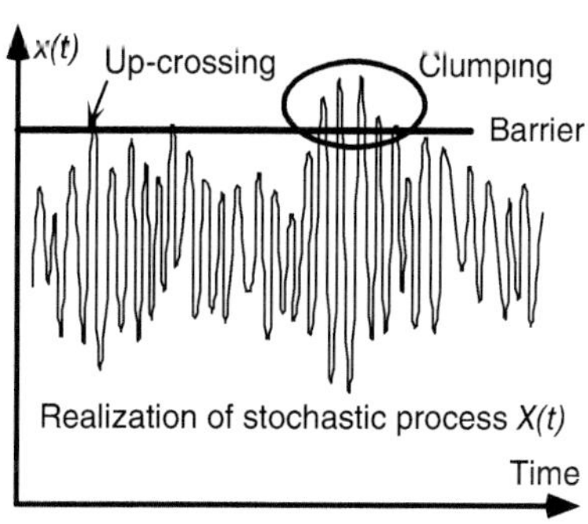

6.4 Stationary stochastic process X(*t*), showing clumping above the barrier.

crossing of a clump is really the failure event. Treating component out-crossing of a clump as an independent out-crossing event, as the theory would suggest, would give a misleading estimate of the probability of failure. The problem is more serious if the processes are not 'narrow-banded' as assumed above.

There have been considerable efforts to extend the asymptotic theory to cases where the out-crossings are not rare or are 'clumped'. An introduction to the problem has been given by Leadbetter *et al.* (1983). There is an extensive theoretical literature dealing with possible solutions. Slightly more accessible expositions have been given by Kordzakhia *et al.* (1999) and Novikov *et al.* (2005).

6.7 Simplifications

For some practical problems, it may be sufficient to by-pass the above out-crossing approach. A plausible and fairly accurate 'rule-of-thumb', known as Turkstra's rule, can be used to combine various load processes using simply the 'maximum' of each load combined with the 'average' values of the others, working on the basis that the combination of two maxima is sufficiently unlikely to be ignored. Turkstra's rule states, simply, that the occurrence of the maximum of one load X_i should be associated with the 'average-point-in-time' values $\bar{X}_j$ of the other load(s). The 'average-point-in-time' loads $\bar{X}_j$ are the loads that would be the loads expected to act on the structure if one were to measure them at any arbitrary time. For most loads, this means rather low load levels. Of the various possible combinations, the maximum (or worst) combination is critical. This may be stated as

$$\max X \approx \max\left(\max X_i + \sum_{j=1}^{n} \bar{X}_j \right), \quad j \neq 1; i = 1, \ldots, n \qquad [6.10]$$

Turkstra's rule is adequate for load combinations where only one extreme load is dominant but not when there are several dominant or near dominant loading systems, such as, for example, wave and wind loadings at sea.

To apply this approach, all maximum and average-point-in-time loads must be defined relative to some time interval, since the 'maximum', for example, will be greater for longer exposure periods. The time interval might be the proposed service life. The extreme load and its statistical extreme value properties may be evaluated from simple combination as dictated by Eqn 6.10 for the given period of interest. This load can then be used in a simple reliability analysis to estimate the probability of failure under the single application of this extreme load for the given service life period. This then provides an estimate of the probability that the design life will be attained. Alternatively, the annual extreme load can be estimated using annual load statistics and the annual failure probability p estimated. Then the failure probability for a number of years, say t_L, (e.g. for the proposed service life 0 to t_L) may be estimated using

$$p_f(t_L) = 1 - \exp(-p.t_L) \approx p.t_L \qquad [6.11]$$

in which independence between years has been assumed (see Eqn 6.4). The last approximation is valid when p is small.

Other forms of simplification based on loading combinations have also been explored. These include the load coincidence method (Wen, 1977; Wen and Chen, 1987). A summary of the method and its extension to deteriorating structures is described by Beck and Melchers (2005).

6.8 Reliability estimation methods, sources of further information and advice, and future trends

Methods to estimate probabilities of failure, such as the Monte Carlo methods and the first order second moment (FOSM) method and its extensions to non-linear limit state functions (second order – SO) and non-Gaussian random variables (reliability methods – RM), will not be described here. A number of well-known texts deal with the details of these (e.g. Thoft-Christensen and Baker, 1982; Madsen *et al.*, 1986) as well as with advanced Monte Carlo methods (e.g. Melchers, 1999; Ditlevsen and Madsen, 1996). There are also numerous review articles for readers to consult.

Methods to estimate the probability of structural failure and the estimation of service life and the theory that underlies it have all reached a high degree of maturity. For this reason most current research effort is devoted to formulating and solving problems with practical applications. These include finite element formulations (Der Kiureghian and Ke, 1988; Lemaire and Mohamed, 2000) and problems with many random variables, for which response surface techniques have been proposed (Faravelli, 1989; Rajashekhar and Ellingwood, 1993; Guan and Melchers, 2001). For details of these, reference might be made to the proceedings of long-running conferences such as the International Conference on the Application of Statistics and Probability (ICASP) in civil engineering, and the International Conference on Structural Safety and Reliability (ICOSSAR).

As indicated above, the basic ideas for service life estimation using probabilistic methods are now well-established. As computational power increases, it will increasingly be the case that the 'brute force' approach of Monte Carlo of Section 6.2 will be applied for practical as well as research problems. It is easy to understand and does not have to be overly concerned with some theoretical issues, such as the asymptotic nature of the solution of Eqn 6.4 or with whether 'clumping' occurs.

Future developments are likely to focus on using these techniques to a greater extent as engineering practitioners become more confident in thinking along probabilistic lines and in starting to apply the type of techniques outlined herein for practical applications. However, some fundamental issues remain, principally in how to deal with uncertainties in general and in translating real-world understanding (including misconceptions, biases and fundamental errors) to robust life-prediction estimation techniques. The issues are not fundamentally different from those recently discussed in relation to ensuring the safety of structures (Melchers, 2007; Brown *et al.*, 2008). This area still represents a significant research challenge – one that is largely outside the domain of conventional engineering and much more aligned with fundamental issues of risk and consequence analysis.

6.9 Conclusions

This chapter has given an overview of the probabilistic methods and approaches that may be applied for the estimation of service life. These are essentially the methods and techniques used for the estimation of the probability of failure of structural systems using the now well-developed theory of structural reliability. Estimates of failure probabilities for a given life expectancy are equivalent to the estimation of the service life associated with a given probability level. An outline has been given of the most general theory that uses stochastic process theory to estimate the out-crossing rate of a vector of stochastic processes from a safe domain. Techniques for solving the problem have been outlined and some simplifications briefly reviewed. Suggestions for further information, reading and likely future directions have also been given.

6.10 References

BECK AT and MELCHERS RE (2005) Barrier failure dominance in time variant reliability, *Probabilistic Engineering Mechanics*, 20 79–85.

BREITUNG, K. and RACKWITZ, R. (1982) Non-linear combination load processes, *J. Structural Mechanics*, 10 (2) 145–166.

BROWN CB, ELMS DG and MELCHERS RE(2008) 'Assessing and achieving structural safety', *Proceedings Institution of Civil Engineers, Structures & Buildings*, 161(SB1) 219–230.

DER KIUREGHIAN A and KE J-B (1988) The stochastic finite element method in structural reliability, *Probabilistic Engineering Mechanics*, 3(2) 83–91.

DITLEVSEN, O. and MADSEN, H.O. (1996) *Structural Reliability Methods*, John Wiley & Sons.

DITLEVSEN, O., OLESEN, R. and MOHR, G. (1987) Solution of a class of load combination problems by directional simulation, *Structural Safety*, 4 95–109.

FARAVELLI L (1989) Response-surface approach for reliability analysis, *J. Engineering Mechanics (ASCE)*, 115(12) 2763–2781.

GRIGORIU, M. (1984) Crossings of non-Gaussian translation processes, *J. Engineering Mechanics Div., ASCE*, 110 (6) 610–620.

GUAN XL and MELCHERS RE (2001) Effect of response surface parameter variation on structural reliability estimates, *Structural Safety*, 23(4) 429–444.

HASOFER, A.M. (1974) 'The upcrossing rate of a class of stochastic processes', in *Studies in Probability and Statistics*, Williams, E. J. (Ed), North-Holland, Amsterdam, 151–159.

HASOFER, A.M., DITLEVSEN, O. and OLESON, R. (1987) Vector outcrossing probabilities by Monte Carlo, *Probabilistic Engineering Mechanics*, 5 122–128.

HOHENBICHLER, M. and RACKWITZ, R. (1986) Asymptotic outcrossing rate of Gaussian vector process into intersection of failure domains, *Probabilistic Engineering Mechanics*, 1 (3) 177–179.

KORDZAKHIA, N, MELCHERS, RE and NOVIKOV, A (1999) 'First passage analysis of a "Square Wave" filtered Poisson process', in *Proceedings Internestional Conference Applications of Statistics and Probability*, (ed) Melchers, RE and Stewart, MG, Balkema, Rotterdam, 35–43.

LEADBETTER MR, LINDGREN G and ROOTZEN H (1983) *Extremes and Related Properties of Random Sequences and Processes*, Springer-Verlag, New York.

LEMAIRE M and MOHAMED A (2000) 'Finite element and reliability: A happy marriage?' *Proceedings 9th IFIP Working Conference, Reliability and Optimization of Struct. Systems*, Ed AS Nowak and MM Szerszen, University of Michigan, Ann Arbor, Mich. 3–14.

MADSEN, H.O., KRENK, S. and LIND, N.C. (1986) *Methods of Structural Safety*, Prentice-Hall.

MELCHERS, R.E. (1992) Load space formulation of time dependent structural reliability, *J. Engineering Mechanics, ASCE*, 118 (5) 853–870.

MELCHERS, R.E. (1995) Load space reliability formulation for Poisson pulse processes, *J. Engineering Mechanics, ASCE*, 121 (7) 779–784.

MELCHERS, R.E. (1999) *Structural Reliability Analysis and Prediction*, Second Edition, John Wiley & Sons.

MELCHERS R.E. (2007) Structural reliability theory in the context of structural safety, *Civil Engineering and Environmental Systems*, 24(1) 55–69.

MOAREFZADEH, M.R. and MELCHERS, R.E. (1996) Sample-specific Linearization in Reliability Analysis of Off-shore Structures, *Structural Safety* 18 (2,3) 101–122.

MORI, Y. and ELLINGWOOD, B.R. (1993) Time-dependent system reliability analysis by adaptive importance sampling, *Structural Safety*, 12 (1) 59–73.

NOVIKOV A, MELCHERS RE, SHINJIKASHVILI E and KORDZAKHIA N (2005) First passage time of filtered Poisson process with exponential shape function, *Probabilistic Engineering Mechanics*, 20 57–65.

RACKWITZ, R. (1993) On the combination of non-stationary rectangular wave renewal processes, *Structural Safety*, 13 (1 + 2) 21–28.

RAJASHEKHAR MR and ELLINGWOOD BR (1993) A new look at the response surface approach for reliability analysis, *Structural Safety*, 12(3) 205–220.

SHINOZUKA, M. (1987) 'Stochastic fields and their digital simulation', in *Stochastic Methods in Structural Dynamics,* Schuëller, G.I. and Shinozuka, M. (Eds), Martinus Nijhoff, The Hague.

THOFT-CHRISTENSEN, P. and BAKER, M.J. (1982) *Structural Reliability and its Applications*, Springer-Verlag.

VENEZIANO, D., GRIGORIU, M. and CORNELL, C.A. (1977) Vector-process models for system reliability, *J. Engineering Mechanics Div., ASCE*, 103 (EM3) 441–460.

WEN, Y.-K. (1977) Statistical combination of extreme loads, *J. Structural Div., ASCE*, 103 (ST5) 1079–1093.

WEN, Y.-K. and CHEN, H.C. (1987) On fast integration for time variant structural reliability, *Probabilistic Engineering Mechanics*, 2 (3) 156–162.

7

Non-destructive testing and evaluation (NDT/NDE) of civil structures rehabilitated using fiber reinforced polymer (FRP) composites

Y. DONG, University of Alaska Fairbanks, USA and
F. ANSARI, University of Illinois at Chicago, USA

Abstract: The chapter begins by discussing the requirement for performance assessment of FRP rehabilitated structures by NDT/NDE methodologies. The general classification of NDT techniques are introduced, followed by the concept of structural health monitoring (SHM) as related to the real-time performance monitoring of the rehabilitated structures. It then explores the suitability of major NDT techniques for defect detection and assessment of FRP composites and FRP rehabilitated concrete structures. The NDT methods reviewed include visual inspections, acoustic emission testing, impact-echo methods, ultrasonics, radiography, ground penetrating radar, microwave based techniques, thermography, optical and laser techniques, and superconducting quantum interference methods. The chapter includes a brief introduction into the basic principles, methodologies, and types of fiber optic sensors, and their applications in FRP composite and FRP rehabilitated concrete structures. Smart structure systems with fiber optic sensors for real-time monitoring of FRP rehabilitated concrete structures are then discussed. A summary of the methods is provided at the end.

Key words: NDT, fiber optic sensor, FRP rehabilitation, reinforced concrete.

7.1 Introduction

Fiber reinforced polymer (FRP) composite systems have been used since the mid-1980s for repairing and strengthening damaged, deteriorated or structurally deficient concrete structures. Current FRP rehabilitation systems fall into two major types. One type consists of factory-manufactured laminates of carbon- or glass-reinforced thermosetting polymers, bonded to the surface of the concrete using an epoxy adhesive. The other type consists of layers of unidirectional sheets or woven/stitched fabrics of dry fibers that are saturated in the field with a thermosetting polymer which simultaneously bonds the FRP laminate to the concrete.

The performance of an FRP rehabilitation system is highly influenced by its integrity, which includes the composite and the adhesive layer in the case of bonded strip, and the composite and the impregnating polymeric resin in the case of dry sheets and fabrics. The interface between the FRP composite and the concrete substrate transfers the loads from the concrete to the FRP composite. The polymer resin plays a critical role in the FRP composite in transferring loads to the fibers and protecting the fibers. Potential defects in FRP composites could also affect the performance of the rehabilitated structures. The presence of such defects as voids, inclusions, debonds, improper cure, and delaminations is almost common during the manufacture and installation of the composite systems. The general effects of potential defects on the rehabilitated structures are discussed by Kaiser and Karbhari (2003).

One of the main challenges of FRP repair/rehabilitation is to assure the structural integrity and long-term reliability of the retrofitted structures. It is recognized that defects in FRP composites need to be identified and evaluated, not just during and immediately after installation of the FRP, but also throughout the service life of the rehabilitated structure. The cause of the damage and its impact on the performance of the structure also need to be assessed. From this perspective, non-destructive testing (NDT) can provide the need by real-time characterization and assessment of the defects and the damages.

Various NDT methods have been proposed and developed for assessing the integrity or quantitatively measuring some characteristic of the materials, components or structures. All NDT methods are based on particular scientific principles, including, mechanical and optical techniques, penetrating radiation, electromagnetic methods, sonic and ultrasonics, and thermal and infrared methods, etc. The various NDT methods and techniques, due to their particular natures, work especially well with certain applications but could be of little or no value at all with other applications. Therefore, choosing the right method and technique is an important part of NDT practices.

Most NDT methods are considered to be localized, as they are used in individual sections of an element or a structure suspected or prone to damage. Moreover, non-destructive testing is often intended to give a snapshot of the damage, and hence results are not usually valid for a given time in the future. In contrast, structural health monitoring (SHM) is intended to detect and locate damage or degradation in structural components without a preconceived notion of the damage location, and to identify damage on a continuous or an intermittent basis. With SHM, it is possible to evaluate the performance, the durability and the service life of FRP-rehabilitated concrete structures.

This chapter explores the suitability of major NDT techniques and the advancement and development of new technologies such as fiber optic sensors for defect detection and performance assessment of FRP rehabilitation of concrete structures.

7.2 Visual inspection

Visual inspection does not usually require special equipment; it is easily implemented provided that the inspector is experienced; and it is inexpensive. Visual inspection is a convenient method to inspect the appearance of FRP-rehabilitated structures in terms of global and local soundness, and can be performed routinely as a means of quality control and damage assessment. Defects such as fabric damage and wrinkles, resin starvation/richness, ply bridging, discoloration due to overheating or lightning strike, and moisture accumulation, may be identified visually and do not require any sophisticated techniques. Once detected, the affected area becomes a candidate for closer inspection.

Visual inspection is often insufficient for locating internal flaws, such as delaminations, debonds, or matrix crazing, in the rehabilitated structures. In addition, tight surface cracks and edge delaminations may not be detected visually. Therefore, visual inspections are usually enhanced by various methods, ranging from low power magnifying glasses to borescopes, and are also supplemented by other NDT methods.

7.3 Acoustic emission (AE) testing

Acoustic emission (AE) is simply the stress waves, in the frequency range of ultrasound usually between 20 KHz and 1 Hz, generated in the materials due to deformation, crack initiation and growth, crack opening and closure, dislocation movement, twining and phase transformation, fiber breakage and delamination. The sources of AE are predominantly damage-related and AE monitoring leads to the prediction of material failure.

AE testing involves recording and evaluating AE signals to identify evolution of damage. This is usually accomplished by direct coupling of piezoelectric transducers on the surface of the structure under test and loading the structure. The output of the piezoelectric sensors (during stimulus) is amplified through a low noise preamplifier, filtered to remove any extraneous noise, and further processed by suitable electronics (see Fig. 7.1). AE tests can non-destructively predict early failure of structures. Further, a whole structure can be monitored from a few locations with additional sensors and while the structure is in operation.

For hybrid structures composed of two or more materials, signals can be initiated in either material or at the interfaces of the materials. In the case of FRP-rehabilitated concrete structures, the AE energy sources may include cracking, plastic deformation, friction of aggregate interlock, and mortar/aggregate debonding in concrete along with fiber/matrix debonding, matrix cracking, delamination and fiber breakage in composite laminates.

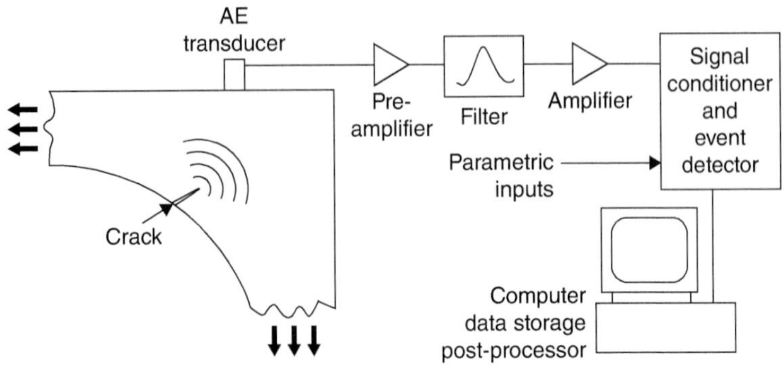

7.1 Principle of acoustic emission process (Huang *et al.* 1998).

Acoustic emission has found wide acceptance for industrial use, such as in proof testing of pressure vessels and tanks (Hellier, 2001). In defect detection of FRP composites, Zheng *et al.* (2000) utilized frequency domain techniques for identifying the damage modes in FRP composites. Their experiments showed that the amplitude range of the signal was not a reliable indication to the type of damage for specifically fabricated specimens because similar damage types may generate signals of different amplitude due to the distance between the damage and the sensor, the properties of the material, and the speed of the loading. In the work of Amoroso *et al.* (2003), AE analysis indicated similar damage effects to the measurement of mechanical tests of FRP composite laminates subjected to impacts.

AE technique was used by Mirmiran *et al.* (1999) to assess the structural integrity of carbon FRP–concrete hybrid columns. It was concluded that rate of change of cumulative AE counts with respect to the applied load correlates well with the degree of damage sustained by the concrete core. Carpinteri *et al.* (2007) used this technique to monitor the creep effects, micro-cracking of the original RC beam, the cracking progression and FRP debonding of FRP-retrofitted RC beams. An AE system was used to monitor the performance of a repaired and FRP-strengthened concrete box beam during load testing (FHWA, 2007). The beam was loaded to failure while the AE system was set up with eight sensors attached to the beam to monitor AE signals from concrete cracking, delaminations in concrete, delaminations in FRP, and/or adhesive bond failure at the interfaces.

It is possible to establish global monitoring of structures using an array of acoustic emission sensors. AE is fast, less labor intensive than competitive evaluation techniques and, in many cases, can be performed without a service shutdown of the structure. Although an AE test detects defects, it does not determine their type and size. AE tests have the capability to determine whether there is a structural problem, approximately where it is, and give a measure of its severity. A complementary inspection method such as ultrasonics may be needed to map out and size the flaws.

Acoustic emissions generated from damaged sections of the structure can be picked up by a wide range of sensor types, including accelerometers, piezoelectric transducers (PZT), and magnetostriction type transducers. However, AE techniques have been limited to localized detection of structural damage, since monitoring of the structure requires placement of multitudes of sensors. Recently, fiber optic AE sensors have been developed to record AE signals. Fiber optic sensors offer advantages over conventional AE transducers. Serial multiplexing of AE signals by a single optical fiber simplifies the logistics involved in dealing with multitudes of cables and leads from the conventional sensors routed from various locations in the structure. A number of novel fiber optic acoustic sensors have been developed (Liu *et al.*, 1990; Shih, 1998; Lim *et al.*, 1999; Chen and Ansari, 2000; Sun *et al.*, 2004). Fiber optic AE techniques have been employed in a number of projects, including detection of damage and micro cracking in concrete and FRP composites (Liang *et al.*, 2004).

7.4 Impact-echo (IE) method

Impact-echo (IE) is a non-destructive test method, specifically developed for concrete and masonry structures. The IE method is based on the use of impact-generated stress (sound) waves that propagate through the structure and are reflected by internal flaws and external boundaries of the structure. The reflected waves are recorded by a displacement transducer placed near the impact point. The record of displacement versus time is transformed into the frequency domain for ease of signal analysis. A study of the time–domain waveform and frequency spectrum makes it possible to locate and detect subsurface flaws (see Fig. 7.2).

The impact-echo method has been employed to detect flaws in concrete materials and members (Sansalone, 1997), and bridge decks (Tawhed and Gassman, 2002). In the FRP retrofit of a concrete bridge, an Olson Instruments impact echo tester was specially modified with an air coupled receiver, and frequency domain analysis was employed to uniquely identify delaminated areas (Maerz *et al.*, 2004). All the major artificially-created delaminations at non-critical locations were detected and some small unplanned delaminations were also identified by the impact-echo testing. Impact-echo techniques have been shown to be fairly effective for locating large voids or delaminations, particularly in plate-like structures (e.g. pavements or bridge decks) where the void tends to be parallel to the structure surface.

Impact-echo is often confused with ultrasonic methods, such as pulse-velocity and pulse-echo. Unlike ultrasonic methods, impact-echo utilizes

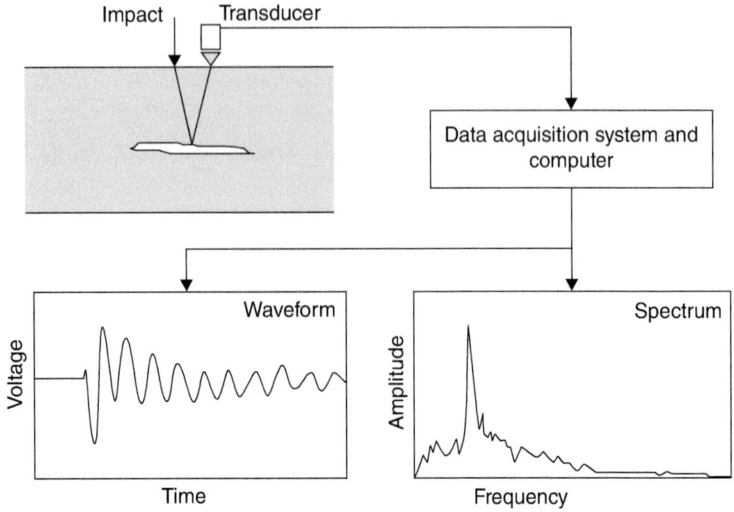

7.2 Schematic diagram of impact-echo method (Sansalone and Streett, 1998).

lower frequency signals (2 to 20 kHz, typically). The use of lower frequencies allows impact-echo to overcome the high signal attenuation and noise frequently encountered with ultrasonic methods. However, smaller cracks and discontinuities are more difficult to detect with this technique, due to the relatively low frequencies involved. Chen and Ansari (1999) developed a distributed fiber optic impact echo sensor in order to simplify the testing process. By distributed sensing, they eliminated the need for moving the sensing transducer from point-to-point.

7.5 Ultrasonics

In ultrasonic tests (UT), high frequency (100 kHz to 4 Hz) stress waves are transmitted into a material to detect defects or changes in material properties. UT actively probes the structure and differs from AE, which listens for emissions from active defects. The most commonly used ultrasonic testing technique is pulse echo, wherein the sound energy is introduced and propagates through the materials in the form of waves. When there is a discontinuity (such as a crack) in the wave path, part of the energy is reflected (echoed) back from the flaw surface. The reflected wave signal is transformed into an electrical signal by the transducer and is displayed on a screen. From the signal, information about the reflector location, size, orientation and other features can sometimes be gained. The measurements are relatively easy to perform with commercially available equipment.

Ultrasonic data can be collected and displayed in a number of different formats. The three most common formats are known as A, B, and C-scans. Each presentation mode (A, B, or C-scan) provides a different way of looking at and evaluating the region of material being inspected. Modern computerized ultrasonic scanning systems can display data in all three presentation forms simultaneously.

Successful applications of the pulse-echo technique have been reported in detecting defects and damage in FRP composites (Scarponi and Briotti, 2000; Doyum and Durer, 2002; Hillger *et al.*, 2004; Hosur *et al.*, 2004; Hsu *et al.*, 2002; Imielinska *et al.*, 2004; Lestari and Qiao, 2005; Ray *et al.*, 2007).

Kundu *et al.* (1999) reported the detection of delaminations between concrete and glass FRP plates by the ultrasonic method. Bastianini *et al.* (2001) used the ultrasonic technique: however, they utilized the amplitude of the reflection signal rather than the time delay for detecting the bond defects. The method was successful in assessment of the bond in FRP strengthened concrete, and masonry. Luprano *et al.* (2006) employed the ultrasonic method for the characterization of the interface between FRP composites and concrete. Their work involved evaluation of the influence of carbon FRP materials and the thickness of the reinforcement on the defects. Ultrasonic test results were compared with the results obtained by infrared thermography.

In the rehabilitation of a concrete bridge with carbon FRP sheets, the acousto-ultrasonic NDT technology has shown the ability to detect and image the delaminations between FRP sheet and concrete substrate (Ekencl *et al.* 2005). The defects, in the form of delaminations, were intentionally formed at the FRP–concrete interface during installation of the FRP sheet to investigate the ability of this technique.

Ultrasonic testing can reveal subsurface flaws in composite materials or at the FRP–concrete interface due to the superior penetrating power of ultrasound. However, these tests require a highly experienced technician or operator to properly acquire and interpret the data. Also, ultrasonic methods, especially those utilizing a scanning bed to obtain C-scans, can be quite expensive.

7.6 Radiography

Radiography involves the use of penetrating gamma or X-radiation and radiographic film to capture images of defects. Radiography requires access to both sides of the structure, with the radiation source placed on one side and the film placed on the other side. Differential absorption of the penetrating radiation by the specimen will produce clearly discernible differences when recorded on radiographic film.

Unlike ultrasonics, radiography cannot capture the volumetric characteristics of defects; however, it does provide a higher resolution of the planar aspects of defects. Defect resolution depends somewhat on the orientation of the defect with respect to the source and the film. Typical discontinuities that are detectable include some delaminations and debondings, depending on the orientation, voids, resin variations, broken fibers, impact damage, and cracks.

Radiography has been used for the characterization and classification of defects in laminated composites (Kim and Achenbach, 1998) and honeycomb structures (Doyum and Durer, 2002). The main disadvantage of this technique is health risk and high equipment cost. The test equipment needs modifications before it can be used to evaluate civil infrastructure in the field.

7.7 Ground penetrating radar (GPR)

Ground penetrating radar (GPR) operates by transmitting electromagnetic waves (in the range of 10~1000 Hz) into the probed material and receiving the reflected pulses as they encounter discontinuities. The discontinuity could be a boundary or interface between materials with different dielectrics or it could be a subsurface object such as a debond or delamination (see Fig. 7.3). The amplitudes of the received echoes and the corresponding arrival times can then be used to determine the nature and location of the discontinuity.

With the advancement in GPR technology, especially the increase in frequency of commercially available GPR antennae and better data processing software, GPR can now be used for subsurface condition assessment in materials consisting of thin layers, such as FRP composites. Careful analysis of GPR waveforms can potentially help detect subsurface debonds

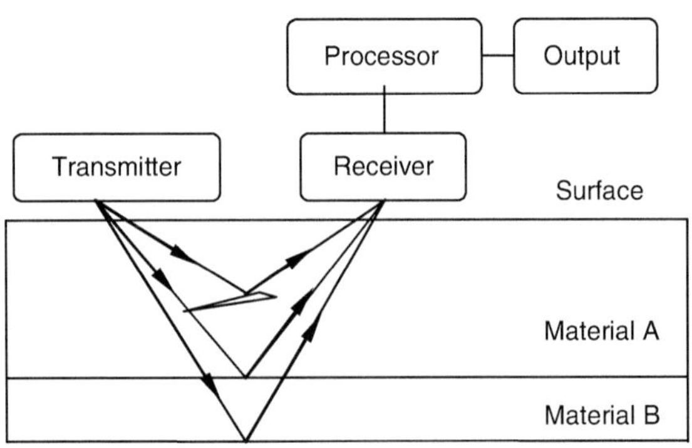

7.3 Principle of testing with the GPR system.

between the wearing surface and the underlying FRP bridge deck, and delaminations within the flanges of the FRP deck.

Compared to other non-destructive techniques such as infrared thermography, ultrasonic or microwave, GPR offers more penetrating power and so can detect concrete defects or deteriorations at greater depths. Results from the literature review show that the lower frequency GPR antenna (1 GHz) cannot detect shallow defects such as debonding in FRP wrapped members, but a higher frequency antenna (2 GHz) can detect those defects (Jackson *et al.*, 2000). On the other hand, a ground coupled 1.5 GHz antenna was found to offer higher penetration capability, which is crucial for testing FRP bridge decks (Halabe *et al.*, 2006). GPR is an excellent tool for detecting water-filled debondings but its performance is not satisfactory for detecting air-filled debondings in FRP wrapped concrete cylinders (Dutta, 2006) and FRP bridge decks (Hing, 2006).

7.8 Microwave-based techniques

Microwave-based techniques use high frequency electromagnetic energy in the range of a few hundred MHz to a few hundred GHz. In this technique, the material or structure is tested by measuring the various properties of the electromagnetic waves scattered by or transmitted through the test article. The ability of microwaves to penetrate inside dielectric materials makes the techniques very suitable for even thick composites.

Microwave techniques have been used to detect debondings, delaminations and voids in stratified media and for determination of the depth of these flaws in FRP-reinforced concrete structures. Important dimensional information, such as spatial resolution, location, and the geometry of a debonded region, can be identified using this method. Defects in the form of air voids between FRP and concrete have been detected by microwaves in the work of Li and Liu (2001), Buyukozturk *et al.* (2003) and Akuthota *et al.* (2004). In AbouKhousa and Qaddoumi's (2004) tests, composite structures were irradiated with electromagnetic waves and the reflected waves were able to capture images of subsurface inclusions in a five layer laminated composite structure of 45.6 mm thickness. Stephen *et al.* (2004) utilized a measurement system consisting of a microwave reflectometer and an automated scanner to detect relatively small debonded regions (smaller than 20 mm × 20 mm) in CFRP-bonded concrete members in a bridge. Ekenel *et al.* (2004) used the same technique as that of Stephen *et al.* (2004) to study the behavior of delaminations under cyclic loading.

Large areas can be scanned by using arrays of sensors. The microwave technique appears to be the best wave propagation method that can potentially overcome some of the noise problems encountered during NDE of composite materials.

7.9 Thermography

The basic principle of thermal inspection consists of measuring or mapping of surface temperatures when heat flows from, to, or through a test object. The most widely used thermographic technique uses an infrared camera to measure radiated electromagnetic energy in the infrared zone, and this is known as infrared thermography (IRT).

Two schemes, passive thermography and active thermography, have been used with IRT. Passive thermographic technique relies on natural heat distributions over the surface of a structure. Passive thermography in most cases is qualitative and is used to detect only anomalies. Abnormal temperature profiles indicate potential problems to be taken care of.

Active infrared thermography drives heat from an external uniform heat source through the test object. Defect-free areas conduct heat more efficiently than areas with defects. Thermal gradients due to the amount of absorbed or reflected heat can be observed by an infrared imaging device. Defects, such as debonding, cracks, impact damage, panel thinning, and water ingress, all could affect the thermal properties. Delamination at the FRP–concrete interface acts as an insulator, causing a higher surface temperature than the areas that are securely bonded. The heat source for active thermography needs to distribute heat uniformly since thermal gradients induced by non-uniform heating can be mistaken for defects. Depending on the external stimulus, different approaches of active thermography have been developed, such as pulse thermography (PT), step heating (SH), lock-in thermography (LT), and vibrothermography (VT) (Maldague, 2000).

Halabe *et al.* (2002) used thermography to inspect glass FRP composites. Delaminations of different sizes were detected in box-section FRP specimens, and between an FRP bridge deck and a wearing surface of 3/8 in. (9.5 mm) thickness. With a more advanced infrared camera, delaminations inserted within the joint between FRP bridge deck modules were able to be detected. Test results showed that infrared thermography was able to detect water-filled delaminations in the top flange of GFRP bridge decks without a wearing surface but were not able to detect the water-filled defects in the top flange after a wearing surface was placed over it (Halabe *et al.*, 2005). On the other hand, air-filled and water-filled debonds between wearing surface and the underlying FRP bridge deck could be readily detected (Halabe *et al.*, 2004). Carbon fibers, due to their higher thermal properties, permit thermographic techniques for inspection of deeper material layers and defects have been identified clearly using phase images (Meola and Carlomagno, 2004).

Thermographic methods have been used to detect debonding between FRP composites and concrete (Hawkins *et al.*, 1998; Jackson *et al.*, 2000;

Mtenga *et al.*, 2001; Nokes and Hawkins, 2001). Besides debonding, defects such as air voids, moisture zones and other types of anomalies are also detected with thermography (Jackson *et al.*, 2000; Hu *et al.*, 2002; Shih *et al.*, 2003). Air voids simulating defects between FRP layers and concrete were even quantitatively sized by infrared thermography (Starnes *et al.*, 2002). When used to characterize the progression of damage in composite strips and at the composite–concrete interface at various stages of loading of FRP rehabilitated bridge decks, the thermographic results supported the experimental observation of the progress of the debond areas along the length of the composite strips with increases in load (Ghosh and Karbhari, 2007). IRT inspections performed on an in-service FRP repaired bridge were able to identify several installation defects and damage to the FRP system in the vicinity of the impact by an over height vehicle (Brown and Hamilton, 2005).

Thus far, the application of infrared thermography for the evaluation of FRP retrofitting has been limited mainly to qualitative efforts. Ghosh and Karbhari (2006) gave a critical review of infrared thermography as a method for non-destructive evaluation of FRP rehabilitated structures.

Laboratory and field-testing results show that infrared thermography is a potentially useful tool for inspection of concrete structures rehabilitated with composite wraps or laminates, and for non-destructive inspection of FRP bridge decks. The technique could be possibly used for several applications such as quality control during pultrusion of new FRP components (in factories), during field construction, and for field inspection of in-service structures. However, as the thickness of the FRP system increases, detection of debonded areas at the FRP/concrete interface becomes increasingly difficult (Brown and Hamilton, 2005).

7.10 Optical and laser techniques

Optical non-destructive testing uses light (often laser light) to detect infinitesimal surface deflections through formation of fringe patterns on the surface of an object. Some of the most applied techniques include optical holography, electronic speckle pattern interferometry (ESPI) and laser shearography. Optical NDT can produce qualitative (defect detection) and quantitative (surface displacement maps or stress/strain analysis) results.

7.10.1 Optical holography

Optical holography is an imaging method that records the amplitude and phase of light reflected from an object as an interferometric pattern on film. A laser beam illuminates the test object and its reflection of the scattered light waves towards the recording device interferes with a reference beam aimed directly at the recording media. The test sample is interferometrically compared in two different stressed states. Stressing can be mechanical, thermal, vibration-induced, etc. The resulting interference pattern contours the deformation undergone by the specimen in between the two recordings. Surface, as well as sub-surface, defects show distortions in the otherwise uniform pattern. In addition, the characteristics of the component, such as vibration modes, mechanical properties, and residual stresses, can be identified through holographic inspection.

7.10.2 Electronic speckle pattern interferometry (ESPI)

Illumination and recording of the speckled reflections from surfaces in a before-and-after type of investigation provide a tool for quantitative analysis of materials. Ansari (1987, 1989) developed a method for detection of microcracks in concrete and for quantitative investigation of the mechanical properties of concrete under uniaxial tension. Although a powerful technique, manual determination of deformations and determination of microcracks required laborious interpretation of speckle generated fringe patterns. For this reason, Ansari and Ciurpita (1987) developed a computer-automated technique for analysis of the fringe patterns. These developments, as well as refinement of computer-based methods, led to the establishment of electronic speckle interferometric techniques (ESPI), also referred to as electro-optic holography or TV holography, in the literature.

ESPI has been used to detect impact damage of thin FRP laminates (Ambu *et al.*, 2006), impact damage in honeycomb composite plates (Kang *et al.*, 2006), and splitting, micro-cracks, unstable zigzag crack extension and also partial delamination inside carbon–carbon composites (Hatta *et al.*, 2005).

7.10.3 Laser shearography

Shearography is an interferometric test method through which damage or defects can be identified on components under load. The test object is illuminated with laser light and observed through a CCD camera equipped with the so-called shearing device (Hung, 2004). The shearing device projects the object image onto the camera chip. The original image of the illuminated surface is recorded via a video image. A second video image is made when the test object is stressed by heating, mechanical loading, changes in pressure, or acoustic vibrations. Changes in the surface contour caused by debonding or delaminations become visible on the video display.

Shearography has been used in production environments for rapid inspection of composite structures. Stressing the structure leads to slight surface deformations that are detected before and during stressing. Display of the computer-processed video image reveals defects as bright and dark concentric circles of constructive and destructive reflected light wave interference. Yang (2006) utilized digital shearography for evaluating defects and delaminations of a CFRP specimen under tensile stressing, a GFRP plate under thermal stressing and a GFRP honeycomb panel under partial vacuum stressing.

7.11 Superconducting quantum interference device (SQUID)

SQUID is a mechanism for measuring extremely weak signals, such as subtle changes of electromagnetic energy field. Due to its inherent high sensitivity and high dynamic range, the SQUID technique is able to detect very small disturbances in a magnetic field produced by structural anomalies located at the surface or inside the material volume, even in the presence of large background magnetic noise.

Ruosi *et al.* (2002) used a highly sensitive high-temperature superconductive superconducting interference device (HTS–SQUID) for testing CFRP panels (3.8 mm thickness) damaged by a concentrated transverse load, by mapping the modulus and phase of the magnetic field. Hatsukade *et al.* (2002) used SQUID for detecting deep lying cracks in the form of hidden slots at various depths within CFRP plates of 20 mm thickness. Their later work showed that SQUID-NDE was also able to detect the hidden slots within multilayer CFRP plates (Hatsukade *et al.*, 2004).

Carr *et al.* (2003) utilized SQUIDs for detecting the regions of heat and impact damage in CFRP plates of 3 mm thickness by inducing an eddy current into the sample by means of wire-wound double-D coils. Hatta *et al.* (2005) utilized a SQUID current-mapping technique to observe the fracture processes of two-dimensionally laminated and three-dimensionally reinforced carbon/carbon composites with pre-cracks and induced-damage near the notches.

The work reported to date indicates that SQUID is a potentially useful NDE technique for testing composite components. However, the equipment that is currently available lacks portability and is not suitable for field use in civil infrastructure systems.

7.12 Fiber optic sensing (FOS) technology

Over the last decade, fiber optic sensor (FOS) technology and instrumentation have become well understood, developed, and accepted widely for composite and smart structure applications in civil engineering. Given their small size, high sensitivity and dielectric properties, fiber optic sensors are attractive sensing devices for non-destructive testing applications. A comprehensive review of the fiber optic sensor applications in civil structures is given elsewhere (Ansari, 2007). A brief review concerning the advancement and development of fiber optic sensing technology for health monitoring of FRP rehabilitated concrete structures is given in this section.

7.12.1 Basic principles

Typical optical fibers are made of silica glass with a core, a cladding to guide the lightwave, and a plastic coating to prevent fracturing of the silica glass due to its high modulus, and allow flexibility and bending of the fiber. Transmission of light through optical fibers can be explained by Snell's law and the concept of total internal reflection. According to Fig. 7.4, as indicated by the refractive index, n, when light travels from the fiber core that has a high refractive index into the cladding with a lower index, the light-wave totally reflects back to the core.

Based on the different sensing principles, various types of fiber optic sensors have been developed. Intensity, spectrometric and interferometric sensors are categorized by their transduction mechanisms, while localized, multiplexed and distributed sensors are categorized by their applications.

Sensors based on intensity modulation pertain to light intensity losses that are associated with straining of optical fibers along any portion of their length (see Fig. 7.5). The advantages of intensity- (or amplitude-) type sensors are the simplicity of construction, and compatibility with multimode fiber technology.

Spectrometric sensors are based on relating the changes in the wavelength of light to the measurand of interest, i.e. strain. An example of such sensors for measuring strains is the Bragg grating sensor (Morey *et al.* 1989) (see Fig. 7.6). These sensors are intended for use as a localized fiber optic sensor. Multiplexed sensors are usually constructed by combining a number of individual sensors for measurement of perturbations over a large structure. The technique of wavelength division multiplexing by using Bragg gratings makes this possible (Kersey and Morey, 1993). In this technique, a broad-band light containing a number of wavelengths within a certain region of the spectrum is employed for scanning a series of Bragg grating

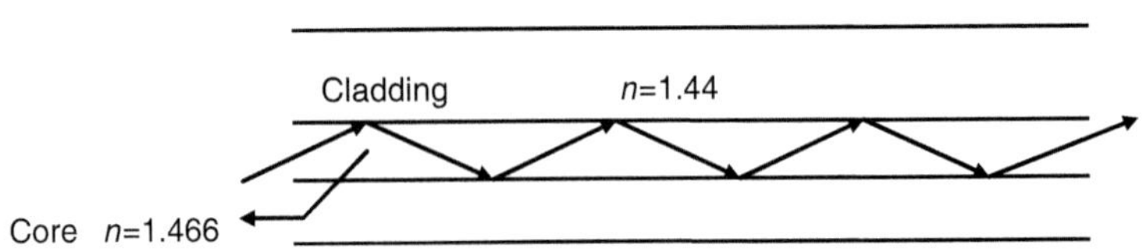

7.4 A typical single-mode optical fiber (Ansari, 1997b).

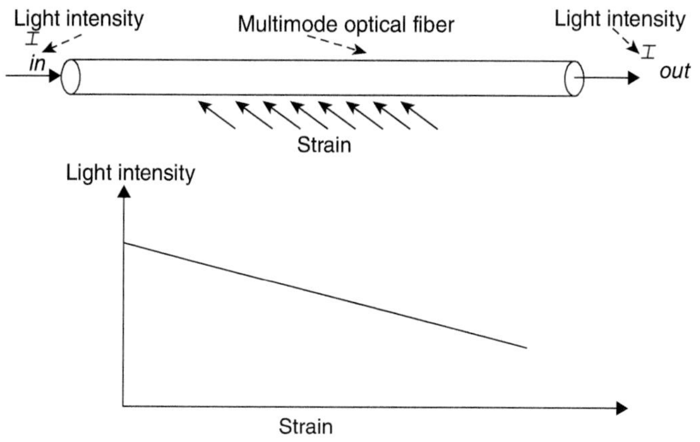

7.5 Optical fiber intensity sensor (Ansari, 1997b).

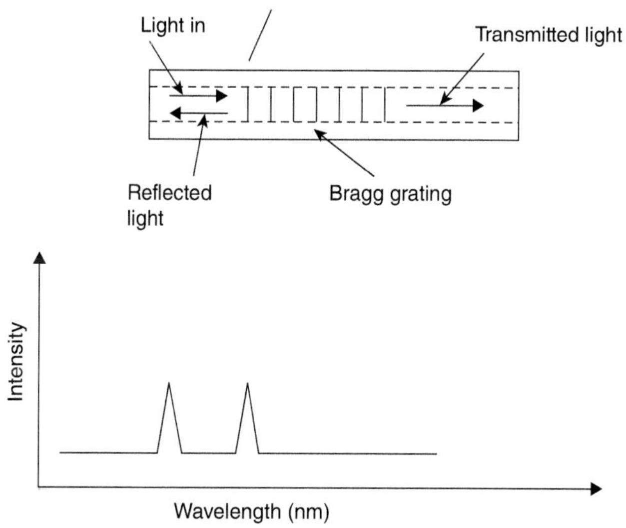

7.6 Strain-induced shift in wavelength for a fiber Bragg grating (Ansari, 1997b).

type sensors. The reflectance wavelength of each Bragg grating is slightly different from the others. The wavelength-shifts of individual sensors are recognized, detected, and then related to the magnitude of strain at specific sensor locations (see Fig. 7.7).

Interferometric sensors can be configured in a number of different ways for sensing purposes (see Fig. 7.8). Interferometric sensors require the interference of light from two identical single-mode fibers, one of which is used as a reference arm and the other is the actual sensor. An exception to

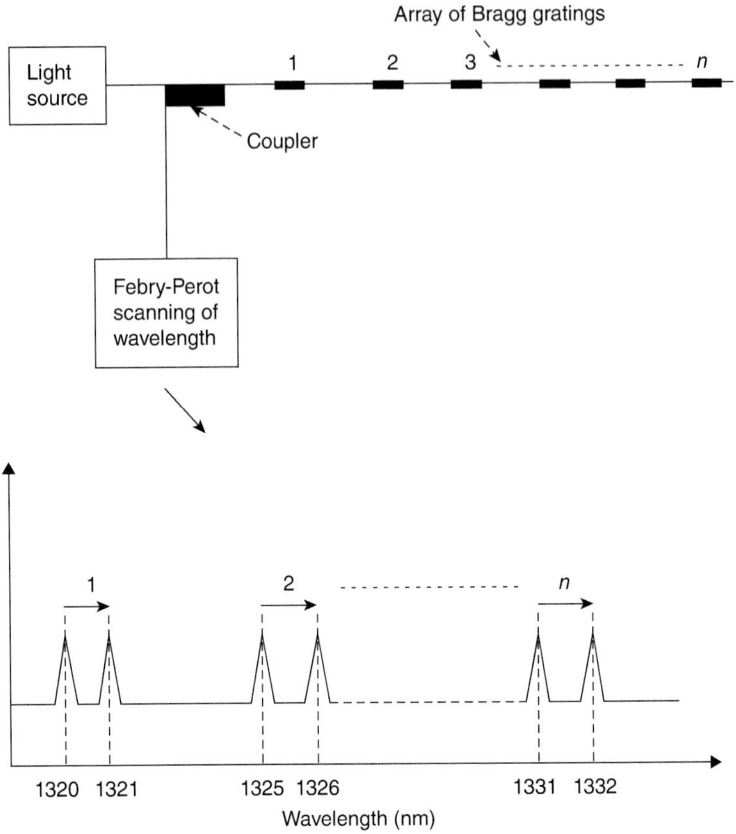

7.7 Multiplexed Bragg grating sensor array (Ansari, 1997b).

a two-arm interferometric sensor is the single-fiber Fabry–Perot type sensor (Claus *et al.*, 1993). In a Fabry–Perot type sensor, the fiber is manipulated with two parallel reflectors (mirrors) perpendicular to the axis of the fiber. The interference of the reflected signals between the two mirrors creates the interference pattern. A Fabry–Perot sensor is only capable of providing localized measurements at the cavity formed by the two mirrors. The interference pattern generated at the output end of the phase sensors is directly related to the intensity of the applied strain field.

Polarimetric sensors take advantage of the polarization characteristic of light for transduction (Rashleigh and Ulrich, 1980; Katsuyami *et al.*, 1981; Ansari and Wang, 1995). Fringe shifts due to external perturbations in polarization-maintaining single-mode fibers are caused by the interference of two mutually perpendicular polarized waves. The advantage of polarimetric sensors is that, unlike their interferometric counterparts, only one fiber is needed for sensing the measurand.

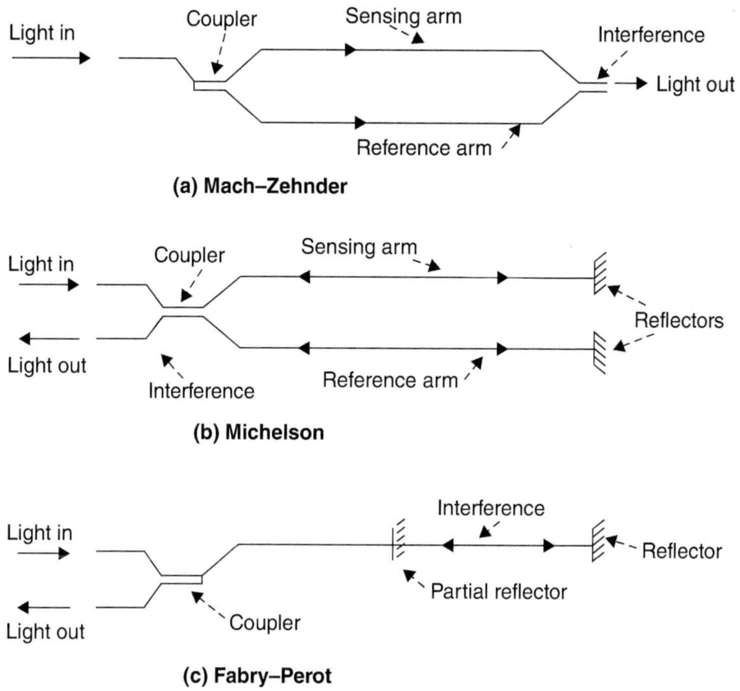

7.8 Fiber optic interferometric sensors (Ansari, 1997b).

Of the sensor types discussed above, intensity type sensors are simple to construct but their sensitivity is rather low. Interferometric sensors offer highest sensitivity but the required components are quite complicated. Polarimetric sensors are based on changes in the propagation constants of the two linearly polarized eigenmodes of a birefringent single mode fiber. For maximum sensitivity in standard measurement setups, both eigenmodes of the sensing fiber have to be excited equally. This requires a tedious tuning procedure and can lead to an unknown sensitivity if the leading fiber experiences external stress due to bending.

Distributed sensors make full use of the optical fibers, in that each element of the optical fiber is used for both measurement and data transmission purposes. The purpose of making measurands by distributed or multiplexed optical fibers is to determine locations and values of measurands along the entire length of the fiber. These sensors are most appropriate for application to large structures owing to their multi-point measurement capabilities. The most widely used distributed sensing technique is based on measurement of propagation time-delays of light traveling in the fiber based on the measurand-induced change in the transmission light. An optical time domain reflectometer (OTDR) is used for this purpose (Tateda and Horiguchi, 1989; Dakin, 1990). A pulsed light signal is transmitted into one end of the fiber, and light signals reflected from a number of partial reflectors along the fiber length are recovered from the same fiber end. By using this concept, it is possible to determine the location of the strain fields by the two-way propagation time-delay and measurement of the reflected time signals (see Fig. 7.9).

More recent developments in distributed sensing involve an interferometric system (Zhao and Ansari, 2001) developed for specific applications in civil structural monitoring. A typical schematic diagram of a white-light distributed sensing system is shown in Fig. 7.10. The system consists of two

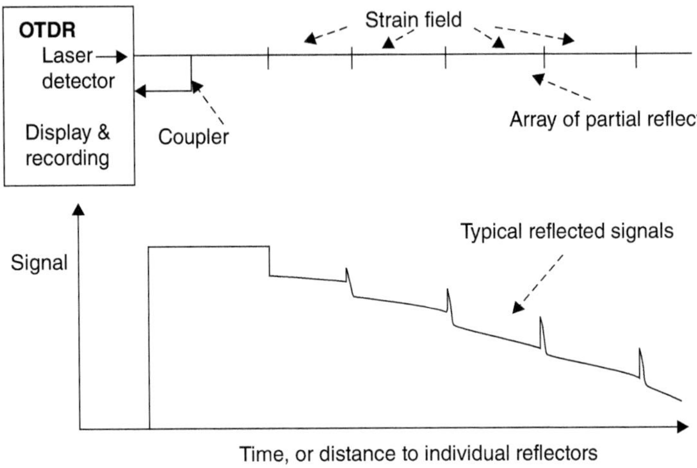

7.9 Distributed sensing through an optical-fiber time-domain reflectometer (Ansari, 1997b).

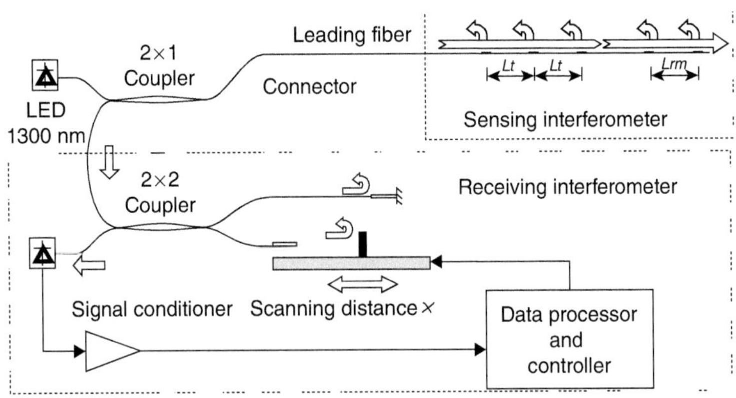

7.10 Schematics of the interferometric distributed sensor (Zhao and Ansari, 2001).

parts: the sensing interferometer module, and the receiving unit. The sensing module comprise a number of individual single-mode fibers of desired gauge lengths. The individual fibers are mechanically connected through ferrules and a portion of beam is reflected when the light wave passes through them. The receiving unit consists of a Michaelson white-light interferometer with a scanning translation stage, signal processing and the system control unit.

Long-gauge structural sensors can provide integrated strain measurements that are not susceptible to the high local strains associated with crack-forming in concrete. These sensors are also well suited for incorporation within rehabilitation and strengthening wraps.

7.12.2 Fiber optic sensors integrated with FRP reinforcements

The state-of-the-art as it pertains to the earlier applications of fiber optic sensors in civil structures is given elsewhere (Ansari, 1997a, b). Only the most recent applications of fiber optic sensors to FRP composites and FRP rehabilitated concrete structures are reviewed here. In some applications, a specially coated heat-resistant optical fiber is embedded within FRP rods to create smart FRPs for self monitoring, or for use as sensors, e.g. for applications in cable stays. A number of investigators have developed innovative methods for embedment of short gauge Bragg and Fabry–Perot sensors in FRP rods (Kalamkarov *et al.*, 2000). On the other hand, by embedment of long gauge interferometric sensors within CFRP and AFRP rods during the pultrusion process, the entire length of the rod becomes a sensor. With this arrangement, it is possible to configure smart tendons for cable stays with arbitrary lengths up to few hundred meters.

Zhou *et al.* (2003) studied the microstructure, mechanical, strain sensing and temperature sensing properties of the FRP-OFBG (FRP integrated with optical fiber Bragg grating) bars used to reinforce concrete beams. Sim *et al.* (2005) reported the tensile strength and pull-out bonding characteristics of hybrid GFRP rods and FBG fiber-optic sensors. The strain acting on the hybrid FRP was measured by the embedded FBG sensor. From the results, it was concluded that hybrid FRP rod can be successfully used for the purpose of reinforcement and smart monitoring in concrete structures. Depending on the measurement requirements, the optical fibers could be terminated prior to or run through the anchorage system (Fig. 7.11). In post-tensioning applications, tendon strains, anchorage slippage, or prestressing losses are able to be monitored by the embedded fiber in the FRP tendons (Ansari, 2005).

Kalamkarov *et al.* (2000) have shown that it is possible to incorporate fiber optic sensors into FRP reinforcements during the pultrusion process,

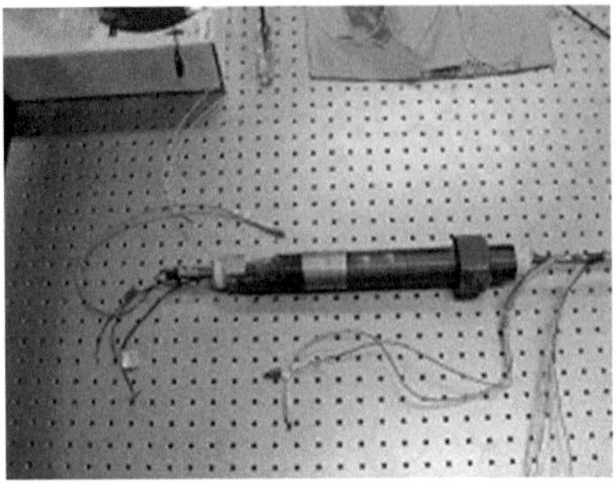

7.11 Instrumented FRP anchorage system with optical fibers (Ansari, 2005).

and the fiber optic sensors can be used for the purpose of monitoring the structural performance. Product specification and manufacturing guidelines for the smart fiber reinforced polymer (FRP) reinforcements incorporating the fiber optic sensors need to be developed before smart FRP reinforcements are widely applied for concrete structures.

7.12.3 Fiber optic sensors monitoring FRP–concrete interfacial bond behavior

In applications that involve monitoring of debonding or strains at interfaces, bare optical fibers can be directly adhered or embedded at the interface. One such application is in monitoring of debonding or peeling of the FRP fabrics from concrete elements, e.g. beams. In such applications, long-gauge fiber optic sensors are adhered to the CFRP fabric during the application of the epoxy resin. This arrangement allows for embedment of the optical fiber within the laminate and the epoxy serves as a protective coating for the sensor. Figure 7.12 shows a four-gauge multiplexed sensor after destructive testing of the retrofitted beam (Zhao *et al.*, 2007). This particular system consisted of four 15 cm long sensors connected in series along the length of the CFRP fabric. All the sensors were still operational following the completion of the destructive test. These sensors were adhered to the fabric from the termination point towards the mid span of the concrete beam in order to detect peeling or debonding due to end-zone stress concentrations. Besides monitoring of fabric deformations, unloading of any of the sensors would be indicative of interface debonding or peeling of concrete associated with a release in strain. Figure 7.13 demonstrates the peeling phenomenon as indicated by the reversal of the strain in sensor segment within 15 cm of the termination point of the FRP fabric.

7.12 Fiber optic sensor embedded within the FRP laminate (Zhao *et al.*, 2007).

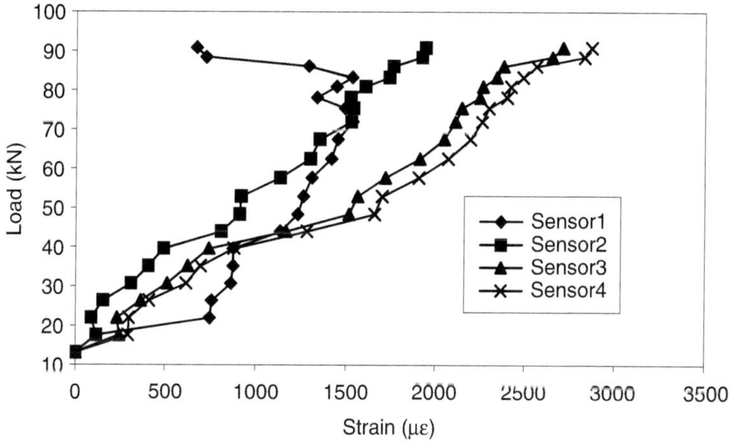

7.13 Typical strains in CFRP as measured by the 15 cm gauge optical fiber sensors (Zhao *et al.*, 2007).

Chan *et al.* (2000) reported the use of multiplexed fiber Bragg grating sensors for monitoring of strain development along the FRP-concrete interface for notched glass FRP-strengthened beams in three-point bending. Results showed that strain measured at the interface deviated noticeably from that at the surface. Embedded optical fibres allowed detection of interfacial debonding between the concrete and composite at load levels that were about 20% less than those of visible separation of the two components.

Fiber optic sensors have also been used at the concrete–FRP interface to measure the concrete strain in order to understand the strain state at a debonded region (Bonfiglioli and Pascale, 2003). Zhu *et al.* (2003) used Mach–Zehnder interferometer sensors at the interface of FRP tubes and inner concrete to measure the internal strain of the structure and the crack-expanding of inner concrete.

7.12.4 Field applications of fiber optic sensors

The combination of advanced composites and fiber optic strain sensors is easy, and offers important advantages over the traditional sensors. Fiber optic sensors can be embedded in the composite material or at the interface between materials in the manufacturing or retrofitting processes.

Fiber optic Bragg grating strain sensors were selected to evaluate the performance and long-term durability of FRP-strengthened concrete beams on the Horsetail Falls Bridge in the state of Oregon, USA. The sensors were installed in pairs – in grooves cut in the concrete and on the cured FRP composites over the concrete (Laylor and Kachlakev, 2000).

Watkins *et al.* (2007) reported using a fiber optic network, consisting of extrinsic Fabry–Perot interferometric (EFPI) sensors, to monitor both static and dynamic load-induced strain of an aging reinforced concrete bridge that was upgraded with FRP wrap and rebar in the state of Missouri, USA. Good agreement with design expectations, finite-element computations, and co-located sensors was demonstrated. A health coefficient parameter was proposed as a single measure of load performance, but would require further development related to loading conditions and aging.

Fiber optic sensors have been extensively employed by ISIS Canada for monitoring of bridges constructed or rehabilitated with FRP composites (Tennyson *et al.*, 2001). Mufti (2002) provides an extensive review of these applications, especially as related to steel-free bridge decks (Mufti, 2005).

7.12.5 Future trends

Fiber optic sensors provide a multitude of capabilities for localized, multiplexed and distributed sensing. Depending on the size of the structure, many of these self-monitoring resident systems could be integrated within the structural system to provide smart capabilities and help monitor aging. Furthermore, fiber optic sensing technologies are especially compatible with concrete and FRP materials. FRP-rehabilitated concrete bridges will greatly benefit from this technology if optical fibres are embedded into the concrete–composite interface to monitor strain distribution and indicate eventual long-term degradation. Embedded sensing allows for determination of the parameters in the future without re-installation and invasive cutting procedures that penetrate the concrete to expose the rebars for installation.

Fiber optic sensing is applicable for remote sensing where a segment of the fiber is used as a sensor gauge and a long length of the same or another fiber conveys the sensed information to a remote station. On-site inspections will be necessary only when this remote sensing system indicates that potential problems have developed, such as excessive loads, structural failure and environmental degradation.

Fiber optic sensing systems must be investigated in the field environment to develop practical protocols and to establish confidence in long-term performance. A current area of research is the field validation of sensing techniques using in-service structures.

7.13 Conclusions

As strengthening and retrofit of concrete structures by FRP composites in the form of externally bonded reinforcement continue to gain acceptance, the need for monitoring the performance of the rehabilitated structures has increased. In addition to the initial condition of the existing structure and the concrete substrate, the overall response of the rehabilitated structure, as well as the condition of the FRP used in the rehabilitation and the integrity of the bond between FRP and concrete substrate, all need to be evaluated, not only during the installation but also throughout the service life of the rehabilitated structures.

A number of NDT techniques have been used in the laboratory for inspection of FRP composites and FRP-retrofitted concrete members. However, most of these methods are not yet fully developed for field application. Currently, there is no single method capable of comprehensively assessing response of FRP-rehabilitated concrete structures both at the global as well as local levels. Methods such as visual inspection are convenient and effective for external defects, but the results are qualitative and subjective. Methods involving acoustic emissions make it possible to monitor entire structures but the test objects need to be loaded and it is often difficult to identify the type and location of a defect. Acoustic impact methods are usually used for local large voids and delaminations within thin structures, whereas ultrasonic techniques provide a viable methodology capable of assessing both surface and deep subsurface discontinuities, but can be constrained by thickness, material interfaces, and the need for contact or use of a coupling agent (often difficult to use under field conditions). Radiographic techniques are of high resolution and sensitivity; they can be used for point detection of defects. At present, their cost and possible health hazards render their applications minimal. Although microwave technique can conduct non-contact, one-sided inspection, the penetration depth is very limited. Ground penetrating radar offers good penetration but it is less sensitive to ambient conditions. Recent advances in thermography provide a tool for rapid inspection, but thermography is typically used for periodic monitoring and not for real-time monitoring. Optical and laser methods are effective for NDT of thin laminate composite materials, but are limited to inspection of defects of sufficient size or shallowness that induce stress concentrations into the surface. SQUID technique can be applied to the detection of deep defects in thick and/or multi-layer FRP composites, though the estimation of the degree of damage requires more quantitative analysis. Currently, nearly all the work on SQUID NDE is financed by government or industrial research grants and promising results are produced. More research needs to be performed to make SQUID applicable for field testing of civil engineering structures.

The recent development of fiber optic sensing systems has made possible the health monitoring of FRP-rehabilitated concrete structures. These systems can provide high-resolution and measurement capabilities that are not feasible with conventional technologies. Fiber optic sensors can be manufactured at a low cost and they offer a number of key advantages, including the ability to multiplex an appreciable number of sensors along a single fiber and interrogate such systems over large distances, strong immunity to electromagnetic interference, and environmental resistance. Integrated into FRP composites or embedded at the FRP–concrete interface, fiber optic sensors provide the ability to measure strain remotely and precisely without imposing any strength degradation on the structure. Thereby, the use of fiber optic sensors in FRP-rehabilitated concrete structures for real-time structural performance monitoring can greatly improve durability and substantially improve the safety of the structure.

In general, there is no single method that provides all the inspection requirements. Combinations of different NDT methods may be necessary. The results of NDT/NDE, supplemented by analytical simulation of structural performance with proper damage models, will truly provide a comprehensive evaluation tool for assessment of the structural condition and service life.

7.14 References

ABOUKHOUSA M and QADDOUMI N (2004), 'Nearfield microwave imaging of subsurface inclusions in laminated composite structures', *16th World Conference on Nondestructive Testing*, Aug. 30–Sept. 03, Montreal, Canada.

AKUTHOTA B, HUGHES D, ZOUGHI R and MYERS J (2004), 'Near-Field microwave detection of disbond in fiber reinforced polymer composites used for strengthening concrete structures and disbond repair verification', *ASCE Journal of Materials in Civil Engineering*, 16(6), 540–546.

AMBU R, AYMERICH F, GINESU F and PRIOLO P (2006), 'Assessment of NDT interferometric techniques for impact damage detection in composite laminates', *Composite Science and Technology*, 66(2), 199–205.

AMOROSO M P, CANEVA C, NANNI F and VALENTE M (2003), 'Acoustic emission performance for damage monitoring of impacted FRP composite laminates', *Proceedings of the Twenty-ninth Annual Review of Progress in Quantitative Nondestructive Evaluation*, Vol. 22 (American Institute of Physics – Vol. CP657), Bellingham, Washington, July 14–19, 1447–1454.

ANSARI F (1987), 'Stress–Strain Response of Microcracked Concrete in Direct Tension', *ACI Materials Journal*, 84(6), 481–490.

ANSARI F (1989), 'Mechanism of Micro Crack Formation in Concrete', *ACI Materials Journal*, 86(5), 459–464.

ANSARI F (1997a), 'Theory and Applicationsof Integrated Fiber Optic Sensors in Structures', *ASCE-EMD, SP, Intelligent Civil Engineering Materials & Structures*, 2–28.

ANSARI F (1997b), 'State-of-the-art in the Applications of Fiber Optic Sensors to Cementitious Composites', *Cement & Concrete Composites*, 19(1), 3–19.

ANSARI F (2005), 'Fiber optic health monitoring of civil structures using long gage and acoustic sensors', *Smart Materials & Structures*, 14(3), 111–117.

ANSARI F (2007), 'Practical Implementation of Optical Fiber Sensors in Civil Structural Health Monitoring', *Journal of Intelligent Material Systems and Structures*, 18(8), 879–889.

ANSARI F and CIURPITA, G (1987), 'Automated Fringe Measurements In Speckle Photography', *Applied Optics*, 26(9), 1688–1692.

ANSARI F and WANG K (1995), 'Rate sensitivity of high birefringent fiber optic sensors under large dynamic loads', *J. Lightwave Engineering*, 13 (1995), 1992–1997.

BASTIANINI F, TOMMASO A D and PASCALE G (2001), 'Ultrasonic non-destructive assessment of bonding defects in composite structural strengthenings', *Composite Structures*, 53(4), 463–467.

BONFIGLIOLI B and PASCALE G (2003), 'Internal Strain Measurements in Concrete Elements by Fiber Optic Sensors', *ASCE Journal of Materials in Civil Engineering*, 15(2), 125–133.

BROWN J R and HAMILTON H R III (2005), *CFRP Repair of Impact-damaged Bridge Girders, Volume II - Inspection of FRP Composite Repairs Using Infrared Thermography*, Florida Department of Transportation Structures Research Report No. 879.

BUYUKOZTURK O, PARK J and AU C (2003), 'NDE of FRP confined concrete using microwaves', *International Symposium Nondestructive Testing in Civil Engineering*, Berlin, Germany, Sept. 16–19.

CARPINTERI A, LACIDOGNA G and PAGGI M (2007), 'Acoustic emission monitoring and numerical modeling of FRP delamination in RC beams with non-rectangular cross-section', *Materials and Structures*, 40(6), 553–566.

CARR C, GRAHAM D, MACFARLANE J C and DONALDSON G B (2003), 'SQUID-based Nondestructive Evaluation of Carbon Fiber Reinforced Polymer', *IEEE Transactions on Applied Superconductivity*, 13(2), 196–199.

CHAN P K C, JIN W, LAU K T, ZHOU L M and DEMOKAN M S (2000), 'Multi-point strain measurement of composite-bonded concrete materials with a RF-band FMCW multiplexed FBG sensor array', *Sensors and Actuators A: Physical*, 87(1–2), 19–25.

CHEN X and ANSARI F (1999), 'Fiber Optic Stress Wave Sensor for Detection of Internal Flaws in Concrete Structures', *Journal of Intelligent Materials Systems and Structures*, 10(4), 274–279.

CHEN Z and ANSARI F (2000), 'Fiber optic acoustic emission sensor for large structures', *J. of Structural Control*, 7(1), 119–129.

CLAUS R O, GUNTHER M F, WANG A B, MURPHY K A and SUN D (1993), 'Extrinsic Fabry–Perot sensor for structural evaluation', in Ansari F, *Applications of Fiber Optic Sensors in Engineering Mechanics*, ASCE-EMD Spect. Pub., ASCE, New York, 60–70.

DAKIN J P (1990), 'Multiplexed and distributed optical fiber sensor systems', in Dakin J P, *The Distributed Fiber Optic Sensing Handbook*, IFS Publications, UK, 3–20.

DOYUM A B and DURER M (2002), 'Defect Characterization of Composite Honeycomb Panels by NDI Methods', *http://www.ndt.net/article/dgzfp02/papers/p36/p36.htm.*

DUTTA S S (2006), *Nondestructive Evaluation of FRP Wrapped Concrete Cylinders using Infrared Thermography and Ground Penetrating Radar*, M.S. Thesis, Department of Civil and Environmental Engineering, West Virginia University, Morgantown, WV.

EKENEL M, STEPHEN V, MYERS J J and ZOUGHI R (2004), 'Microwave NDE of Reinforced Concrete Beams Strengthened with CFRP Laminates Containing Surface Defects and Tested under Cyclic Loading', *16th World Conference on Nondestructive Testing*, Montreal, Canada, August–September.

EKENEL M, GALATI N, MYERS J J, NANNI A and GODÍNEZ V (2005), 'Acousto–Ultrasonic Technology for Nondestructive Evaluation of Concrete Bridge Members Strengthened by Carbon Fiber-Reinforced Polymer,' *Transportation Research Record*, 1928, 245–251.

FHWA (2007), *Bridge Acoustic Emission Local Area Monitoring System. http://www.tfhrc.gov/hnr20/nde/acoustic.htm.*

GHOSH K K and KARBHARI V M (2006), 'A critical review of infrared thermography as a method for non-destructive evaluation of FRP rehabilitated structures', *International J. of Materials and Product Technology*, 25(4), 241–266.

GHOSH K K and KARBHARI V M (2007), 'Evaluation of strengthening through laboratory testing of FRP rehabilitated bridge decks after in-service loading', *Composite Structures*, 77(2), 206–222.

HALABE U B, BANGALORE G, GANGARAO H V S and KLINKACHORN P (2002), 'Infrared Scanning of FRP Composite Members', *Proceedings of the Twenty-ninth Annual Review of Progress in Quantitative Nondestructive Evaluation*, Vol. 22 (American Institute of Physics–Vol. CP657), Bellingham, Washington, July 14–19, 1003–1010.

HALABE U B, HING C L, KLINKACHORN P and GANGARAO H V S (2006), 'Subsurface Defect Detection in FRP Bridge Decks using Ground Penetrating Radar', *Proceedings of the 33rd Annual Review of Progress in Quantitative Nondestructive Evaluation*, Vol. 26 (American Institute of Physics – Vol. CP894), Portland, Oregon, July 30–August 4, 1443–1450.

HALABE U B, ROY M, KLINKACHORN P and GANGARAO H V S (2005), 'Detection of Air and Water Filled Subsurface Defects in GFRP Composite Bridge Decks using Infrared Thermography', *Proceedings of the 32nd Annual Review of Progress in Quantitative Nondestructive Evaluation*, Vol. 25 (American Institute of Physics – Vol. CP820), Bowdoin College Brunswick, Maine, July 31–August 5, 1632–1639.

HALABE U B, VASUDEVAN A, GANGARAO H V S, KLINKHACHORN P and LONKAR, G (2004), 'Subsurface Defect Detection in FRP Composites using Infrared Thermography', *Proceedings of the 31st Review of Progress in Quantitative Nondestructive Evaluation*, Vol. 24 (American Institute of Physics – Vol. CP760), Colorado School of Mines, Golden, Colorado, July 25–30, 1477–1484.

HATSUKADE, Y, KASAI, N, TAKASHIMA, H and ISHIYAMA, A (2002), 'Non Contact SQUID – NDT Method using a Ferrite Core for Carbon Fiber Composites', *Superconductor Science and Technology*, 15, 1728–1732.

HATSUKADE Y, KASAI N, MARUNO Y, ISHIYAMA A and TANAKA S (2004), 'Detection of Deep Lying Defects in Carbon Fiber Composites using SQUID – NDE System', *Physica C*, 412, 1484–1490.

HATTA H, HASAN A, HATSUKADE M S, WAKAYAMA S, SUEMASU H and KASAI, N (2005), 'Damage Detection of C/C composites using ESPI and SQUID Techniques', *Composites Science and Technology*, 65(7–8), 1098–1106.

HAWKINS G F, JOHNSON E and NOKES J (1998), 'Typical manufacturing flaws in FRP retrofit applications', *Concrete Repair Bulletin*, pp.14–17.

HELLIER C J (2001), *Handbook of Nondestructive Evaluation*, New York: McGraw-Hill.

HILLGER W, HENRICH R, MEIER R, AHRHOLDT M and ROHLING H (2004), 'Non-contact Ultrasonic Imaging Techniques for Composite Components', *16th World Conference on Nondestructive Testing*, Montreal, Canada, Aug. 30–Sept. 5.

HING C L (2006), *Nondestructive Evaluation of Fiber Reinforced Polymer Bridge Decks using Ground Penetrating Radar and Infrared Thermography*, Ph.D. Dissertation, Department of Civil and Environmental Engineering, West Virginia University, Morgantown, WV.

HOSUR M V, VAIDYA U K, ULVEN C and JEELANI S (2004), 'Performance of Stitched/Unstitched Woven Carbon/Epoxy Composites under High Velocity Impact Loading using Ultrasonics', *Composite Structures*, 64(3–4), 455–466.

HSU D K, BARNARD D J, PETERS J J, DAYAL V and KOMMAREDDY V (2002), 'Nondestructive Inspection of Composites and their Repairs', *Proceedings of 6th Joint AAA/DOD Aging Aircraft Conference*, San Francisco, CA, Sep. 16–19, 1–11.

HU C W, SHIH J K C, DELPAK R and TANN D B (2002), 'Detection of Air Blisters and Crack Propagation in FRP Strengthened Concrete Elements using Infrared Thermography,' *Inframation - The Thermographer's Conference*, Orlando, FL, Sep. 29–Oct. 2.

HUANG M, JIANG L, LIAW P K, BROOKS C R, SEELEY R and KLARSTROM D L (1998), 'Using Acoustic Emission in Fatigue and Fracture Materials Research', *JOM-e, Member Journal of TMS*, 50 (11).

HUNG M (2004), 'Shearography and Applications in Nondestructive Evaluation', *16th World Conference on NDT*, Aug. 30–Sep. 3, Montreal, Canada.

IMIELINSKA K, CASAINGS M, WOJTYRA R, HARAS J, LE CLEZIO, E and HOSTEN B (2004), 'Air Coupled Ultrasonic C Scan Technique in Impact Response Testing of Carbon Fiber and Hybrid Composites', *Journal of Materials Processing Technology*, 157–58, 513–522.

JACKSON D, ISLAM M and ALAMPALLI S (2000), 'Feasibility of Evaluating the Performance of Fiber Reinforced Plastic (FRP) Wrapped Reinforced Concrete Columns using Ground Penetrating Radar and Infrared Thermography Techniques', *Structural Materials Technology – An NDT Conference*, Feb. 28–Mar. 3, Atlantic City, NJ, 390–395.

KAISER H and KARBHARI V M (2003), 'Identification of potential defects in the rehabilitation of concrete structures with FRP composites', *International Journal of Materials and Product Technology*, 19(6), 498–520.

KALAMKAROV, A L, D O MACDONALD, S B FITZGERALD and A V GEORGIADES (2000), Reliability assessment of pultruded FRP reinforcements with embedded fiber optic sensors, *Composite Structures*, 50(1), 69–78.

KANG K, CHOI M, KIM K, CHA Y, KANG Y, HONG D and YANG S (2006), 'Inspection of Impact Damage in Honeycomb Composite Plate by ESPI, Ultrasonic Testing, and

Thermography', *12th A-PCNDT 2006 – Asia-Pacific Conference on NDT*, Nov. 5–10, Auckland, New Zealand.

KATSUYAMI T, MATSUMURA H and SUGAMUME T (1981), 'Low-loss single polarization fibers', *Electronics Letters*, 17, 473.

KERSEY A D and MOREY W W (1993), 'Multiplexed Bragg grating fiber-laser strain sensor system with modelocked interrogation', *Electronic Letters*, 29, 112.

KIM N and ACHENBACH J D (1998), 'Quantitative characterization of multiple delaminations in laminated composites using the Compton backscatter technique', *Journal of Nondestructive Evaluation*, 17(2), 53–65.

KUNDU T, EHSANI M, MASLOV K I and GUO D (1999), 'C-scan and L-scan generated images of the concrete/GFRP composite interface', *NDT & E International*, 32, 61–69.

LAYLOR H M and KACHLAKEV D I (2000), *Fibre-reinforced Composites for Strengthening Bridges in Oregon, http://onlinepubs.trb.org/onlinepubs/trnews/rpo/rpo.trn208.pdf.*

LESTARI W and QIAO P (2005), 'Experimental Investigation of Damage Detection on Composite Plates using Wave Analysis', *Nondestructive Evaluation and Health Monitoring of Aerospace Materials, Composites and Civil Infrastructure IV*, March 8–10, San Diego, CA.

LI J and LIU C (2001), 'Non-contact Detection of Air Voids Under Glass Epoxy Jackets Using a Microwave System', *Subsurface Sensing Technologies and Applications*, 2(4), October, 411–423.

LIANG Y, SUN C and ANSARI F (2004), 'Acoustic Emission Characterization of Damage in Hybrid Fiber Reinforced Polymer Rods'. *ASCE, Journal of Composites for Construction*, 8(1), 70–78.

LIM T K, ZHOU Y, LIN Y, YIP Y M and LAM Y L (1999), 'Fiber optic acoustic hydrophone with double Mach–Zehnder interferometers for optical path length compensation', *Optics Communications*, 59(4–6), 301–308.

LIU K, FERGUSIN S M and MEASURES R M (1990), 'Fiber optic interferometric sensor for the detection of acoustic emission within composite materials', *Optical Letters*, 15 (22), 1255–1257.

LUPRANO V A M, TUNDOT A, TATI A, GRINZATO E, MARINETTI S and BISON P G (2006), 'Non-destructive Defect Characterization in Civil Structures Reinforced By Means Of FRP', (*http://www.ndt.net/article/ecndt2006/doc/We.1.6.1.pdf*).

MAERZ N, GALECKI G and NANNI A (2004), 'Experimental Nondestructive Testing of FRP Materials, Installation, and Performance', *16th World Conference on Nondestructive Testing*, Montreal, Canada. August 30–September 3.

MALDAGUE X (2000), 'Applications of Infrared Thermography in Nondestructive Evaluation', *Trends in Optical Nondestructive Testing* (invited chapter), P. Rastogi and D. Inaudi., ed., Elsevier, Science, Amsterdam 591–609. (Paper also available at *http://www.ndt.net/v06n04.htm*).

MEOLA C and CARLOMAGNO G M (2004), 'Recent Advances in the Use of Infrared Thermography', *Measurement Science Technology*, 15, 27–58.

MIRMIRAN A, SHAHAWY M and EL ECHARY H (1999), 'Acoustic emission monitoring of hybrid FRP–concrete columns', *Journal of Engineering Mechanics*, 125(8), 899–905.

MOREY W W, MELTZ G and GLENN, D H (1989), 'Fiber optic Bragg grating sensors', *Proc. SPIE Fiber Optic and Laser Sensors*, 1169, 98.

MTENGA P V, PARZYCH J and LIMERICK R L (2001), 'Quality Assurance of FRP Retrofit using Infrared Thermography', *Proceedings of ASCE Structures Congress*, Washington DC, USA.

MUFTI A (2002), 'Structural Health Monitoring of Innovative Canadian Civil Engineering Structures', *An International Journal – Steel and Composite Structures*, 1(1), 89–103.

MUFTI A (2005), 'Structural Health Monitoring of Innovative Bridge Decks', *Structure and Infrastructure Engineering: Maintenance, Management, Life-Cycle Design and Performance Journal*, 1(2), 119–133.

NOKES J P and HAWKINS G F (2001), 'Infrared Inspection of Composite Reinforced Concrete Structures', Ohio Department of Transportation, OH. USA. *http://www.dot.state.oh.us/research/2001/Structures/14688-FR.pdf.*

RASHLEIGH S C and ULRICH R (1980), 'High birefringence in tension-coiled single-mode fiber', *Optical Letters*, 5, 354–356.

RAY B C, HASAN S T and CLEGG D W (2007), 'Evaluation of Defects in FRP Composites by NDT Techniques', *J. of Reinforced Plastics and Composites*, 26 (12), 1187–1192.

RUOSI A, VALENTINO M, LOPRESTO V and CAPRINO G (2002), 'Magnetic Response of Damaged Carbon Fibre Reinforced Plastics Measured by a HTS-SQUID Magnetometer', *Composite Structures*, 56, 141–149.

SANSALONE M J (1997), 'Impact-echo: The complete story', *ACI Structural Journal*, 94(6), 777–786.

SANSALONE M J and STREETT W B (1998), 'The Impact-Echo Method', NDTnet, *The online Journal of Nondestructive Testing & Ultrasonics*, 3(2).

SCARPONI C and BRIOTTI G (2000), 'Ultrasonic technique for the evaluation of delaminations on CFRP, GFRP, KFRP composite materials', *Composites: Part B*, 31(3), 237–243.

SHIH J K C, TANN D B, HU C W, DEPAK R and ANDREOU E (2003), 'Remote sensing of air blisters in concrete–FRP bond layer using IR thermography', *International Journal of Materials and Product Technology*, 19(1/2), 174–187.

SHIH S T (1998), 'Wide-band polarization-insensitive fiber optic acoustic sensors', *Optical Engineering*, 37(3), 968–976.

SIM J, MOON D, HONGSEOB O H, PARK C and PARK S (2005), 'Hybrid FRP rod for reinforcement and smart-monitoring in concrete structures', in JF Chen and JG Teng, *Proceedings of the International Symposium on Bond Behaviour of FRP in Structures (BBFS 2005)*, 393–400.

STARNES M A, CARINO N J and KAUSEL E A (2002), 'Quantitative Infrared Thermography for Quality Control of Concrete Structures Strengthened with FRP Composites', *Proceedings, ACI International Conference*, American Concrete Institute, Detroit, MI.

STEPHEN V, KHARKOVSKY S, NADAKUDUTI J and ZOUGHI R (2004), 'Microwave Field Measurement of Delaminations and their Growth in CFRP Concrete Members in a Bridge', *16th World Conference on Nondestructive Testing*, Montreal, Canada, August–September.

SUN C, LIANG Y and ANSARI F (2004), 'Serially Multiplexed Dual-point Fiber Optic Acoustic Emission Sensor', *IEEE, OSA, Journal of Lightwave Engineering*, 22(2), 487–493.

TATEDA M and HORIGUCHI T (1989), 'Advances in optical time domain reflectometry', *IEEE J. Lightwave Technology*, 7(8), 1217–1223.

TAWHED FW and GASSMAN S L (2002), 'Damage assessment of concrete bridge decks using impact-echo method', *ACI Materials Journal*, 99(3), 273–281.

TENNYSON R C, MUFTI A A, RIZKALLA S, TADROS G and BENMOKRANE B (2001), 'Structural Health Monitoring of Innovative Bridges in Canada with Fibre Optic Sensors', *Journal of Smart Materials and Structures*, 10(3), 560–572.

WATKINS S E, FONDA J W and NANNI A (2007), 'Assessment of an Instrumented Reinforced-concrete Bridge with Fiber-reinforced-polymer Strengthening', *Optical Engineering*, 46(5).

YANG L (2006), 'Recent Developments in Digital Shearography for Nondestructive Testing', *Material Evaluation*, 64(7), 703–709.

ZHAO M, DONG Y, ZHAO Y and ANSARI F (2007), 'Monitoring Bond in FRP Retrofitted Concrete Structures', *Journal of Intelligent Material Systems and Structures*, 18(8), 753–890.

ZHAO Y and ANSARI F (2001), 'Quasi-distributed fiber-optic strain sensor: Principle and experiment', *Applied Optics*, 40(19), 3176–3181.

ZHENG G, BUCKLEY M A, KISTER G and FERNANDO G F (2000), 'Blind Deconvolution of Acoustic Emission Signals for Damage Identification in Composites', *Nondestructive Evaluation of Aging Materials and Composites IV*, SPIE, 3993, Newport Beach, CA, March 8–9, 47–57.

ZHOU Z, OU J P and WANG B (2003), 'Smart FRP-OFGB bars and their application in reinforced concrete beams. *Proceedings of the First International Conference on Structural Health Monitoring and Intelligent Structure*', 13–15, Nov. 2003, Tokyo, Japan: 861–866.

ZHU Q, FANG R, CAO Z and ZENG W (2003), 'Measurement of internal strain of FRP concrete structure by fiber optic sensor', in Shen G X, Cha S, Chiang F P and Mercer C R, *Optical Technology and Image Processing for Fluids and Solids Diagnostics, Proceedings of the SPIE*, Volume 5058, 538–542.

8 Structural health monitoring and field validation of civil engineering structures

C. SIKORSKY, California Department of Transportation, USA, and V. M. KARBHARI, University of Alabama in Huntsville, USA

Abstract: With the increasing age of infrastructure and the use of new technologies, there is a need to derive, on a continuous basis, knowledge about the actual condition of a structure, or system, with the aim of not just knowing that its performance may have deteriorated, but rather to be able to pinpoint the area of degradation and more importantly to assess remaining performance levels and life. This has led to the development of the field of structural health monitoring (SHM) as a means of validating current and assessing potential future performance characteristics of structures and systems.

Key words: structural health monitoring, sensors rehabilitation, damage, capacity.

8.1 Introduction

Civil infrastructure forms the social and economic backbone of the civilized world, encompassing fixed systems that comprise our basis for transportation of goods and services, collection and supply of water and sewage, and for the transmission of power. Irrespective of the design approach, materials, construction methods and quality control, all these systems deteriorate with time, and a major challenge for all owners and operators is the determination of existing capacity, remaining service life, and functional operability of these systems. In addition to the concerns related to changes faced over periods of use, there is the important issue of the actual capacity of an 'as-built' structure and its correlation to the 'as-designed' functionality. In a number of historical cases, the differences between design and *in-situ* response were estimated through engineering judgment and use of extremely conservative sets of systems level factors of safety. The combination of a deteriorating civil infrastructure and limitations on available resources necessitates maximizing or extending the service life of structures.

The aging infrastructure is gradually becoming a global concern. This is especially true in the case of advanced nations, wherein a large fraction of critical civil infrastructure systems were built decades ago and are now either considered structurally deficient or are approaching that state. This reality, coupled with the extensive use of new materials and innovative designs, has greatly increased the need for the owners and operators to have a clear and quantitative picture of the condition of the structure. In addition, to prevent catastrophic failure of structures and ensure safety of the public, owners and operators also need a system that can provide early warning of unsafe conditions, ensuring a comprehensive shift to condition-based, rather than time-based, inspection.

Currently, visual inspection, and use of methods such as localized load-testing (load rating) are the predominant nondestructive evaluation techniques used for inspection of structural integrity. Previous research has shown that such inspections, in so far as bridges are concerned, have limited accuracy and efficiency (FHWA, 2001) and, in addition, there is the possibility that sudden, catastrophic failure could occur in between inspections. Furthermore, rapid evaluation of structural conditions after major events such as earthquakes is not possible. In such cases, critical life-lines are often closed until such a time when visual inspection can be completed. Since there are generally a limited number of qualified inspectors in any agency, this often results in undue delay in opening critical life-lines.

In an attempt to increase the service life of deteriorated and under-strength structures, of late there has been an increased use of new advanced materials such as fiber reinforced polymer (FRP) composites in major load-carrying components in bridge structures. While the use of these materials has proved to be an efficient means of enhancing service life and safety of some structures, there is limited information regarding the long-term performance of such materials and rehabilitated systems, which results in a level of concern regarding reliability over extended time periods of service or under extreme events.

There is thus a need to derive, on a continuous basis, knowledge about the actual condition of a structure, or system, with the aim of not just knowing that its performance may have deteriorated, but rather to be able to pinpoint the area of degradation and more importantly to assess remaining performance levels and life. This has led to the development of the field of structural health monitoring (SHM) as a means of validating current, and assessing potential future, performance characteristics of structures and systems.

8.2 Structural health monitoring (SHM)

Over the past two decades, a significant level of research effort has focused on the development of algorithms to locate and quantify damage in a structure using non-destructive methods. Similarly, work has focused on systems identification algorithms as well. More recently, the area of structural health monitoring (SHM) has gained popularity as a complementary technology to system identification and non-destructive damage detection methods. In the past decade, the use of strategies for long-term structural health monitoring has been mentioned as a potential solution to the above challenges. Long-term structural health monitoring refers to the practice of using an integrated system of sensors, data acquisition devices, data transmission and processing devices, and corresponding algorithms, to continuously monitor the condition of a structure (in this case a bridge) over an extended period of time. Through monitoring of structural responses to external excitations, the SHM system can provide the owner with quantitative information about changes in structural response, which can be correlated to changes in the structural condition of the bridge, and this information can be later used to estimate structural reliability and load capacity. Combined with an appropriate set of structural degradation models, the remaining life of the structure can then be predicted. Such systems may be able not only to provide early warning of possible unsafe conditions, but also to help bridge owners decide whether a repair or a replacement should be made. Thus, limited maintenance budgets can be optimized while at the same time improving the safety of the public transportation system.

Structural health monitoring systems can be classified either as model-driven or data-driven. In a model-driven approach, a physical model of the structure is developed, typically using finite elements. Experimental results from the health monitoring system are used to update this model, and then locate and quantify damage. A data-driven approach requires a model as well, but the model is based more on statistical theory than structural theory. There are similarities between these approaches, however. Both approaches rely on a data acquisition system for data collection and transmission. The data acquisition system can be permanently mounted to the structure or be mobile, depending on user requirements and needs. Typically, these systems collect data on a continuous or periodic basis and transmit the data to a common point, either via a wireless or hard-wired transmission line. Most health monitoring systems collect data that is compared with results from some numerical model. The weakness here is that not all systems attempt to numerically update a physical model of the structure using experimental data from the structure in service. While some systems incorporate a system identification or non-destructive damage evaluation algorithm to rapidly process the data, these are the exception, not the rule. It is also unclear how structural capacity and remaining service life can be estimated using a data-based model.

Part of the problem can be attributed to poor definitions or a lack thereof. Structural health monitoring has been defined by some as 'the use of *in-situ*, non-destructive sensing and analysis of structural characteristics, including the structural response, for detecting changes that may indicate damage or degradation' (Housner *et al.*, 1997). Unfortunately, such a definition fails to consider the effect of damage on capacity and remaining service life. Intrinsically, owners and operators of modern civil engineering systems need knowledge of the integrity and reliability of the network, structural system, and/or components in real time, such that they can not only evaluate the state of the structure but also assess when preventive actions need to be taken. Thus, what is needed is an efficient method to collect data from a structure in-service and process the data to evaluate key performance measures such as serviceability, reliability and durability. For the current work, the definition by Housner *et al.* (1997) is modified, and structural health monitoring is defined as 'the use of in-situ, non-destructive sensing and analysis of structural characteristics, including the structural response, for the purpose of estimating the severity of damage/deterioration and evaluating the consequences thereof on the structure in terms of response, capacity, and service life'. Essentially, a SHM system then must have the ability to collect, validate, and make accessible, operational data on the basis of which decisions related to service life management can be made (Karbhari, 2009). While this task may not have been possible a decade ago, recent progress in (a) sensor technology, (b) methods of damage identification and characterization, (c) computationally efficient methods of analysis, and (d) data communication, analysis and interrogation, have made it possible for one to consider SHM as a tool, not just for operation and maintenance but also for the eventual development of a true reliability and risk-based methodology for design.

There are at least three major weaknesses, from the perspective of a bridge owner, associated with the current focus of structural health monitoring. First, the cost to install and maintain the infrastructure associated with a database system is enormous. Second, a database system is not mobile and precludes rapid evaluation of structural safety. Third, the ability of a data-based monitoring system to evaluate safety is limited to the events seen by the monitoring system. While the issue of evaluation of bridge capacity and estimation of remaining service has been investigated by several researchers, no one, to date, has presented a methodology by which capacity and remaining service life could be obtained. Even more frustrating to a state bridge engineer is the lack of consensus among researchers and a failure to evaluate differing systems. An owner needs a system that can evaluate the ability of the structural system to function as designed, and estimate the remaining service life of that system. That is, a Level IV damage evaluation method as defined by Rytter (1993).

The objective of this chapter is to provide examples of SHM as a method of non-destructive field validation of structures using available techniques to assess the condition of a structural system (using bridges as examples) in-service, based on four specific attributes. First, the system must be mobile and utilize the minimum number of sensors possible. Second, it must be able to rapidly locate and quantify damage on a global basis. Third, it must identify damage mechanisms locally. Lastly, the system must be able to estimate structural capacity and remaining service life. Readers interested in a more in-depth treatment of the principles of SHM *vis-a-vis* civil structures are referred to a recent work edited by Karbhari and Ansari (2009).

8.3 Theoretical framework of a structural health monitoring system

This section summarizes the theoretical framework of an example structural health monitoring system. First, the theory supporting the system identification scheme is presented. Next, the scheme to detect damage is summarized, including damage localization, severity estimation and identification of the stiffness loss of the structure. Lastly, a theoretical link between structural capacity and damage is presented. A time domain decomposition (TDD) technique is used to extract the modal parameters as given by Kim *et al.* (2002). (Obviously other techniques may be used to extract modal parameters.)

As shown in Fig. 8.1, a structure having a flaw of magnitude which is small in comparison to the original structure can be described using field-measured mode shapes Φ_i^* and eigenfrequencies ω_i^* that are different from those measured in the undamaged state (Fig. 8.1b), using only the frequency information from the flawed structure. Then the identified model with the effective parameters (Fig. 8.1b) will have eigenfrequencies close to ω_i^* (for example in the least squares sense) of the flawed model (Fig. 8.1a) but the mode shapes of the two structures will be different in the neighborhood of the flaw. It is also reasonable to assume that the identified elastic properties of the flawless structure are close to those of the flawed structure. The differences in the mode shapes of the identified structure and the measured mode shapes of the existing structure can then be exploited to localize the flaw. A systems identification methodology to identify effective elastic properties of a structure can be developed following Stubbs and Kim (1996).

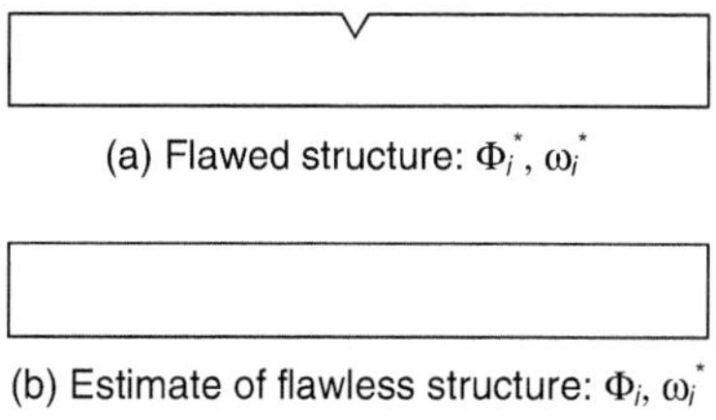

8.1 Flawed structure and estimate of flawless structure.

Consider a linear skeletal structure with NE members and N nodes. Suppose k_j^* is the unknown stiffness of the jth member of the structure for which M eigenvalues are known. Also, suppose k_j is a known stiffness of the jth member of a finite element (FE) model for which the corresponding set of M eigenvalues are known. Then, relative to the FE model, the fractional stiffness change of the jth member of the structure, α_j, and the stiffnesses are related according to:

$$k_j^* = k_j(1 + \alpha_j) \quad [8.1]$$

Similarly, the fractional mass change of the jth member of the structure, β_j, and the masses are related according to:

$$m_j^* = m_j\,(1 + \beta_j) \quad [8.2]$$

The fractional stiffness change and the fractional mass change of NE members may be obtained using:

$$Z = F\alpha - G\beta \quad [8.3]$$

as suggested by Stubbs and Osegueda (1990) where $\boldsymbol{\alpha}$ is a $NE \times 1$ matrix containing the fractional changes in stiffness between the FE model and the structure, β is a $NE \times 1$ matrix containing the fractional changes in mass between the FE model and the structure, $\mathbf{Z}$ is a $M \times 1$ matrix containing the fractional changes in eigenvalues between the two systems, $\mathbf{F}$ is a $M \times NE$ stiffness sensitivity matrix relating the fractional changes in stiffnesses to the fractional changes in eigenvalues, and $\mathbf{G}$ is a $M \times NE$ mass sensitivity matrix relating the fractional changes in masses to the fractional changes in eigenvalues.

The $M \times NE$, $\mathbf{F}$ matrix can be determined as follows: first, M eigenvalues are numerically generated from the initial FE model; second, the stiffness of the first member of the FE model is modified by a known amount; third, the corresponding set of M eigenvalues are numerically generated for the modified FE model; fourth, the fractional changes between the M initial eigenvalues and M eigenvalues of the modified structure are computed; fifth, each component of the first column of the $\mathbf{F}$ matrix (i.e. the $M \times 1$, $\mathbf{F}$ matrix) is computed by dividing the fractional changes in each eigenvalue by the magnitude of the modification at member one; and finally, the $M \times NE$, $\mathbf{F}$ matrix is generated by repeating the entire procedure for all NE members. The $M \times NE$, $\mathbf{G}$ matrix can be determined in a similar manner.

Using the above rationale as a basis, the following seven-step algorithm can be used to identify a given structure:

(i) Select a target structure (e.g. a post-damage state of the structure) for which sufficient eigenfrequencies that can be used to identify the associated flawless structure are available. (Note that the mode shapes of the damaged structure in defining the target structure are ignored.)

(ii) Select an initial finite element (FE) model of the structure, utilizing all possible knowledge about the design and construction of the structure.

(iii) As outlined above, compute the sensitivity matrices of the FE model.

(iv) As outlined above, compute the fractional changes in eigenvalues between the FE model and the target structure.

(v) Solve Equation 8.3 to estimate fractional changes in mass and stiffness.

(vi) Update the FE model using the results in Step (v) and Equations 8.1 and 8.2.

(vii) Repeat steps (iv)–(vi) until $\mathbf{Z} \cong 0$, which indicates that the effective parameters of the structure have been identified.

In the problem of interest here, the dynamic responses of the structure in the time domain represent the physical world data and the modal parameters represent the pattern space. The feature space is represented by indicators that are a function of measurable pre-damage and post-damage modal parameters. These indicators can be selected in such a manner that they reflect internal structure in the data. The function of the decision algorithm is to partition the data space into D_n clusters (decision spaces). In this study, $n = 2$ and the decision spaces correspond to the cases: (a) a structure is not damaged at a given location, and (b) a structure is damaged at a given location. For each instance, the indicator of damage will fall into one of the two categories. The damage indicator DI_{ij}, used herein, and which forms the basis of the feature space (in the pattern recognition sense) is given by (Stubbs *et al.*, 2000) as:

$$DI_{ij} = \frac{k_j}{k_j^*} \cong \left[\frac{\Phi_i^{*T} C_{jo} \Phi_i^* + \Phi_i^{*T} C \Phi_i^*}{\Phi_i^T C_{jo} \Phi_i + \Phi_i^T C \Phi_i} \right] \frac{\Phi_i^T C \Phi_i}{\Phi_i^{*T} C \Phi_j^*} \qquad [8.4]$$

where the scalars k_j and k_j^*, respectively, are parameters representing the material stiffness properties of the undamaged and damaged jth member of the structure, Φ_i is the ith modal vector, C is the system stiffness, and C_{jo} is the matrix containing only geometric quantities (and possibly terms containing Poisson's ratio). There are two important characteristics of the indicator DI_{ij} given by Equation 8.4: first, the expression attempts to express the changes in stiffness at a specific location in terms of measurable pre-damage and post-damage mode shapes (Φ_i and Φ_i^*) and second, the term C_{jo} on the right hand side can be determined from a knowledge of only the topology of the structure. Thus, for each damage location j, there are as many DI_{ij}s available as there are mode shapes. As noted above, in the context of pattern recognition, the latter values of DI_{ij} define the feature space. A convenient form of the damage index DI_{ij} for a single location, if several modes (MN) are used, can be described by:

$$DI_j = \frac{\sum_{i=1}^{MN} (\Phi_i^{*T} C_{jo} \Phi_i^* + \Phi_i^{*T} C \Phi_i^*) \Phi_i^T C \Phi_i}{\sum_{i=1}^{MN} (\Phi_i^T C_{jo} \Phi_i + \Phi_i^T C \Phi_i) \Phi_i^{*T} C \Phi_i^*} \tag{8.5}$$

The final step in damage localization is classification. The act of classification assigns an object to one of a number of possible groups on the basis of observations made on the objects. In this study, the objects are members of the structure. Each and every member of the structure falls into one of two groups: undamaged elements and damaged elements. Finally, the observations made on the objects are the DI_{ij}s. The criteria for damage localization used herein utilizes statistical reasoning. In the approach used here, the values $DI_1, DI_2, DI_3, \ldots, DI_{NE}$ for each element are considered as realization of a random variable. From these values, a normalized damage indicator is given by

$$z_j = \frac{DI_j - \mu_{DI}}{\sigma_{DI}} \tag{8.6}$$

where μ_{DI} and σ_{DI} represent the mean and standard deviation of the damage index, DI_j, respectively. Finally, the classification of a structural element (or part thereof) is accomplished using the following set of rules: let H_o be the hypothesis that member j of the structure is not damaged and let H_1 be the hypothesis that member j of the structure is damaged. Then, for example, the following decision rules, with 98% confidence level, may be used to assign damage to member j: (i) choose H_1 if $z_j = 2$ and (ii) choose H_o if $z_j < 2$. Obviously, other levels of confidence can be selected.

Note that in Equation 8.4 the indicator of damage is the ratio of the undamaged stiffness to the damaged stiffness. Such a number exists for each potentially damaged member. Herein, the damage can be expressed as the fractional change in stiffness of an element:

$$\alpha_j = \frac{k_j^* - k_j}{k_j} = \frac{1}{DI_j} - 1 \tag{8.7}$$

Thus if there is no damage, $\alpha_j = 0$; if there is damage, $\alpha_j < 0$. Note that if $\alpha_j = -1$, all stiffness capacity is completely lost. Having stiffness parameters for the baseline as-built structure, location of damage, and the severity of damage, the stiffness properties of the existing structure can be obtained from:

$$k_j^{(\text{existing})} = k_j^{(\text{baseline})}[1 + \alpha_j] \tag{8.8}$$

Note that if there is no damage at location j, the stiffness properties of the baseline and the existing structure are the same.

While this technique can locate and quantify damage severity, some means is required to relate this damage information to either structural demand or capacity. In other words, can the structure continue to carry the design load, given the presence of damage? Relying on previous work in the area of continuum damage mechanics, a theoretical link between damage indicators and structural demand can be formulated. Let A_0 be the initial area of the undamaged member. After that section has incurred some damage, a certain part of the section is cracked or 'lost' (Kachanov, 1986). Let us denote the lost area as A. It is important to note the difference between local damage characteristics that can be detected visually, as opposed to an analytical expression for damage. A local damage characteristic such as a crack or delamination can be identified visually. However, that damage characteristic will also cause some change in the dynamic characteristics of the bridge. The analytical expression for damage developed here is an average damage indicator for the area in question. Therefore, this analytical expression for damage can be used to locate a visible sign of damage, or more importantly warn of impending damage that has yet to manifest itself. Thus, the value $A_0 - A$ can be interpreted as the actual area (or material) of the section. For the case of isotropic damage, cracks and voids are equally distributed in all directions. The damage indicator can be taken as a scalar. Here we define the damage indicator as ω:

$$\omega = \frac{A}{A_0} \quad \text{where} \quad 0 \le \omega \le 1 \tag{8.9}$$

For an undamaged material, $\omega = 0$, and $\omega = 1$ when the structure has lost all capacity and failed. In order to determine structural demand, let us subject the section to uniaxial tension or an actual stress σ_a.

$$\sigma_a = \frac{P}{A_0 - A} = \frac{P}{A_0(1-\omega)} = \frac{\sigma}{(1-\omega)} \tag{8.10}$$

where σ is the nominal stress and σ_a is the stress related to the damaged member. The underlying assumption here is that the strain response of the body is modified by damage only through the actual stress. We also assume that the rate of damage growth is determined primarily by the level of actual stress (Kachanov, 1986). Based on these assumptions, the elastic strain of a damaged material is

$$\varepsilon = \frac{\sigma_a}{E'} = \frac{\sigma}{E(1-\omega)} \tag{8.11}$$

where E' represents the damaged modulus. Reformulating Equation 8.11 yields:

$$\psi = (1-\omega) = \frac{1}{E}\frac{d\sigma}{d\varepsilon} = \frac{E'}{E} \qquad [8.12]$$

Based on this formulation, damage may be estimated by measuring the elastic response. In previous work, Stubbs and Kim (1996) demonstrated that relative to the FE model, the fractional stiffness change of the jth member of the structure, α_j, and the stiffnesses are related according to:

$$k_j^* = k_j(1 + \alpha_j) \quad \text{for} \quad -1.0 \leq a \leq 0.0 \qquad [8.13]$$

where k_j^* represents the stiffness of the damaged member and k_j is the original undamaged member. It should be noted that for $\alpha_j = 0$, no damage is present. Therefore,

$$1+\alpha_j = \frac{k_j^*}{k_j} = \frac{\dfrac{c_j E_j^* I_j^*}{L_j^3}}{\dfrac{c_j E_j I_j}{L_j^3}} = \frac{E_j^* I_j^*}{E_j I_j} \qquad [8.14]$$

If we accept the notion that damage is not due to gross changes in geometric properties, then the length (L) of the damaged and undamaged members are numerically equivalent. Note also that if a large section of the bridge had disappeared, similar to a punching shear failure, a visual inspection would be sufficient to locate the damage and therefore no need would exist for an inspection using an NDE technique such as this. Therefore, we can reduce Equation 8.14 to

$$1+\alpha = \frac{E_j^* I_j^*}{E_j I_j} \qquad [8.15]$$

Substituting Equation 8.12 into Equation 8.15 yields

$$1+\alpha = \frac{E_j^* I_j^*}{E_j I_j} = (1-\omega) \qquad [8.15a]$$

or

$$\alpha = -\omega \qquad [8.15b]$$

Based on this argument, it may be hypothesized that the damage indicator ω as defined in damage mechanics (Kachanov, 1986) is the same as the damage severity estimation determined from this modal-based nondestructive damage detection scheme. Therefore, the estimation of damage severity (α) may be used to compute the actual load that a 'damaged' structure experiences. For example, the flexural stress in an undamaged beam element is given by

$$\sigma = \frac{Mc}{I} \qquad [8.16]$$

and for a damaged element, the flexural stress can be determined by substituting Equation 8.16 into 8.10 to yield

$$\sigma_a = \frac{Mc}{I(1+\alpha)} = \frac{\sigma}{(1+\alpha)} \quad \text{where} \quad -1.0 \le \alpha \le 0.0 \qquad [8.17]$$

Using the Caltrans Bridge Design Specifications (2000), the dead load moment demand for the bridge deck slab can be expressed as:

$$M_{DL} = \frac{0.8wS^2}{8} \qquad [8.18]$$

Utilizing Equations 8.17 and 8.18 it is obvious that for a damaged member, the flexural moment due to dead load (M^*_{DL}) is given by

$$M^*_{DL} = \frac{0.8wS^2}{8(1+\alpha)} \qquad [8.19]$$

where w is distributed dead load and S is the effective span distance. Similar equations can be formulated to determine the live load demand for a damaged structure (M^*_{LL}).

$$M^*_{LL} = \frac{1.3(S+2)P}{32(1+\alpha)} \qquad [8.20]$$

where $P = 71$ kN (16 kips) for an HS-20 loading. Note also that Equation 8.20 includes a 30% increase for impact loads. The load combination, $Grp(1)$ used to design the bridge is defined in the design specification as:

$$Grp(1)^* = 1.3\left[M^*_{DL} + \frac{5}{3}M^*_{LL}\right] \qquad [8.21]$$

A ratio of demand to capacity (*D/C*) can then be computed to quickly identify where demand exceeds capacity in flexure. A demand to capacity ratio of unity or less is deemed acceptable:

$$\frac{D}{C} = \frac{Grp(1)^*}{\varphi M_u} \qquad [8.22]$$

where $M_u = [A_s f_y (d - 0.5^*\text{a})]$ and $\phi = 0.9$. In addition, since the ability to estimate capacity is possible, this could result in improved estimates of structural reliability as demonstrated by Park *et al.* (1997).

8.4 Example applications highlighting field validation and assessment

The concepts of structural health monitoring have been applied to numerous structures including bridges, buildings, historical monuments and other systems and there is a wide range of literature that provides details on completed and ongoing applications. A recent publication (Karbhari and Ansari, 2009) provides a reference on both SHM technologies and applications in civil infrastructure systems and hence this will not be repeated in this chapter. Rather, this chapter provides two examples to illustrate how the technique can be used in both post-event analysis (wherein a reinforced concrete bridge was evaluated for damage after an earthquake) and to assess the case wherein the initial deterioration cannot be stopped but is reduced (a structure rehabilitated with the sole purpose of providing a small extension of service life). It is emphasized that the latter case provides significant challenges to engineers both due to risk assessment and cost–benefit, and that SHM provides mechanisms for enabling these that were heretofore not available. Both structures discussed were monitored using a mobile monitoring system.

8.4.1 Damage detection after an earthquake

The Hector Mine earthquake occurred on 16 October 1999, with an epicenter in the Mojave Desert approximately 76 km south east of Barstow. The Lavic Road Bridge, which passes over I-40 approximately 64 km east of Barstow, was reportedly the only major structure in the vicinity of the epicenter to experience significant structural damage. The structure was completed in 1968 and contains two spans of 36 m and 37.5 m respectively. The superstructure is a 2.1 m deep reinforced concrete triple box girder, which includes a 10.4 m wide deck and four 0.2 m wide webs spaced at 2.7 m. The bridge is supported approximately at mid-span by a 1.5 m diameter column, which in turn is supported on a spread footing. As part of a routine procedure, modal tests were performed on the structure in September 1999. The occurrence of the earthquake in October 1999 provided an opportunity to quantify the impact of the event on the structure using the results of a second modal test. The impact of the earthquake on the stiffness properties of the bridge is presented here by comparing the results of system identification before and after the earthquake. Analysis indicated that the average decrease in stiffness of the deck, column, and abutments were, respectively, approximately 30%, 20%, and 50%. Figure 8.2 shows the change in concrete modulus for the deck and column since monitoring began. If deterioration is assumed to be linear, then the remaining useful life of the structure can be estimated using Fig. 8.2. Although fairly simple,

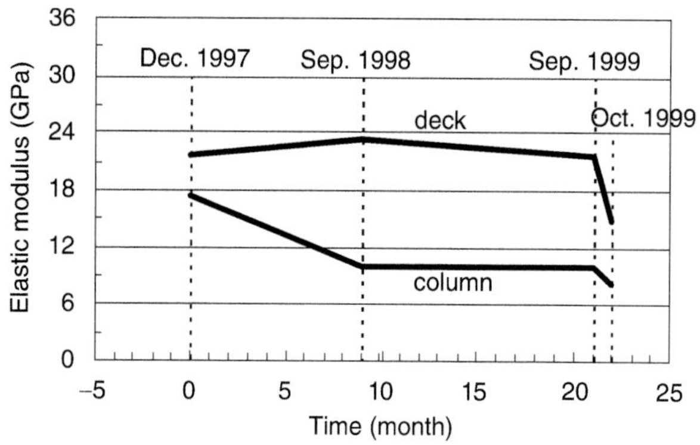

8.2 Changes in elastic modulus of the deck and column.

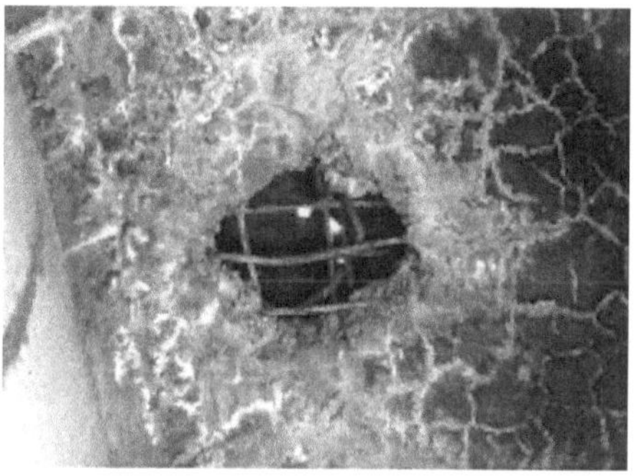

8.3 Punching shear in deck.

the methodology provides ease in post-event assessment and enables the evaluation of need for repair and its consequent cost versus remaining service life.

8.4.2 Assessment of rehabilitation in the presence of continuing deterioration

A 103.6 m long, five span, two-lane highway bridge (spanning a canal that is part of the California Aqueduct System,) that was built in 1964 had shown significant signs of structural distress in the form of significant cracking in the deck, shear cracking of girders, and even punching shear in the deck (as in Fig. 8.3). The bridge superstructure consisted of five, cast in-place, continuous, reinforced concrete T-girders, monolithically connected to the bents, with spans of lengths of 17.6 m, 22.8 m, 22.8 m, 22.8 m, and 17.6 m, respectively. The 0.16 m thick reinforced concrete deck spanned transversely between the 2.2 m center-to-center spaced girders with expansion provided in Span #3 in the form of a hinge, whereas the abutments are a fixed diaphragm. It should also be noted that the bridge was retrofitted in the past decade to resist larger magnitude earthquakes. As part of the retrofit, the top 51 mm of the concrete deck were removed and replaced with a polymer concrete overlay of equivalent thickness. Unfortunately, cores taken from the bridge deck indicated that the deck may not have been cleaned properly prior to the application of the polymer concrete. Therefore, the structural integrity of the concrete deck and polymer overlay was also questionable. Detailed investigations clearly showed the lack of capacity and accelerating deterioration of the bridge which would be cause for it, at minimum, to be posted and perhaps even closed. Since the bridge served a major artery and a reroute would have added significant distance and led to additional congestion in the area, a rehabilitation scheme was selected so as to enable life extension till such a point when additional funds could be developed to build a new bridge.

To reverse the effects of deterioration and strengthen the bridge, a rehabilitation scheme using fiber reinforced polymer (FRP) composites was developed, based on detailed analysis for punching shear, and a pattern of external FRP reinforcement in both the longitudinal and transverse directions of deck soffits was installed by a contractor (Sikorsky *et al.*, 2001). Examples of the use of both impregnated unidirectional fabric and adhesively bonded pultruded strips are shown in Figures 8.4a and 8.4b, respectively. Dynamic vibrational tests were performed on the bridge to determine the potential changes in the stiffness properties of the bridge, as well as evaluate the impact of those changes on the strength of the bridge. The change in stiffness properties were evaluated for each span of the bridge, as well as for the entire bridge.

In order to enable use of SHM, a finite element (FE) model for the bridge was developed to create a baseline model of the bridge. This enabled the use of theoretical modal analysis to determine the theoretical eigenfrequencies and eigenvectors, as well to optimize the instrumentation layout for the modal test. The schematic of the FE model is shown in Fig. 8.5. Data were collected at 44 sensor locations on the bridge, using Kistler 8390A2 triaxial accelerometers for all acceleration measurements. These devices were attached directly to response points in orientations matching global cartesian response directions. Data from the accelerometers and impact hammer were acquired and processed on a 16-channel SigLab 20–42 DSP analyzer manufactured by DSP Technologies. Time data were transferred to a laptop computer for further analysis. The modal parameters were extracted from the time histories using ME'Scope, a commercially available modal analysis software package. In addition, data were collected at deck locations above the piers to help identify mode shapes.

8.4 (a) Example of strengthening through the use of field impregnated (wet layup) unidirectional fabric. (b) Example of strengthening through the use of adhesively bonded pultruded strips.

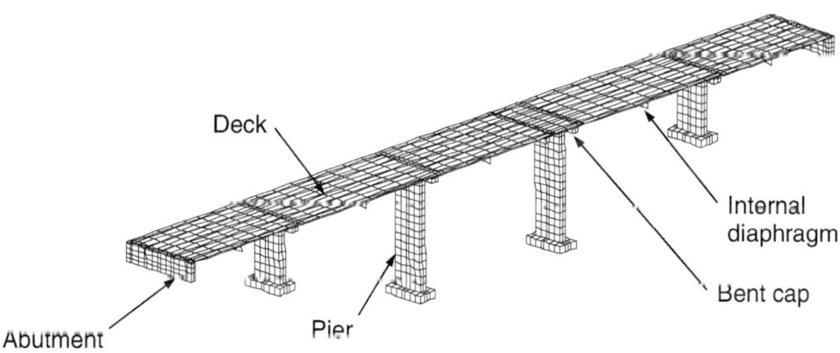

8.5 Schematic of the finite element model.

Modal tests were performed at annual intervals. The experimental frequencies and mode shapes from the first test (conducted prior to rehabilitation), and from tests conducted annually after the rehabilitation are summarized in Table 8.1 and Figures 8.6a–8.6c. The effective moduli of the entire bridge superstructure over this period of time, using systems identification are listed in Table 8.2. The trend showed an initial relative increase in stiffness α of the deck due to the FRP rehabilitation of up to 7.6%,

Table 8.1 Measured frequencies for the structure (in Hz)

Mode	Prior to rehabilitation	After rehabilitation	
		Year 1	Year 2
1	3.950	4.072	4.197
2	4.428	4.501	4.445
3	6.007	6.249	6.157
4	6.788	6.985	6.273

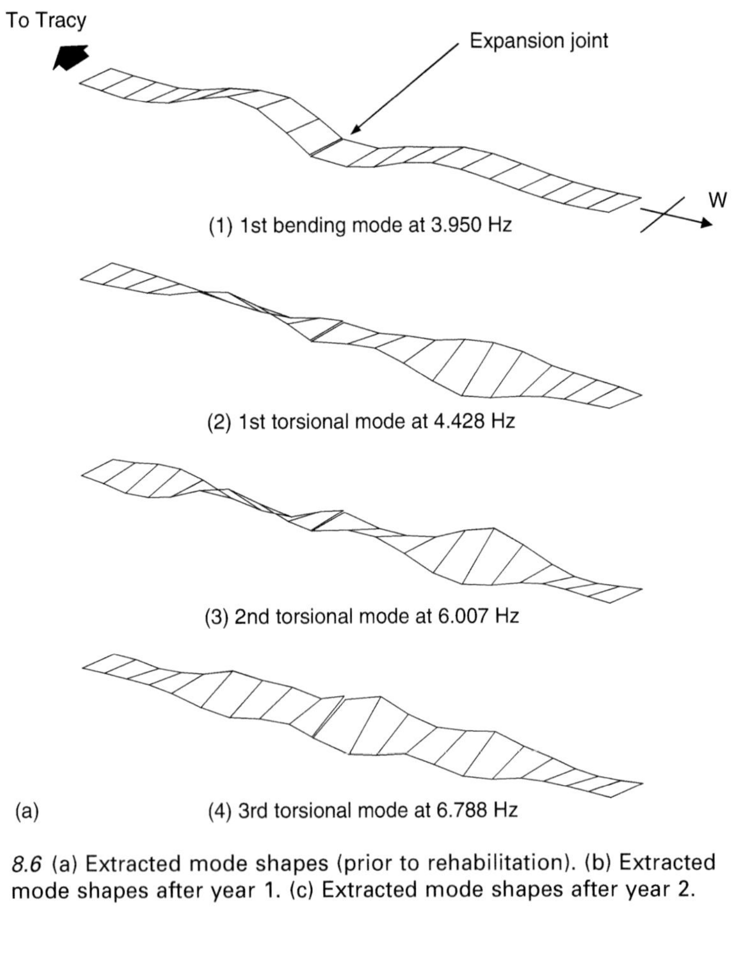

8.6 (a) Extracted mode shapes (prior to rehabilitation). (b) Extracted mode shapes after year 1. (c) Extracted mode shapes after year 2.

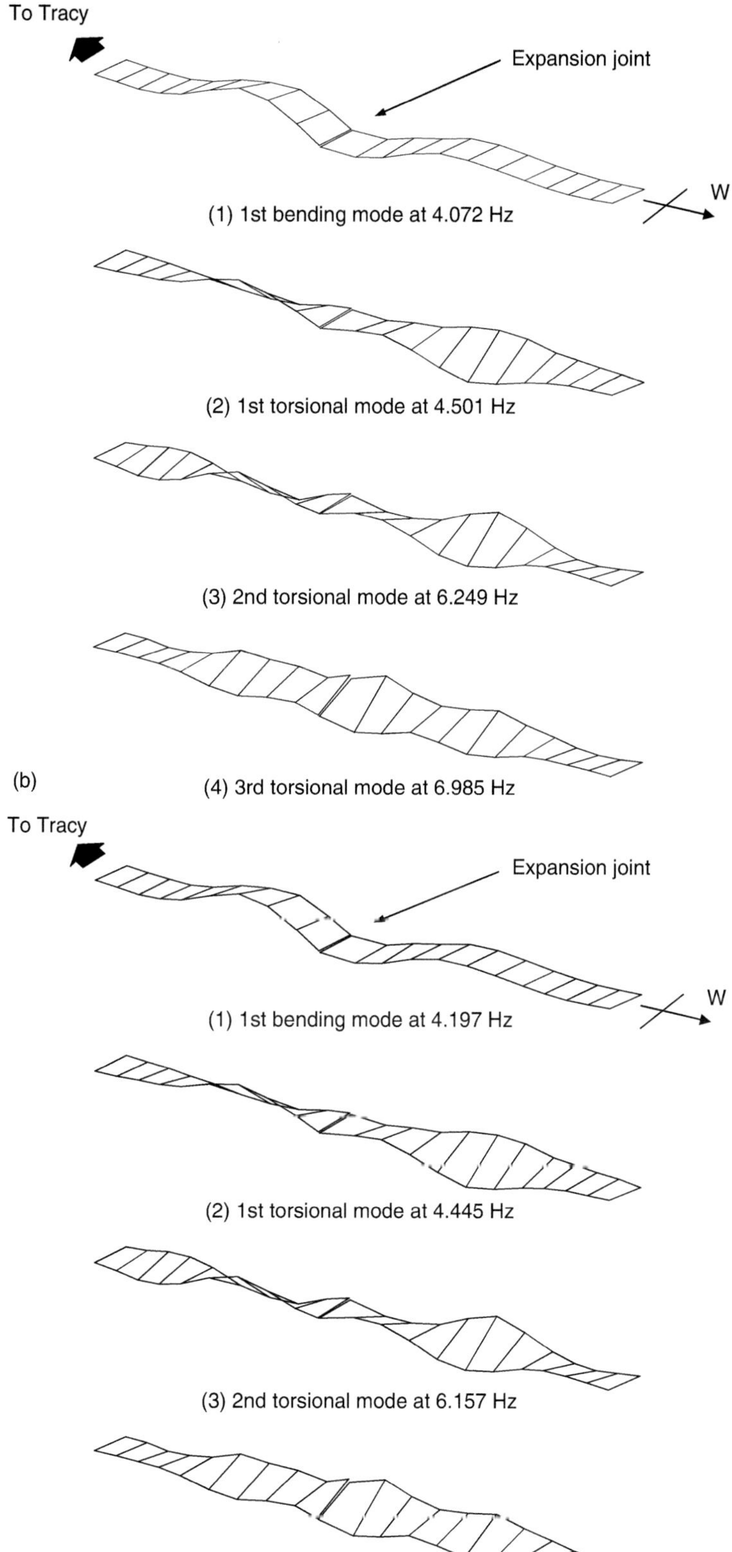

8.6 Continued

Table 8.2 Results using successive system identification

	Prior to rehabilitation	Year 1	Year 2
Effective modulus of superstructure GPa (ksi)	9.2 (1333)	9.9 (1433)	9.5 (1379)
Percent change from initial	–	+7.6	+3.3

Table 8.3 Progression of damage and demand/capacity ratios for slab

	Damage index		*D*/*C* ratio	
Span	Prior to rehab	Year 2 after rehab	Prior to rehab	Year 1 after rehab
as-built	0.000	0.000	0.96	0.96
1	0.040	0.068	1.00	0.83
2	0.241	0.406	1.27	1.29
3	0.118	0.199	1.09	0.96
4	0.162	0.273	1.15	1.06
5	0.032	0.054	1.00	0.81

followed by a decrease in the next year due to continuing damage, albeit at a much slower rate than before, allowing for a continued positive effect of the rehabilitation in terms of service life extension. A summary of changes in α and slab demand/capacity, D/C, ratios over this period is shown in Table 8.3.

Based on a review of the results, several statements can be made regarding the condition of the bridge and its ability to support legal loads (HS-20). First, the experimental frequencies increased as a result of the use of externally bonded FRP, which indicates a relative increase in stiffness. Based on the results in Table 8.3, it can be seen that the deck slab is able to support legal loads, except in Span #2. However, SHM showed that, after the rehabilitation, the weak link in the system continued to be the girders wherein the demand/capacity (D/C) ratios exceeded 1.0 for Spans 2, 3 and 4. It should be noted that while frequency data-based analysis indicated an increase in stiffness after the rehabilitation was completed, the results of the damage evaluation using the mode shapes indicated that damage was continuing, albeit at a significantly reduced rate in all five spans. Figure 8.7 shows some of the damage as related to the FRP and adjacent areas and an overview of damage locations in shown in Figure 8.8.

Use of both frequency and mode shape-based data enabled a comprehensive assessment of damage progression and, in this case, enabled the

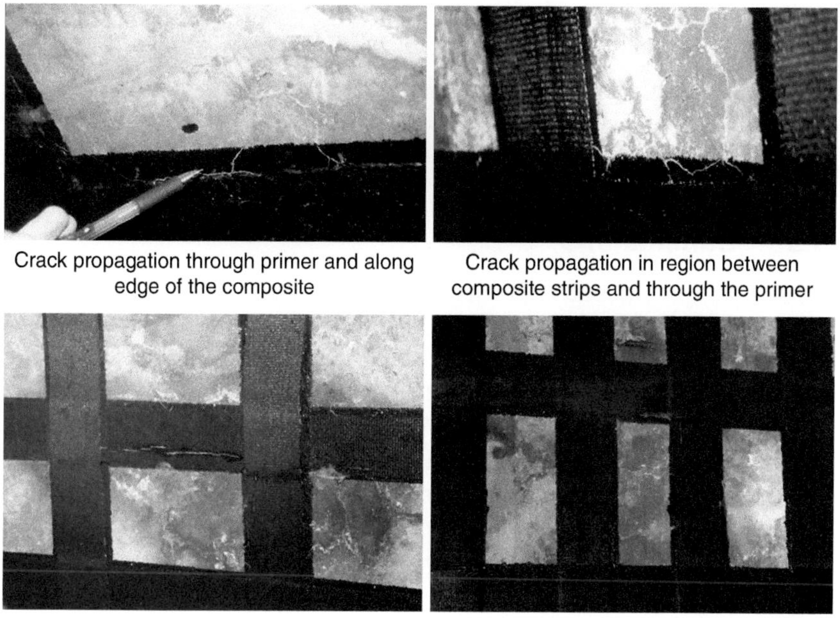

Crack propagation through primer and along edge of the composite

Crack propagation in region between composite strips and through the primer

Split in composite

Occurrence of damp patches on soffit

8.7 Damage in the FRP and adjacent areas over time.

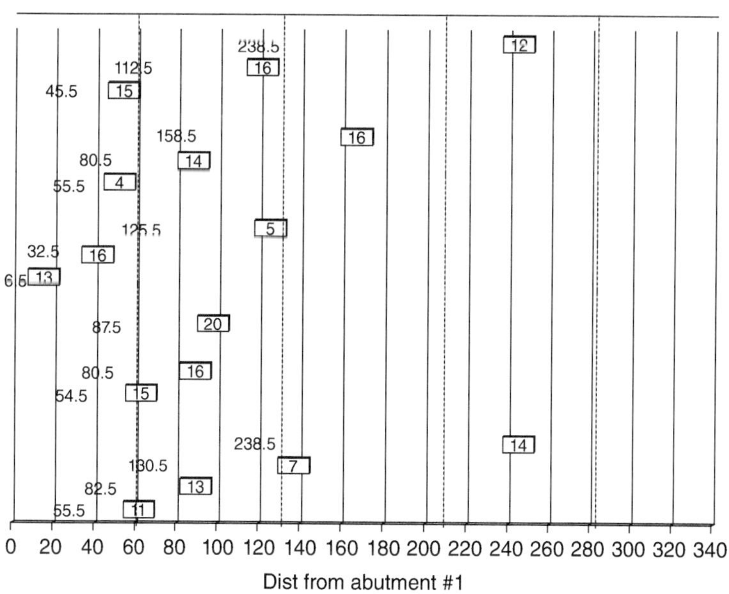

8.8 Overall identification of damage locations after rehabilitation.

engineers to provide for adequate service life extension at minimal cost. The use of SHM was shown to allow for a lower level of rehabilitation for the period of life extension required, emphasizing its use not just in assessment of need but also in field validation and assessment of real-time cost–benefit analysis to allow for managing risk.

8.5 Conclusions

While SHM provides an important diagnostic tool, it should be emphasized that, if used appropriately, it can also provide a means of assessing the level of rehabilitation needed for specific periods of extension of service life and can also be used not only to validate rehabilitation but also to follow the changes in the structure over time. When combined with analytic and computational tools such as FE analysis, field-based data collection through an appropriately designed sensor system can provide a crucial aspect of prognosis as well. Efficient prognostic tools, when integrated with appropriately designed sensor-based monitoring systems and diagnostic output, can thus be used to enable real-time estimates of damage, residual capacity and remaining service life of civil structures. This will provide the owner with sufficient information not only to assess the level and time for maintenance actions but also allow for real-time planning for replacement of structures. It should be noted that there are differences between mere health monitoring systems and management systems, in that the latter use monitoring data to detect damage and then evaluate how the damage/deterioration affects the overall response and use of the structure. A true SHM system thus must contain five aspects (i) monitoring of appropriate structural response characteristics, (ii) mapping of damage areas of the original state through FE analysis (or other appropriate tools) to assess the state of health of the structure, (iii) identification of specific mechanisms of damage causing the changes in response over time, (iv) development of hypothetical scenarios to assess how these mechanisms could continue deterioration, thereby enabling assessment of prognostic parameters related to remaining capacity, service life and risk based on preset thresholds for performance in much the same way as damage tolerance methodologies are used in the composites-based aerospace sector, and (v) decision tools to enable real-time decision analysis based on risk and economic criteria.

8.6 References

CALTRANS (2000), *Bridge Design Specifications*. LFD Version, State of California, Department of Transportation, Sacramento, CA.

FHWA (2001), *Reliability of Visual Inspection for Highway Bridges*, Volume I: Final Report, US Department of Transportation, Federal Highway Administration.

HOUSNER, G.W., BERGMAN, L.A., CAUGHEY, T.K., CHASSIAKOS, A.G., CLAUS, R.O., MASRI, S.F., SKELTON, R.E., SOONG, T.T., SPENCER, B.F., and YAO, J.T.P. (1997), 'Structural control: Past, present, and future,' *ASCE Journal of Engineering Mechanics*, 123(9), 897–971.

KACHANOV, L.M. (1986), *Introduction to Continuum Damage Mechanics*, Martinus Nijhoff Publishers, Dordrecht, The Netherlands.

KARBHARI, V.M. (2009), 'Introduction: Structural Health Monitoring – A Means to Optimal Design in the Future,' in *Structural Health Monitoring of Civil Infrastructure Systems*, V. M. Karbhari and F. Ansari, CRC/Woodhead Publishing Ltd, Oxford.

KARBHARI, V.M. and ANSARI, F. (2009), *Structural Health Monitoring of Civil Infrastructure Systems*, CRC/Woodhead Publishing Ltd, Oxford.

KIM, B.-H., STUBBS, N., and SIKORSKY, C. (2002), 'Local Damage Detection Using Incomplete Modal Data.' *20th IMAC*, 435–441.

PARK, S., STUBBS, N. and SIKORSKY, C.S. (1997), 'Linkage of nondestructive damage evaluation to structural system reliability', in *Proceedings of Smart Materials and Structures 1997*, SPIE, Bellingham, WA, pp. 234–245.

RYTTER, A. (1993), *Vibration Based Inspection of Civil Engineering Structures*, Ph.D. Thesis, University of Aalborg, Denmark.

SIKORSKY, S., STUBBS, N., KARBHARI , V. and SEIBLE, F. (2001), 'Capacity Assessment of a Bridge Rehabilitated Using FRP Composites,' *Proceedings of the 5th International Conference on Fibre-Reinforced Plastics for Reinforced Concrete Structures, FRPRCS-5*, Volume 1, Cambridge, July 2001, 137–146.

STUBBS, N. and J.T. KIM (1996), 'Damage localization in structures without baseline modal parameters,' *AIAA Journal*, 34, 1644–1649.

STUBBS, N. and R. OSEGUEDA (1990), 'Global Non-destructive Damage Evaluation in Solids.' *The International Journal of Analytical and Experimental Modal Analysis*, 5 (2), 67–79.

STUBBS, N., PARK, S., SIKORSKY C. and CHOI, S. (2000), 'A Global Non-destructive Damage Assessment Methodology for Civil Engineering Structures.' *International Journal of Systems Science*, 31 (11), 1361–1372.

9

Databases and knowledge-based systems for service life estimation of fiber reinforced polymer (FRP) rehabilitated civil engineering structures

V. M. KARBHARI, University of Alabama in Huntsville, USA

Abstract: While a substantial amount of research has been conducted using FRP materials, one of the biggest stumbling blocks to its increased usage is the lack of readily available information related to material characteristics, durability in various environments, successful applications, and for the purposes of design. This chapter summarizes thoughts outlining the need for the development of an integrated knowledge system or 'virtual designer' that would simultaneously serve as a centralized repository for information sets and would also enable data and knowledge to be accessed in a manner amenable to designers. Beyond the immediate use of such a system in design, it is also possible that it would serve as a means for further science-based development of materials and structural concepts for civil infrastructure applications, and therefore serve as a catalyst for the increased use of FRP materials.

Key words: FRP, composites, databases, knowledge base, durability, design.

9.1 Introduction

Fiber reinforced polymer (FRP) composites are increasingly being considered for use in a variety of applications pertaining to civil infrastructure. The attractiveness of this class of materials derives from its light weight, tailorable performance characteristics, corrosion resistance and potentially high durability. Over the past decade, these materials have been used for seismic retrofit of columns, externally bonded rehabilitation of deteriorating and deficient concrete slabs and girders, and even in the form of all-composite bridge decks, and rapidly constructible building structures. In a number of cases, the enhanced environmental durability of FRP materials has resulted in the implementation of solution strategies not possible with conventional materials. While a substantial amount of research has been conducted using FRP materials, one of the biggest stumbling blocks to their increased usage is the lack of readily available information for design. Unlike traditional construction materials such as steel and concrete, the use of FRP is still new and hence there is a lack of information related to material characteristics, durability in various environments, successful applications, and designs. In recent years, a number of researchers have reported results pertaining to the characterization of various forms of FRP composites used in civil infrastructure, others have reported on results of short- and long-term durability testing under a range of exposure conditions, others have reported on tests conducted at the level of components and systems, and still others have reported on the development of models. However, each of these has, for the most part, been conducted in isolation, resulting in both a lack of integration among these results, and difficulty in obtaining data in a form readily useable for designers. This chapter summarizes thoughts outlining the need for the development of an integrated knowledge system, or 'virtual designer', that would simultaneously serve as a centralized repository for information sets and would also enable data and knowledge to be accessed in a manner amenable to designers. Beyond the immediate use of such a system in design, it is also possible that it would serve as a means for further science-based development of materials and structural concepts for civil infrastructure applications, and therefore serve as a catalyst for the increased use of FRP materials.

It should be noted that the development of such a schema is not new and similar considerations have been raised in the past for conventional materials. Frohnsdorf and Skalny (1983) first discussed the use of databases and models for further development of cements, and an overview of knowledge bases and models was provided in 1980 by Frohnsdorf and Clifton. Extensive databases on structural shapes are available for steel, as are models for corrosion-induced deterioration and degradation in various environments. While there is still substantial areas of research related to steel and concrete, there is no lack of data for these materials. The same cannot be said of FRP composites.

9.2 Knowledge systems

Knowledge systems represent more than merely the replacement of a paper-based system of information with one that is computer based. Rather they represent a means of enabling the integration of various sources of information to accelerate both fundamental science-based and applied engineering-based development of technology. For the purposes of this chapter, a knowledge system is understood to include not only the information itself, but also the mechanism for its storage, retrieval, processing. transfer, and in the case of simulations, for the generation of additional knowledge. The system thus has multiple components, any, or all, of which can be accessed by the user. Following Rasdorf (1985), knowledge can be classified into four major types, (i) design facts, (ii) procedures, which

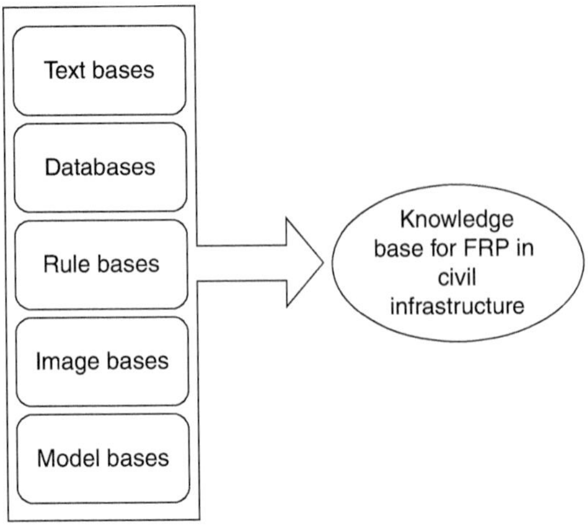

9.1 Classification of knowledge.

includes algorithms and simulations, (iii) judgments, consisting of heuristical knowledge and that derived on the basis of experience, and (iv) meta-knowledge. However, if one is to consider the type of information that could be stored electronically, knowledge can be further differentiated, as shown in Fig. 9.1, into (i) text bases, (ii) databases, (iii) rule bases, (iv) image bases, and (v) models, or model bases. Each of these is briefly discussed below.

Text bases provide a means for storage of information in narrative form. Thus this class includes archival papers and reports, and essentially is the same as a common database used for searching within collections of archival literature. While this form of data storage is crucial for purposes of education, the extraction of data relevant to design is unwieldy unless the search object is exactly the same as one described in the literature. Within this set, knowledge is not provided in a uniform context. It may be available as a combination of textural data, numerical information, equations, and images. However, it serves as an important source of information regarding prior work and provides a description of procedures, protocols and methodology. There are a number of journals dedicated to the science and application of FRP composite materials, including those that emphasize the use of FRP in civil infrastructure. These provide a substantial base for the novice and expert to gain knowledge about specific aspects. While they provide a source of information to the designer and user/owner, the knowledge is primarily in the form of information that cannot be directly manipulated or used further in design through electronic means. Thus the transfer of this information to a new design is purely through the human user.

Westbrook (1985) defined *databases* as a self-describing collection of integrated files. For the purposes of the present discussion, this set represents the source for design information. Characteristics of different FRP materials would be available therein, as would the effect of various environmental conditions. Since a large amount of data could be stored, similar sets can be correlated, allowing for better statistical descriptions of data. It is noted that the structure of this database would have to be carefully developed, as would the criteria for validation of data prior to their inclusion. Since information from this class of knowledge base will be used directly in design, it is essential that the data is evaluated prior to its inclusion, and that the user is given an indication of the level of reliability of the information. This aspect is, perhaps, the most difficult, since the discrepancy in results could be due to the use of varying material sets, changes in test protocols, or even just in the manner of reporting data. For example, various researchers have reported conflicting results pertaining to the effects of freeze–thaw cycling on the bond between FRP composites and concrete, with some suggesting that this is perhaps the most deleterious environment and others suggesting that it has no effect whatsoever. It is crucial that the user be aware of the contradictions in data available in such cases.

An extensive database of material characteristics related to FRP composites used in the aerospace/defense sectors already exists in the form of MIL-HDBK-17, Vol. 2 (2002). While this provides useful information, the materials are generically not of the same form nor are processed in similar fashion to those used in civil infrastructure applications. However, various data sets related to FRP used in civil infrastructure drawn from research conducted at universities and research institutes already exist although they have not been synthesized and integrated into a common database. An example of these data is shown in Fig. 9.2. The development of a validated set of data following pre-defined protocols for testing, statistical analyses, and exposure would enable diverse sets of data from different laboratories to be collected in a centralized database which could be made available to all users. This would, in effect, accelerate the use of FRP in civil infrastructure through both availability of data and through the creation of a confidence level in the validity of data.

Rule bases essentially consist of sets of heuristic knowledge gleaned from experts and hence they complement the information provided in databases. This form of knowledge has been shown to be extremely valuable in the development of expert systems for medical diagnosis since it allows for the use of 'if-then-else' type scenarios. One of the benefits of this type of knowledge base is that, if structured appropriately, it can serve as a guide for routine design, as well as for the rapid selection (or deselection) of concepts or materials. This, then, serves as both a design tool for the novice and as a means of instruction. Further, it provides a means of capturing the

Test Period Months	Environment	Sample No.	Thickness (mm)	Width (mm)	Tensile Modulus (GPa)	Tensile Strength (MPa)
0	DI Water, 22.7°C	1	1.26	12.79	130.45	2769.23
		2	1.26	12.72	144.12	2861.06
		3	1.27	12.81	134.96	2752.88
		4				
		5				
		Average				
		St. Dev.				
		COV				
2	DI Water, 22.7°C	36				
		37				
		38				
		39				
		40				
		Average				
		St. Dev.				
		COV				
6	DI Water, 22.7°C	44				
		45				
		46				
		47				
		48				
		Average				
		St. Dev.				
		COV				
12	DI Water, 22.7°C	52				
		53				
		54				
		55				
		56				
		Average				
		St. Dev.				
		COV				
18	DI Water, 22.7°C	60				
		61				
		62				
		63				
		64				
		Average				
		St. Dev.				
		COV				
24	DI Water, 22.7°C	68				
		69				
		70				
		71				
		72				
		Average				
		St. Dev.				
		COV				

Test Period Months	Environment	Sample No.	Thickness (mm)	Width (mm)	Tensile Modulus (GPa)	Tensile Strength (MPa)
0	DI Water 37.8°C	1	1.26	12.79	130.45	2769.23
		2	1.26	12.72	144.12	2861.06
		3	1.27	12.81	134.96	2752.88
		4	1.25	12.74	155.95	2603.10
		5	1.27	12.70	138.47	2851.60
		Average	**1.26**	**12.75**	**140.79**	**2767.57**
		St. Dev.	0.01	0.05	9.84	103.77
		COV	0.01	0.00	0.07	0.04
2	DI Water 37.8°C	84	1.28	13.09	140.25	2742.16
		85	1.27	13.00	141.80	-
		86	1.26	12.80	139.80	2686.33
		87	1.25	12.94	144.70	2947.18
		88	1.26	12.86	139.46	2800.20
		Average	**1.27**	**12.94**	**141.20**	**2793.97**
		St. Dev.	0.01	0.11	2.15	112.22
		COV	0.01	0.01	0.02	0.04
6	DI Water 37.8°C	92	1.26	13.09	-	-
		93	1.25	13.00	-	-
		94	1.29	13.07	-	-
		95	1.28	13.19	-	-
		96	1.26	12.91	-	-
		Average	**1.27**	**13.05**	-	-
		St. Dev.	0.01	0.10	-	-
		COV	0.01	0.01	-	-
12	DI Water 37.8°C	100	1.28	13.01	134.91	2662.97
		101	1.27	13.07	139.91	2747.73
		102	1.26	12.72	137.00	2507.02
		103	1.27	12.75	131.63	2777.10
		104	1.28	13.10	134.57	2293.55
		Average	**1.27**	**12.93**	**135.60**	**2597.67**
		St. Dev.	0.01	0.18	3.08	199.80
		COV	0.01	0.01	0.02	0.08
18	DI Water 37.8°C	108	1.27	13.18	134.40	-
		109	1.27	13.01	126.72	-
		110	1.28	13.08	141.02	2469.62
		111	1.27	12.43	131.48	2687.44
		112	1.26	12.89	135.89	2569.45
		Average	**1.27**	**12.92**	**133.90**	**2575.50**
		St. Dev.	0.01	0.29	5.30	109.04
		COV	0.00	0.02	0.04	0.04
24	DI Water 37.8°C	116	1.26	12.81	127.42	2653.78
		117	1.26	12.92	-	-
		118	1.26	12.87	126.19	2031.27
		119	1.27	13.17	136.01	2410.93
		120	1.27	12.95	146.01	2686.55
		Average	**1.26**	**12.94**	**133.91**	**2445.63**
		St. Dev.	0.00	0.14	9.17	302.36
		COV	0.00	0.01	0.07	0.12

Notes: "-" indicates no data obtained for that mechanical property.
Control signifies 22.7°C Air at 30% Relative Humidity.

9.2 Example of database for material characteristics of prefabricated strips used in rehabilitation after environmental exposure.

knowledge of 'subject experts' built through years of experience, which often is lost in the transition between personnel. An example of this is in the deselection of materials for use in specific environments wherein the application of simple rules would essentially result in the deselection of a large number of material choices based on the expected type and severity of exposure, as well as the anticipated service life. Another form of rule base is in the development of heuristic knowledge that is often used in design guidelines. In many cases the guidelines are based on consensus within a committee and do not necessarily relate to specific data sets but rather relate to overall factors prescribed or recommended for use, irrespective of the differences between various materials available. An example of this is provided in Table 9.1.

Table 9.1 Reduction factors recommended for use in ACI-440.2R-08 (2008)

Environmental reduction factor for various FRP systems and exposure conditions

Exposure conditions	Fiber type	Environmental reduction factor C_E
Interior exposure	Carbon	0.95
	Glass	0.75
	Aramid	0.85
Exterior exposure (bridges, piers, and unenclosed parking garages)	Carbon	0.85
	Glass	0.65
	Aramid	0.75
Aggressive environment (chemical plants and wastewater treatment plants)	Carbon	0.85
	Glass	0.50
	Aramid	0.70

Sustained plus cyclic service load stress limits in FRP reinforcement

	Fiber type		
Stress type	GFRP	AFRP	CFRP
Sustained plus cyclic stress limit	$0.20f_{fu}$	$0.30f_{fu}$	$0.55f_{fu}$

Recommended additional reduction factors for FRP shear reinforcement

$\psi_f = 0.95$	Completely wrapped members
$\psi_f = 0.85$	Three-side and two-opposite-sides schemes

Advances in image processing and data storage have made it fairly easy and cost-efficient to store large amounts of visual data in the form of *image bases*. From a design perspective, it is important to be able to view previous designs, steps in the construction process, and even aspects related to failure. Knowledge regarding characteristics of materials as provided in databases, while important, does not often provide the complete picture without inclusion of aspects related to failure. The ability to store micrographs, and images of failure mechanisms (Fig. 9.3, for example), greatly assists in the provision of essential information to the user. It is, however, essential that image bases are not constructed in isolation but that their contents are directly linked to pertinent portions of other knowledge bases. For example, while an image base may be able to highlight the difference in failure mode of two E-glass/epoxy systems used in seismic retrofit depending on the use of glass or aramid transverse threads (Zhang *et al.*, 2001), or even due to minor changes in sizing on the fibers (Zhang *et al.*, 2003), the essential

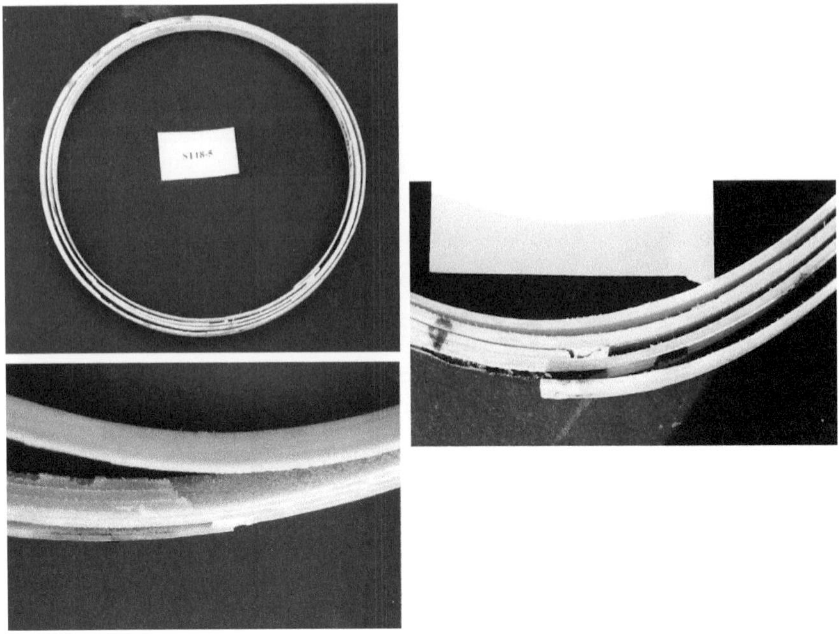

9.3 Failure in adhesively bonded FRP jackets subjected to freeze–thaw.

difference is not clear unless it is linked to actual data (in the corresponding database) showing the effect of deterioration in properties.

The content of *model bases* are mathematical models and simulations. In the context of this application, these could be models describing the deterioration of materials under specific environmental conditions, or models pertaining to the design/analysis of entire structures. Our knowledge at both the materials and structural levels has enabled the construction of sets of equations to represent the response of complex systems. This aspect of the knowledge base is perhaps the most powerful and pervasive since it enables the development of contextual knowledge from an integration of the mathematical representation of knowledge over a range of possible scales. One could essentially think of this as being the ultimate repository and generator of information. Simulation models have specific advantages since they allow for the investigation of various alternatives, as well as the effect of variation over a range of variables. In its final form, the model base would make it possible to investigate the implications of choice of constituent materials, configuration and processing method, not just on the durability of the resulting material but also on the reliability and structural integrity of the component and/or system formed from those materials.

9.3 The fiber reinforced polymer (FRP) virtual knowledge engine

The use of FRP composites in civil infrastructure necessitates the consideration not only of materials level information but also information at the systems level, especially related to interactions with conventional materials such as concrete and steel. Further, since the advantages of these materials include the potential of prefabricating components for entire systems that can be rapidly assembled, aspects related to assembly and construction have to be considered as well. Thus, any knowledge-based system being developed for such an end application has to necessarily consider aspects ranging from the level of material constituents (fiber and resin) to those of assembly and construction.

The virtual engine conceptually provides the nexus for web-based synthesis and sharing of experiments, simulations, data and heuristics, as well as for focused test beds. It thus serves as more than just a linear knowledge-based tool, providing the kernel for science-based advances in the multidisciplinary progress of materials and structural level advances. In essence, it serves to accelerate the development and insertion of FRP composites in intended applications. The overall structure is shown schematically in Fig. 9.4 and the integration is shown in Fig. 9.5. Based on the specific needs of the application, the overall structure can be thought of as being anchored in the system's design component, which includes the definition of the design envelope and of system functionality. A potential user would thus use this core to define and refine detailed requirements for an application. Once this is completed, the separate structural constructs (each a knowledge base of its own) related to materials, sensing, sustainability, and logistics (which would include aspects of transportation and construction) can be brought to bear on the problem at hand. As shown schematically in Fig. 9.5, this would enable the systems level needs to be defined at one level and then acted upon, using resource tools as required.

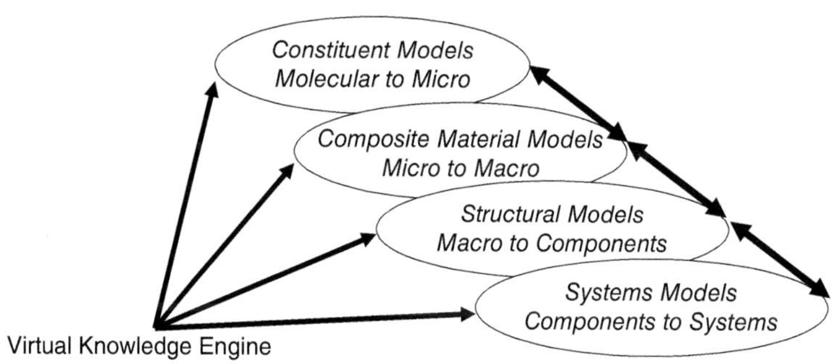

9.4 Schematic showing overall purview of the knowledge engine.

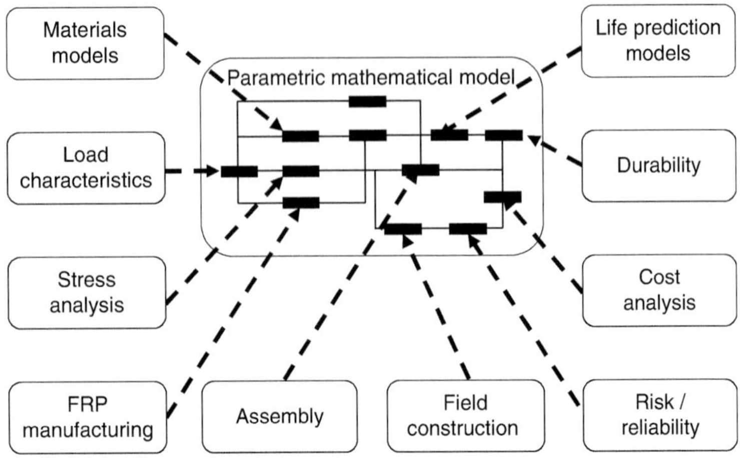

9.5 Integration of models in the simulation tool.

9.4 Application to the durability-based design of FRP strengthened systems

While many components of the knowledge engine exist, they have not been integrated into a single system. However, it is possible to show the application of such a scheme in the durability-based design of strengthening systems for girders following the structure presented by Wilcox (2008). This section reports on the results of an investigation aimed at the consideration of statistical variation in the response of prefabricated FRP pultruded strips used in infrastructure rehabilitation within a load and resistance factor design (LRFD) framework. The explicit incorporation of effects of aging and continued deterioration of the concrete structure (such as through corrosion) are included through databases and models.

Data related to durability of three prefabricated unidirectional carbon/epoxy FRP strips and the manufacturer-recommended high solids content epoxy adhesives, denoted as material systems A, B, and C, respectively, were obtained through extensive testing. Equations based on an Arrhenius type model were developed to predict change in performance with time, based on time–temperature superposition principles wherein performance at time t (based on immersion in water at 23°C) is predicted in terms of the percentage retention of initial characteristics as

$$P(t) = \frac{P_o}{100}[100 - A \cdot \ln(t)] \qquad [9.1]$$

where P_o is the original (or unexposed/control) value of a mechanical property, t is the time at which the predicted value is queried beyond the unex-

Table 9.2 Values of degradation constant, *A*, for the materials considered

	FRP strip				Adhesive	
Material system	Tensile strength	Tensile modulus	Flexural strength	SBS strength	Tensile strength	Tensile modulus
A	0.63	0.69	0.42	2.64	4.64	6.67
B	2.52	0.78	4.89	5.63	5.89	4.44
C	3.66	0.92			0.81	0.01

posed ($t = 0$) state, and A is a material property-based constant. Values of A are given in Table 9.2 for the prefabricated strips and the corresponding adhesives. The constant A determines the behavior of the degradation trend line and therefore controls the property retention values determined by each equation. Examination of this constant affords a statistic for comparison of the degradation trends within each material property and between properties as well. Large variation is identified within the material properties for tensile strength, flexural strength and short beam shear strength (SBS) in the prefabricated strips, and in both the tensile strength and tensile modulus for the adhesives. Table 9.2 could be considered part of the knowledge base for the materials and a review indicates that variability of results precludes any realistic combination of systems or creation of an overlaying case which prescribes the behavior of all three material systems. Table 9.2 also shows that there is a need to consider degradation in modulus, in contrast to its neglect in the ACI guidelines (ACI 440, 2008). Especially in the case of adhesives, the tensile modulus is not immune to degradation. This emphasizes the difference between models and rule bases described in the previous section. Factors for degradation in salt and alkali (concrete leachate) environments were determined following the procedure outlined by Abanilla *et al.* (2006) through use of a modification factor to be applied as a product of the degradation constant, A, in Equation 9.1, and are given in Table 9.3. It is important to note that the use of Equation 9.1, with factors from Tables 9.2 and 9.3, enables the designer to anticipate deterioration based on intended period of service rather than through a single arbitrary number that is used as a factor irrespective of intended service life.

To illustrate the difference between designing with differently determined characteristic material properties, examples of rehabilitation were considered using both ACI 440 (2008) as a design guide and using degradation-based material properties substituted wherever possible, thereby providing a comparison of efficacy between model-based design and rule-based design. The capacity of the sample girders was varied in a number of

Table 9.3 Modification factors for other environments

	Salt water		Alkali	
Material system	Tensile strength	Tensile modulus	Tensile strength	Tensile modulus
A	0.99	0.97	0.93	0.96
B	0.90	0.97	0.82	0.98
C	0.67	0.96	0.84	0.93

Table 9.4 Girder details

	Girder #5	Girder #20
Span length (m)	16.76	22.86
Girder spacing (m)	2.33	2.59
Top slab depth (cm)	17.78	16.82
T-beam web width (cm)	30.48	40.64
T-beam web depth (cm)	106.68	152.4
Effective flange width (cm)	223.42	242.57
Reinforcing steel (#-size)	6- #36 bars	9- #36 bars
Streel strength (MPa)	275	275
Bottom cover (cm)	5.08	5.08
Concrete strength (MPa)	22.4	22.4

ways to show different strengthening situations. The original capacity of the as-designed girders was calculated as a control, giving a target moment capacity for the differing strengthening cases. To show the effects of environmental aging on reinforced concrete structures, the amount of reinforcing steel was decreased and then FRP was used for strengthening as appropriate. The FRP is also considered to degrade over time in some cases, using the degradation equations based on factors in Table 9.2 to provide comprehensive insight into the behavior at the systems level and to enable the owner to conduct a risk and economic analysis for the decision regarding amount of rehabilitation to be conducted.

Two typical girders from a set studied previously for development of a LRFD framework for rehabilitation (Atadero, 2006) were used, with details as in Table 9.4. Strengthening of the example girders was carried out according to the procedure prescribed in ACI 440 (2008), using a sectional analysis. Two cases of steel deterioration were considered in this study. In the first case, a specific amount of loss in steel cross-section is assumed, requiring rehabilitation using externally bonded FRP, following which no further corrosion is assumed to occur. In the second case, further corrosion of steel is included after rehabilitation, resulting in continuing loss of capacity over time. Cases of incremental steel loss are used as a function of loss of capacity at the level of 10, 20, and 30%. The consideration of both cases is important to enable investigation of response over time, since use of FRP does not necessarily stop corrosive activity within the reinforced concrete girder. The model used to classify the rate at which corrosion of steel occurs was proposed by Vu and Stewart (2000).

Two design formats were considered in this study, (i) using ACI 440 (2008) recommended FRP properties, and (ii) using characteristics that incorporate time-based deterioration. A typical sectional analysis is conducted and the required number of FRP strips is determined analytically. Then, a comprehensive reliability analysis is conducted for each girder, accounting for variation of selected parameters in the girders (including steel strength and girder size) to determine a reliability index, β, through which a comparison of strengthened girders can be made. The reliability index provides a comparison statistic that is independent of the varying strip statistical properties discussed previously relating to manufacturer strip design and quality control, and uses relevant statistical distributions to fully examine the girders and the selected parameters. The reliability index is calculated using the procedure listed in NCHRP Report 368 (Nowak, 1999). The loads used for reliability analysis are based on the mean loads found for bridge girders of all sizes, as characterized by Nowak (1999) in the formulation of load and resistance factors for the new AASHTO LRFD Bridge Design Specifications (AASHTO, 2004; Nowak, 1999). A summary of loads used in this study after application of bias factors is listed in Table 9.5.

The procedure used to calculate β for the examples in this research is a hybrid reliability approach using a FORM method found in NCHRP Report 368 (Nowak, 1999) and Monte Carlo Simulation (MCS) as used by Plevris *et al.* (1995) – details are provided in Wilcox (2008). Since the application of FRP strips to a concrete girder does not give an immediate connection to a statistical distribution of the resistance offered by the FRP reinforced concrete, MCS was employed to bridge this gap. Using descriptions of all the variables involved in the resistance of an FRP-strengthened girder obtained from the literature and relevant design codes, a total resistance distribution for the strengthened girders can be found. The results from the tests conducted on the 'control' FRP strips are used to characterize statistical distribution of pertinent FRP properties, which are then used to simulate the strengthening of girders so that a statistical description of resistance can be found and used for reliability. Variability of properties was included in the simulations, including steel strength and bar size, concrete strength, geometric dimensions of girders and FRP thickness, strength and modulus. A large number of simulations

Table 9.5 Summary of loads used for the girder examples in reliability analysis after application of bias and other factors, following Atadero (2006). Static live load moments reflect M_{LM+IM} with application of appropriate girder distribution factors

Parameter	Girder 5	Girder 20
Dead load moments		
μ_{DL}	240.10 [177.08]	542.02 [399.77]
σ_{DL}	24.01 [17.71]	54.20 [39.98]
Wearing load moments		
μ_{W}	63.05 [46.50]	97.37 [71.82]
σ_{W}	15.76 [11.63]	24.34 [17.96]
Distribution factor	0.64	0.74
Live load bias factor	1.325	1.315
Impact factor	1.1	1.1
Static live load moments		
μ_{LL}	691.60 [510.10]	1724.35 [1271.81]
σ_{LL}	124.49 [91.82]	310.38 [228.93]
Total load moments		
μ_{Q}	994.74 [733.68]	2363.75 [1743.41]
σ_{Q}	127.76 [94.23]	316.01 [233.08]

Note: Loads/moments reported as: kN-m [kip-ft]

(two million) were used to both verify that resistance is of a Lognormal distribution (as suggested by Nowak (1999)), and to determine distribution parameters for the Lognormal distribution (mean and standard deviation).

Sample results are shown in Tables 9.6–9.8. The impact of the characteristic design value can be seen in Cases 1, 3, and 4 of each girder and material system. In general, the β for case 1 is the lowest, as the characteristic design value is penalized the most for that case. Similarly, the β for Case 2 should be the highest because it essentially uses the mean as the design value. Logically, the β for Cases 3 and 5 usually falls between the two. The results are not always this straightforward; due to the range of convergence for each case and number of pultruded strips, the trends may deviate slightly. Notice that β falls faster for the case with continuous steel degradation than for the case with continuous FRP degradation, implying that the steel degradation model used predicts more rigorous corrosion of the steel than experimental data shows for the FRP. When including FRP material degradation in the examples, the results show that more material must be applied to the girders to maintain a reliability index above the target. This could be avoided if a design life of the retrofit or strengthening is chosen before design takes place. Retrofits are not typically meant to last for 50 years, especially if steel degradation and FRP degradation continue through-

Table 9.6 Results for girder 5 and material system A

a	b	c	d	e	f	g	h	l	j	k
Overall degradation										
Girder & material ID	Case ID #	Characteristic design value	Environmental factor, CE	FRP time, if applicable, (years)	Steel time, if applicable, (years)	Steel c/s loss, if applicable, (%)	# of plys, sectional analysis	Satisfies LRFR loads	# of plys, reliability analysis	Reliability index, β
5-A	1	CE($\mu - 3\sigma$)	0.85	–	–	0	1	Yes	1	5.03
5-A	1	CE($\mu - 3\sigma$)	0.85	–	–	10	1	Yes	3	3.64
5-A	1	CE($\mu - 3\sigma$)	0.85	–	–	20	2	Yes	**5**	**2.69**
5-A	1	CE($\mu - 3\sigma$)	0.85	–	–	30	4	Yes	**5**	**1.82**
5-A	2	CE(μ)	1.00	–	–	0	1	Yes	1	5.29
5-A	2	CE(μ)	1.00	–	–	10	1	Yes	2	3.53
5-A	2	CE(μ)	1.00	–	–	20	2	Yes	**5**	**2.94**
5-A	2	CE(μ)	1.00	–	–	30	4	Yes	**5**	**2.01**
5-A	3	CE($\mu - 3\sigma$)	1.00	–	–	0	1	Yes	1	5.48
5-A	3	CE($\mu - 3\sigma$)	1.00	–	–	10	1	Yes	3	3.77
5-A	3	CE($\mu - 3\sigma$)	1.00	–	–	20	2	Yes	**5**	**2.89**
5-A	3	CE($\mu - 3\sigma$)	1.00	–	–	30	4	Yes	**5**	**1.93**
5-A	4	CE($\mu - 3\sigma$)	1.00	–	0	–	1	Yes	1	5.48
5-A	4	CE($\mu - 3\sigma$)	1.00	–	10	–	1	Yes	1	4.83
5-A	4	CE($\mu - 3\sigma$)	1.00	–	20	–	1	Yes	1	4.44
5-A	4	CE($\mu - 3\sigma$)	1.00	–	30	–	1	Yes	1	4.24
5-A	4	CE($\mu - 3\sigma$)	1.00	–	40	–	1	Yes	1	4.03
5-A	4	CE($\mu - 3\sigma$)	1.00	–	50	–	1	Yes	1	3.69
5-A	5	CE(μ)	1.00	0	–	–	1	Yes	1	5.29
5-A	5	CE(μ)	1.00	10	–	–	1	Yes	1	5.23
5-A	5	CE(μ)	1.00	20	–	–	1	Yes	1	5.10
5-A	5	CE(μ)	1.00	30	–	–	1	Yes	1	5.02
5-A	5	CE(μ)	1.00	40	–	–	1	Yes	1	4.97
5-A	5	CE(μ)	1.00	50	–	–	1	Yes	1	4.95
5-A	6	CE(μ)	1.00	0	0	–	1	Yes	1	5.29
5-A	6	CE(μ)	1.00	10	10	–	1	Yes	1	4.72
5-A	6	CE(μ)	1.00	20	20	–	1	Yes	1	4.38
5-A	6	CE(μ)	1.00	30	30	–	1	Yes	1	4.15
5-A	6	CE(μ)	1.00	40	40	–	1	Yes	1	3.97
5-A	6	CE(μ)	1.00	50	50	–	1	Yes	2	3.63

Note: BOLD refers to the maximum number of strips available and the reliability index achieved using that maximum.

Table 9.7 Results for girder 5 and material system B

a	b	c	d	e	f	g	h	l	j	k
				Overall degradation						
Girder & material ID	Case ID #	Characteristic design value	Environmental factor, CE	FRP time, if applicable, (years)	Steel time, if applicable, (years)	Steel c/s loss, if applicable, (%)	# of plys, sectional analysis	Satisfies LRFR loads	# of plys, reliability analysis	Reliability index, β
5-B	1	CE($\mu-3\sigma$)	0.85	–	–	0	1	Yes	1	5.08
5-B	1	CE($\mu-3\sigma$)	0.85	–	–	10	1	Yes	2	3.58
5-B	1	CE($\mu-3\sigma$)	0.85	–	–	20	2	Yes	**5**	**2.90**
5-B	1	CE($\mu-3\sigma$)	0.85	–	–	30	4	Yes	**5**	**2.04**
5-B	2	CE(μ)	1.00	–	–	0	1	Yes	1	5.10
5-B	2	CE(μ)	1.00	–	–	10	1	Yes	3	3.64
5-B	2	CE(μ)	1.00	–	–	20	2	Yes	**5**	**3.07**
5-B	2	CE(μ)	1.00	–	–	30	4	Yes	**5**	**2.16**
5-B	3	CE($\mu-3\sigma$)	1.00	–	–	0	1	Yes	1	5.09
5-B	3	CE($\mu-3\sigma$)	1.00	–	–	10	1	Yes	2	3.56
5-B	3	CE($\mu-3\sigma$)	1.00	–	–	20	2	Yes	**5**	**2.98**
5-B	3	CE($\mu-3\sigma$)	1.00	–	–	30	4	Yes	**5**	**2.11**
5-B	4	CE($\mu-3\sigma$)	1.00	–	0	–	1	Yes	1	5.09
5-B	4	CE($\mu-3\sigma$)	1.00	–	10	–	1	Yes	1	4.40
5-B	4	CE($\mu-3\sigma$)	1.00	–	20	–	1	Yes	1	4.39
5-B	4	CE($\mu-3\sigma$)	1.00	–	30	–	1	Yes	1	4.18
5-B	4	CE($\mu-3\sigma$)	1.00	–	40	–	1	Yes	1	3.76
5-B	4	CE($\mu-3\sigma$)	1.00	–	50	–	1	Yes	1	3.69
5-B	5	CE(μ)	1.00	0	–	–	1	Yes	1	5.10
5-B	5	CE(μ)	1.00	10	–	–	1	Yes	1	5.04
5-B	5	CE(μ)	1.00	20	–	–	1	Yes	1	4.99
5-B	5	CE(μ)	1.00	30	–	–	1	Yes	1	4.96
5-B	5	CE(μ)	1.00	40	–	–	1	Yes	1	4.92
5-B	5	CE(μ)	1.00	50	–	–	1	Yes	1	4.89
5-B	6	CE(μ)	1.00	0	0	–	1	Yes	1	5.10
5-B	6	CE(μ)	1.00	10	10	–	1	Yes	1	4.26
5-B	6	CE(μ)	1.00	20	20	–	1	Yes	1	4.07
5-B	6	CE(μ)	1.00	30	30	–	1	Yes	1	3.86
5-B	6	CE(μ)	1.00	40	40	–	1	Yes	1	3.53
5-B	6	CE(μ)	1.00	50	50	–	1	Yes	2	3.88

Note: BOLD refers to the maximum number of strips available and the reliability index achieved using that maximum.

Table 9.8 Results for girder 20 and material system C

a	b	c	d	e	f	g	h	l	j	k
					Overall degradation					
Girder & material ID	Case ID #	Characteristic design value	Environmental factor, CE	FRP time, if applicable (years)	Steel time, if applicable, (years)	Steel c/s loss, if applicable, (%)	# of plys, sectional analysis	Satisfies LRFR loads	# of plys, reliability analysis	Reliability index, β
20-C	1	CE($\mu - 3\sigma$)	0.35	–	–	0	1	Yes	1	4.17
20-C	1	CE($\mu - 3\sigma$)	0.35	–	–	10	1	Yes	7	3.76
20-C	1	CE($\mu - 3\sigma$)	0.35	–	–	20	4	Yes	**7**	**1.87**
20-C	1	CE($\mu - 3\sigma$)	0.35	–	–	30	**7**	**No**	**7**	**1.06**
20-C	2	CE(μ)	1.00	–	–	0	1	Yes	1	4.27
20-C	2	CE(μ)	1.00	–	–	10	1	Yes	6	3.82
20-C	2	CE(μ)	1.00	–	–	20	4	Yes	**7**	**2.09**
20-C	2	CE(μ)	1.00	–	–	30	**7**	**No**	**7**	**1.21**
20-C	3	CE($\mu - 3\sigma$)	1.00	–	–	0	1	Yes	1	4.33
20-C	3	CE($\mu - 3\sigma$)	1.00	–	–	10	1	Yes	7	3.51
20-C	3	CE($\mu - 3\sigma$)	1.00	–	–	20	4	Yes	**7**	**1.52**
20-C	3	CE($\mu - 3\sigma$)	1.00	–	–	30	**7**	**No**	**7**	**1.17**
20-C	4	CE($\mu - 3\sigma$)	1.00	–	0	–	1	Yes	1	4.33
20-C	4	CE($\mu - 3\sigma$)	1.00	–	10	–	1	Yes	1	3.81
20-C	4	CE($\mu - 3\sigma$)	1 00	–	20	–	1	Yes	1	3.64
20-C	4	CE($\mu - 3\sigma$)	1 00	–	30	–	1	Yes	2	3.51
20-C	4	CE($\mu - 3\sigma$)	1 00	–	40	–	1	Yes	4	3.62
20-C	4	CE($\mu - 3\sigma$)	1 00	–	50	–	2	Yes	5	3.71
20-C	5	CE(μ)	1.00	0	–	–	1	Yes	1	4.27
20-C	5	CE(μ)	1.00	10	–	–	1	Yes	1	4.15
20-C	5	CE(μ)	1.00	20	–	–	1	Yes	1	4.08
20-C	5	CE(μ)	1.00	30	–	–	1	Yes	1	4.07
20-C	5	CE(μ)	1.00	40	–	–	1	Yes	1	4.04
20-C	5	CE(μ)	1.00	50	–	–	1	Yes	1	4.00
20-C	6	CE(μ)	1.00	0	0	–	1	Yes	1	4.27
20-C	6	CE(μ)	1.00	10	10	–	1	Yes	1	3.83
20-C	6	CE(μ)	1.00	20	20	–	1	Yes	1	3.59
20-C	6	CE(μ)	1.00	30	30	–	1	Yes	3	3.79
20-C	6	CE(μ)	1.00	40	40	–	1	Yes	3	3.87
20-C	6	CE(μ)	1.00	50	50	–	2	Yes	6	4.09

Note: BOLD refers to the maximum number of strips available and the reliability index achieved using that maximum.

out that entire duration without any remediation. If a design life is chosen and the properties of the FRP at that time period are used during initial design, an appropriate level of rehabilitation could be selected.

While the process described in the example was conducted in a step-wise fashion, it was based on the use of existing databases for the FRP materials and the concrete girders, and was augmented through the use of deterioration models for FRP and steel. These were integrated to develop a design tool which could, given pre-existing criteria for thresholds on use of material and β, in essence have been constructed to form a decision-based tool for durability-based design.

9.5 Conclusions

FRP composites have been shown to have significant advantages in specific cases of use in civil infrastructure, both from intrinsic performance and logistical perspectives. However, design knowledge is not easily available in forms amenable for use, and hence the creation of a knowledge-based system is hypothesized as being an enabling tool for the accelerated development of FRP materials, components, and fabrication technologies that would accelerate optimized development in these areas. The development of an integrated tool is hypothesized as being of immense value, both for design and to assess risk and economics of use.

9.6 Acknowledgements

The concepts outlined in this paper were developed as a result of discussions held during the *NSF Workshop on Performance and Design of Fiber Reinforced Polymer Composites at Very Cold Temperatures*, held at the University of Alaska on August 13–14, 2004.

9.7 References

AASHTO, AMERICAN ASSOCIATION OF STATE HIGHWAY TRANSPORTATION OFFICIALS (2004). *AASHTO LRFD Bridge Design Specifications, SI Units*, 3rd edition; AASHTO: Washington D.C., USA.

ABANILLA, M.A., LI, Y. AND KARBHARI, V.M. (2006), 'Durability Characterization of Wet Layup Graphite/Epoxy Composites Used in External Strengthening,' *Composites B*, 37(2/3), pp. 200–212.

ACI 440.2R-08 (2008), *Guide for the Design and Construction of Externally Bonded FRP Systems for Strengthening Concrete Structures*, ACI, Farmington Hills, MI, USA.

ATADERO, R.A. (2006), *Development of Load and Resistance Factor Design for FRP Strengthening of Reinforced Concrete Structures*, Ph.D. dissertation; University of California, San Diego, USA.

FROHNSDORF, G. AND CLIFTON, J.R. (1980), *Cements Research Progress*, Chapter 13, American Ceramic Society.

FROHNSDORF, G. AND SKALNY, J. (1983), Cement in the 1990s: Challenges and Opportunities, *Philosophical Transactions of the Royal Society of London*, A, 310, pp. 17–30.

MIL-HDBK-17 (2002), *Department of Defense Handbook, Composite Materials Handbook, Volume 2: Polymer Matrix Composites Materials Properties*, USA.

NOWAK, A.S. (1999) *Report 368: Calibration of LRFD Bridge Design Code*, National Academy Press: Washington D.C, USA.

PLEVRIS, N., TRIANTAFILLOU, T.C. and VENEZIANO, D. (1995), Reliability of RC Members Strengthened with CFRP Laminates, *Journal of Structural Engineering*, 121, pp 1037–1044.

RASDORF, W.J. (1985), 'Perspectives on Knowledge in Engineering Design,' in *Proceedings of the International Computers in Engineering Conference*, Vol. 2, American Society of Mechanical Engineers, Boston, MA, pp. 249–253.

VU, K.A.T. STEWART, M.G. (2000), Structural Reliability of Concrete Bridges Including Improved Chloride-induced Corrosion Rates, *Structural Safety*, 22, pp 313–333.

WESTBROOK, J.H. (1985), 'Some Considerations in the Design of Properties Files for a Computerized Materials Information System,' in *The Role of Data in Scientific Progress*, Editor: P.S. Glaeser, *Proceedings of the 9th International CODATA Conference*, pp. 241–245.

WILCOX, P. (2008), *Reliability Based Assessment of FRP Rehabilitation of Reinforced Concrete Girders*, Thesis for Master of Science in Structural Engineering, University of California San Diego, June.

ZHANG, S., KARBHARI, V.M. AND REYNAUD, D. (2001), 'NOL-ring Based Evaluation of Freeze and Freeze–Thaw Exposure Effects on FRP Composite Column Wrap Systems,' *Composites: Part B*, 32, pp. 589–598.

ZHANG, J.S., KARBHARI, V.M., ISLEY, F. AND NEUNER, J. (2003), 'Fiber-Sizing-Based Enhancement of Materials Durability for Seismic Retrofit,' *ASCE Journal of Composites for Construction*, 7(3), pp. 194–199.

10

Pipeline rehabilitation systems for service life extension

M. NAJAFI, University of Texas at Arlington, USA

Abstract: With the continued aging of existing pipelines and underground infrastructure, the requirement for renewal and service life extension of these assets is also increasing. The needs for improving quality of life, reduction of inconvenience to society, and less impact on the environment make trenchless renewal methods a sustainable choice for pipeline owners and public agencies. Since the 1980s, a number of methods have become available to renew *in-situ* existing and deteriorated underground pipelines. These methods belong to the family of 'trenchless technology' methods. Trenchless technology alternatives can install a new pipe or renew an existing and deteriorated pipe with minimum surface or subsurface excavation and disruption. Trenchless renewal methods can prolong life of the pipe, reduce operation and maintenance costs (by eliminating inflow/infiltration and cost of emergency repairs, and improve quality of service). These methods can achieve short-term and long-term savings over repairs and eventual replacement costs with open-cut replacement and repair methods, which often involve enormous social costs. The objectives of this chapter are to present an overview of pipeline deterioration mechanisms and how trenchless renewal methods can be applied to stop leaks, resist corrosion and abrasion, and install a new pipe in place of the existing and deteriorated pipe to provide a new design life. An overview of current and emerging design concepts also are presented.

Key words: pipeline deterioration, trenchless renewal methods (TRMs), decision support system (DSS), pressure and gravity pipes, partially and fully deteriorated design concepts.

10.1 Introduction

Pipeline systems require constant maintenance and can become impaired for a number of reasons. A comprehensive study performed by WRc (Serpente, 1993) concludes that the concept of measuring the 'rate of deterioration' of sewers is unrealistic, but deterioration is more influenced by random events in a pipe life-span (a storm or an excavation nearby) and severe defects do not always lead immediately to collapse. Pipes are prone to certain types of failures based on the type of material, physical design, age, functionality and external and internal environment. Distress and collapse of a pipe are the result of the complex interactions of various mechanisms that occur within and around the pipeline (McKim *et al.*, 2000). The impact of the deterioration of the pipeline system depends upon its size, complexity, topography and service. While it is almost impossible to predict when a pipe will collapse, it is feasible to estimate whether a pipe has deteriorated sufficiently for collapse to be likely.

Pipelines can have defects classified as built-in or long-term. Built-in defects are generated during pipeline construction and represent conditions that affect the performance of pipes after installation. Long-term defects are caused as a result of the deterioration process. Construction-related or built-in defects can be offsets in alignment, joints loosely fitted or loosened by vibrations, flattened or ovaled pipes, sags due to settlement, stresses caused by dynamic loadings of backfill, removal of trench sheathing and pilings, overburden compaction, etc. Joints can experience construction defects, such as pinching of rubber gaskets, misalignment of gaskets, and squeezing due to 'overshoving' of one pipe into another. A structural failure can be a crack, break, split, cavitation of the pipe opening, or separation at a joint (Najafi, 2005).

Examples of causes of long-term pipeline deterioration are sulfate corrosion due to pipe gases, excessive hydraulic flows, structural failures, leaks and infiltrations, and erosions. Bacteria in the wastewater stream convert the sulfates to hydrogen sulfides which, when released into the pipe air space, become oxidized into sulfuric acid. The sulfuric acid is reactive to some pipe materials, making it corrode. Severe corrosion can jeopardize the structural integrity of a pipe or manhole, and lead to collapse. A condition of pipeline deterioration that occurs over an extended time period and is not a result of construction practice is considered a long-term deterioration. Proper maintenance of pipelines is essential to keep the pipeline in good condition. As a safety warning, it should be noted that hydrogen sulfide, a common byproduct of sewage, is a rapid-acting systemic poison that causes respiratory paralysis with consequent asphyxia at high concentrations. Inhalation of high concentrations may cause coma after a single breath and may be rapidly fatal. Prolonged exposure may cause pulmonary edema (NIOSH, 1977).

The state of the surrounding soil is of fundamental importance in assessing the structural condition of a pipe. The main factors that affect the rate of ground loss include pipe defect size, hydraulic conditions (water table, and frequency and magnitude of surcharge), and soil properties (cohesive or non-cohesive soil). Severe defects (larger than 4 in. or 100 mm), high water table (above pipe level), frequent and high magnitude of hydraulic surcharge, and soil types (silts, fine sands, etc.) can also have serious effects on ground loss. Loss of side support will allow the side of the pipe to move outward when loaded vertically, and collapse will be likely to occur once the pipe deformation exceeds 10%. Uneven loading of pipes due to joint displacement also accelerates the pipe deterioration process.

10.1.1 Modes of pipeline deterioration

Pipeline deterioration is a complex process; many factors are responsible for their deterioration and failure – structural, hydraulic, environmental, functional, age of the pipe, quality of initial construction, etc. The intensity of structural failures depends on the size of the defect, soil type, fluid parameters, ground-water level and fluctuations, corrosion, method of pipe construction, and loading on the pipe. Hydraulic failures in gravity pipes can be caused by infiltration and inflow (I/I) problems. These I/I problems reduce the planned hydraulic capacity of gravity pipes, leading to system backups with increasing potential for collapse. Figure 10.1 illustrates various kinds of internal and external forces acting on a pipe.

Pipe breakage is likely to occur when the environmental and operational stresses act upon pipes whose structural integrity has been compromised by corrosion, degradation, inadequate installation or manufacturing defects. Pipe breakage types were classified by O'Day *et al.* (1986) into three categories: (i) circumferential breaks, caused by longitudinal stresses; (ii) longitudinal breaks, caused by transverse stresses (hoop stresses); and (iii) split bell, caused by transverse stresses on the pipe joint. This classification may be complemented by an additional breakage type, viz. holes due to corrosion or impact. Circumferential breaks due to longitudinal stress are typically the result of one or more of the following occurrences: (i) thermal contraction (for example, due to low temperature of fluid in the pipe and the pipe surroundings) acting on a restrained pipe, (ii) bending stress (beam

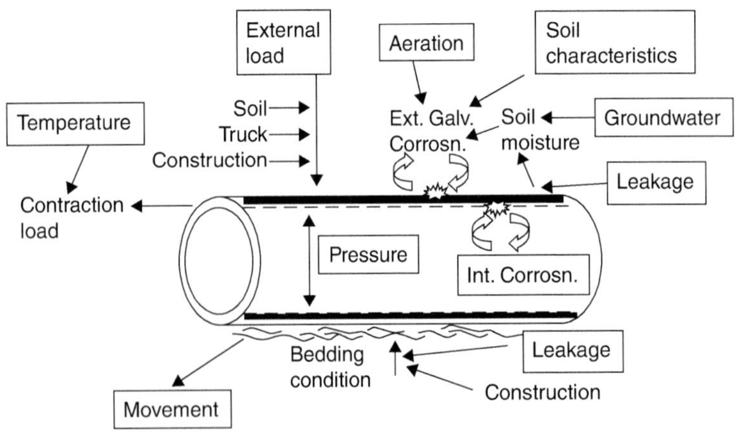

10.1 Pipeline interactions leading to failure (O'Day *et al.*, 1986).

Table 10.1 Typical defects in gravity pipes (Davies *et al.*, 2001)

Defect	Description
Longitudinal cracks and fractures	May occur at springing level as well as at the crown and invert. A result of excessive 'crushing' or 'ring' stress.
Tension cracks	Cracks are diagonal and spread from the point of overload, which is often a hard spot beneath the pipe.
Circumferential cracks and fractures	Relative vertical movement of successive lengths of pipe causing cracks and/or fractures due to excessive shear or bending stresses. Most likely to occur near joints.
Broken pipes	Occurs when pieces of a cracked or fractured pipe visibly move from their original position. Normally represents a further stage in deterioration of a cracked or fractured pipe and is a very serious defect.
Socket bursting	Excessive pressure inside the joint due to the expansion of the jointing material may cause a bursting failure of the socket.
Deformed pipes	Occurs when a longitudinally cracked or fractured pipe loses the support of the surrounding ground.

failure) due to soil differential movement (especially in clayey soils) or large voids in the bedding near the pipe (resulting from leaks, I/I, etc.), (iii) inadequate trench and bedding practices, and (iv) third-party interference (e.g. accidental breaks). Table 10.1 lists the most typical types of defect found in gravity pipes.

Pipeline corrosion and failure is an environmental, safety, economic, and operational issue. These factors are the basis for the trenchless renewal of pipelines and they also have significant influence on the decision to renew the pipeline. In selecting an appropriate renewal method, it is important to consider such factors as renewal method durability, reliability, strength to carry loads (hydrostatic and soil/traffic), social impacts, constructability and bypassing requirements, as well as the suitability of the specific method for the type of application, and availability of qualified contractors. The aim of the selection process is to consider all factors in order to arrive at the most cost-effective, technically viable solution.

10.2 Extending service life

Service life extension can be broadly defined as any technology that can be applied to an existing, aging, and deteriorated infrastructure system to increase its useful life. In the pipeline industry, the threshold for new service design life is commonly set at 50 years, although some methods and pipe products under certain conditions may provide a service life of up to 100 years.

Since the 1980s, a number of methods have become available to renew existing and deteriorated underground pipelines *in-situ*. These belong to the family of 'trenchless technology' methods. Trenchless technology methods include all the methods that can install a new pipe or renew an existing deteriorated pipe with minimum surface or subsurface excavation and disruption. Trenchless renewal methods not only can prolong the life of a pipe, reduce operation and maintenance costs (by eliminating inflow/infiltration and cost of emergency repairs) and improve quality of service, but also can achieve economic short-term and long-term savings in repairs and eventual replacement costs when compared with 'open-cut', which often involves enormous social costs.

The term 'renewal methods' refers to all the methods that can extend the service life of a pipeline beyond its original design life. Renewal methods are used in this chapter to refer collectively to 'rehabilitation,' 'upgrade,' 'renovation,' and 'replacement' of a deteriorated pipeline using trenchless technology methods. The term 'repair' refers to all the methods that can fix localized defects, without adding to the service life of the pipeline (Najafi, 2005). Repair methods are not included in this chapter.

Figure 10.2 is a conceptual plot of life expectancy. The decision to renew a pipeline is a complicated issue and requires much technical and managerial information. While age of the pipeline is the most important factor, other parameters such as failure or break rates, hydraulic capacity, quality of service, consequences of failures, environmental issues, location of the pipeline, budgetary and regulatory concerns and so on, need to be considered. The lack of knowledge of the condition of pipes and remaining service life, present difficulty in terms of determining which trenchless renewal to use and when to use these methods in order to ensure pipeline sustainability and make the best asset management option at the lowest cost.

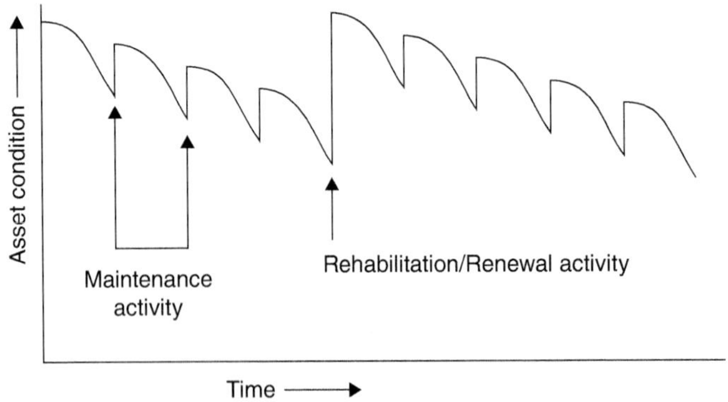

10.2 Asset deterioration and renewal, modified from AMSA (2002).

The maintenance of assets needs to be balanced between planned and unplanned maintenance schedules so that the cost of maintenance activities is minimized. Figure 10.2 also illustrates deterioration and renewal strategies for an asset that result in the effective management and operation of the asset. This figure demonstrates several maintenance activities during the life of the asset, with continued deterioration, until the point that a major renewal/rehabilitation/replacement is required. After the renewal process, the asset starts a new life with a new cycle of maintenance activities and deterioration processes (AMSA, 2002).

10.3 Trenchless renewal methods (TRMs)

Many trenchless renewal methods are available, or are currently under development. These methods can be used to replace, rehabilitate, upgrade, repair or renovate (collectively called renewal) existing pipelines. They also can be used to replace and enlarge existing pipes. The basic trenchless renewal methods can be categorized into the following types (Fig. 10.3):

(i) Cured-in-place pipe (CIPP)
(ii) Sliplining (SL)
(iii) Modified sliplining (MSL)
(iv) In-line replacement (ILR)
(v) Close-fit pipe (CFP)
(vi) Spray-in-place pipe (SIPP)
(vii) Thermoformed pipe (ThP)
(viii) Sewer manhole renewal (SMR).

The choice of trenchless pipeline renewal method depends on the physical conditions of the existing pipeline system, such as pipeline length, type, material, size, type and number of manholes, service connections, and the nature of the problem or problems involved. The problems with the existing pipe may include structural or non-structural infiltration or inflow, exfiltration or outflow, pipe breakage, joint settlement, joint or pipe misalignment and capacity problems. Other features of the renewal systems, such as applicability to a specific project, constructability, cost factors, and life expectancy should also be considered. Figure 10.3 presents the main divisions of pipeline renewal methods.

10.3.1 Cured-in-place pipe (CIPP)

The CIPP method involves the insertion of a resin-impregnated fabric tube into an existing pipe by use of water inversion or winching. The fabric is a polyester material, fiberglass reinforced or similar. Usually hot water or steam is used for the inversion process. Hot water, steam, or ultraviolet

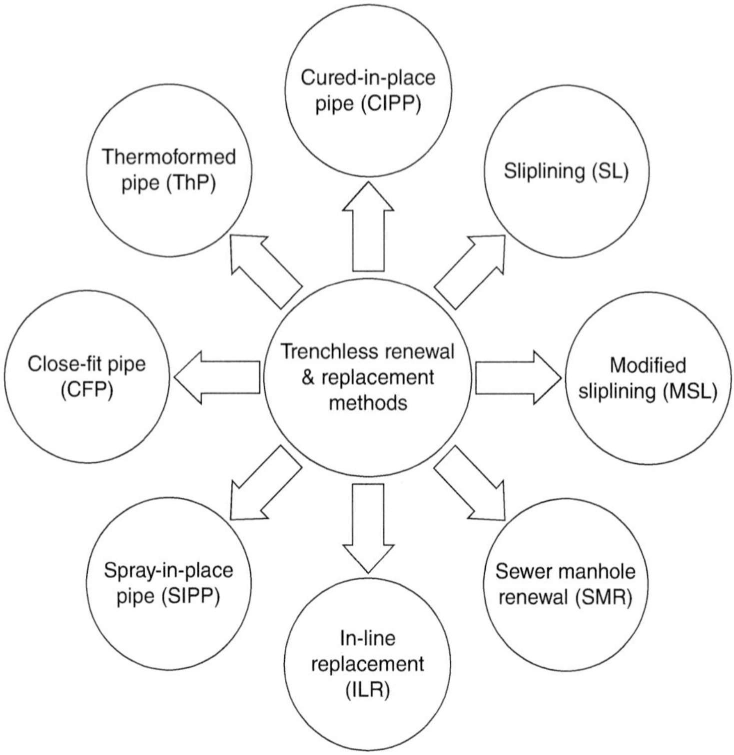

10.3 Basic trenchless renewal methods (Najafi, 2010).

Table 10.2 Main characteristics of CIPP methods (Najafi, 2005)

Method	Diameter range in. (mm)	Max inst. ft (m)	Liner material	Applications
Inverted in place	4–108 (100–2700)	3000 (1000 m)	Thermoset resin/Fabric composite	Gravity and pressure pipelines
Winched in place	4–54 (100–1500)	1000 (300 m)	Thermoset resin/Fabric composite	Gravity and pressure pipelines

(UV) light can be used for curing. The pliable nature of the resin-saturated fabric prior to curing allows installation around curves, filling of cracks, bridging of gaps, and maneuvering through pipe defects. CIPP can be applied for structural or non-structural purposes. Table 10.2 presents main characteristics of CIPP methods. Figure 10.4 illustrates CIPP installation process.

10.4 CIPP winch-in-place installation method (courtesy of Insituform Technologies).

Table 10.3 Main characteristics of sliplining methods (Najafi, 2005)

Method	Diameter range in. (mm)	Max inst. ft (m)	Liner material	Applications
Segmental	4–158 (100–4000)	1000 (300)	HDPE, PVC, GRP	Gravity and pressure pipelines
Continuous	4–63 (100–1600)	1000 (300)	HDPE, PVC	Gravity and pressure pipelines

10.3.2 Sliplining (SL)

Sliplining is mainly used for structural applications when the existing pipe does not have joint settlements or misalignments. In this method, a new pipeline of smaller diameter is inserted into the existing pipe and usually the annulus space between the existing pipe and the new pipe is grouted. This installation method has the merit of simplicity and is relatively inexpensive. However, there can be a significant loss of hydraulic capacity. Table 10.3 presents the main characteristics of sliplining methods. Figures 10.5 and 10.6 illustrate the two variations of the sliplining method, namely *segmental* sliplining and *continuous* sliplining.

10.3.3 Modified sliplining (MSL)

Modified sliplining (MSL) methods include methods whereby pipe sections or plastic strips are installed in close-fit with existing pipe, and the annular

10.5 Segmental sliplining method (courtesy of Midwest Society for Trenchless Technology).

10.6 Typical continuous sliplining process.

space is grouted. There are three variations of the MSL method: panel lining (PL), spiral wound lining (SWL) and formed-in-place pipe (FIPP).

Panel linings can be used to structurally renew large diameter (more than 48 in. or worker-entry) pipes. This method can accommodate different shapes, such as non-circular pipelines. The main type of material for this method is fiberglass. Figure 10.7 illustrates a panel-lining method.

The spiral wound method uses a layered composite PVC liner and cementitious grout to renew the existing pipe. The combination of the

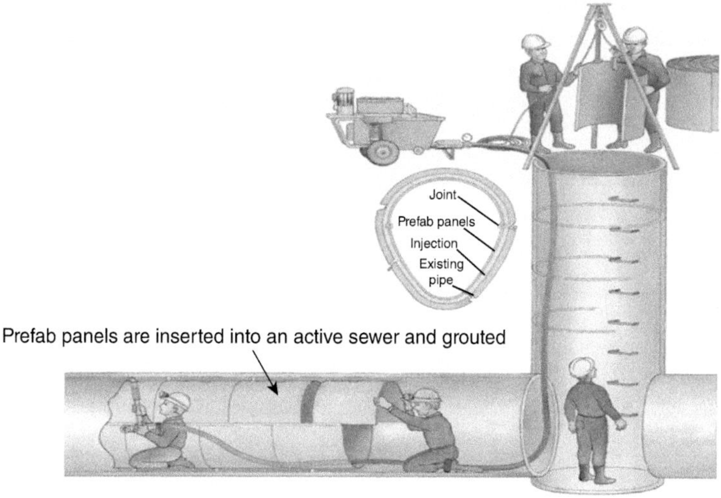

10.7 Panel lining process (courtesy of Channel Line).

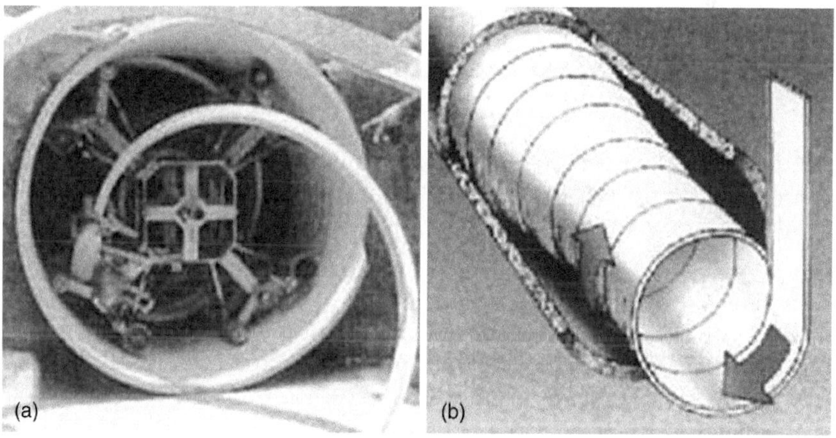

10.8 (a) Spiral wound installation machine, (b) expanding spiral winding (courtesy of Danby of North America, Inc.).

ribbed profile of the PVC liner and the fluid nature of the grout produces a highly integrated structure with the PVC liner 'tied' to the existing pipe through the grout. For worker-entry pipes, the structural strength of the renewed pipe is determined by the grout characteristics.

Formed-in-place pipe applications include renewal of wastewater and stormwater pipes, and culverts for diameters ranging from 48 inches to 2 feet, regardless of the shape and material of the host pipe. Figure 10.8

Table 10.4 Main characteristics of modified sliplining method (Najafi, 2005)

Method	Diameter range in. (mm)	Max inst. ft (m)	Liner material	Applications
Panel lining	More than 48 in. (1200)	Varies	GRP	Gravity pipelines
Spiral wound	4–100 (100–2500)	1000 (300)	PE, PVC	Gravity pipelines
Form-in-place pipe	48–144 (1200–3600)	Varies	PVC, HDPE	Gravity pipelines

Table 10.5 Main characteristics of in-line replacement method (Najafi, 2005)

Method	Diameter range in. (mm)	Max inst. ft (m)	Liner material	Applications
Pipe bursting	4–140 (100–3500)	750 (200)	PE, clay, PVC, GRP, DI	Gravity and pressure pipelines
Pipe removal	Up to 36 (900)	300 (100)	PE, PVC, GRP, DI	Gravity and pressure pipelines

illustrates a schematic diagram for the spiral wound process. Table 10.4 presents the main characteristics of modified sliplining methods.

10.3.4 In-line replacement (ILR)

When the capacity of a pipeline is found to be inadequate, then in-line replacement should be considered. There are two categories representing in-line replacement: pipe bursting and pipe removal. Pipe bursting, as the name implies, uses a hammer to break the existing pipe and force particles into the earth while a new pipe is pulled and/or pushed in its place simultaneously. Pipe removal, can also be performed by the use of horizontal directional drilling (HDD) (also called pipe reaming), horizontal auger boring (HAB) or microtunneling (MT) (also known as pipe eating) equipment. In this method, the existing pipe is broken into small pieces and is taken out by means of a slurry (for HDD or MT methods) or auger (for HAB method). Table 10.5 presents the main characteristics of the inline replacement method. Figures 10.9 and 10.10 illustrate pneumatic pipe bursting and pipe reaming methods.

When using the horizontal directional drilling (HDD) equipment to remove the existing pipe (pipe reaming), the HDD rig with minor modifica-

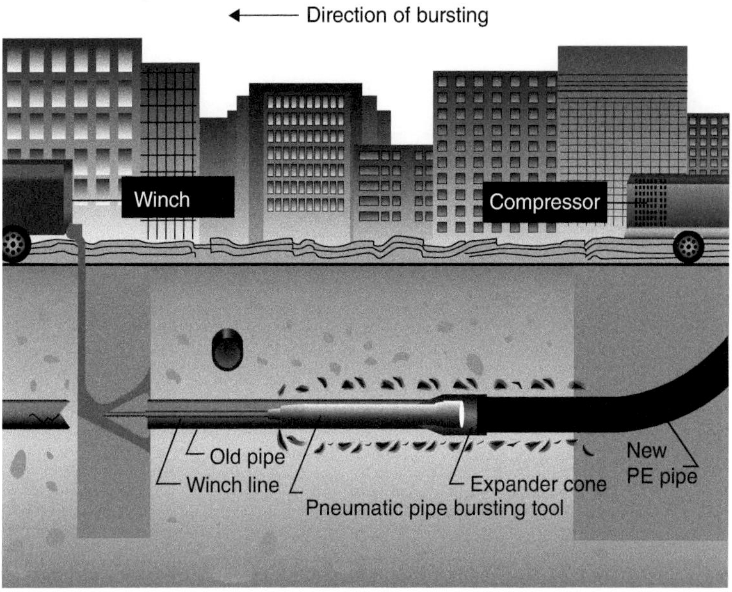

10.9 Pneumatic pipe bursting method (courtesy of TT Technologies).

10.10 Pipe reaming method (courtesy of Nowak Pipe Reaming).

tion pushes the drill rods through the existing pipe and connects the rods to a special reamer (see Fig. 10.10). Then, the new pipe string is attached to the reamer via a swivel and towing head. As the drill rig rotates and simultaneously pulls back the drill rod, the existing pipe is ground up and replaced by the new pipe.

10.3.5 Close-fit pipe (CFP)

This type of trenchless pipeline renewal temporarily reduces the cross-sectional area of the new pipe before it is installed, then expands it to its original size and shape after placement to provide a close fit with the existing pipe. The method can be used for both structural and non-structural purposes. Lining pipe can be reduced on-site or in the manufacturing plant and reformed by heat and/or pressure or naturally. There are three versions of this approach: fold and formed (F&F), drawdown (DD) and rolldown (RD). Table 10.6 presents the main characteristics of close-fit lining methods.

Table 10.6 Main characteristics of close-fit methods (Najafi, 2005)

Method	Diameter range in. (mm)	Max inst. ft (m)	Liner material	Applications
Fold & formed	4–30 (100–800)	700 (210)	HDPE, PVC	Gravity and pressure pipelines
Drawdown	3–60 (75–1500)	1000 (300)	HDPE, MDPE	Gravity and pressure pipelines
Rolldown	3–24 (75–600)	1000 (300)	HDPE, MDPE	Gravity and pressure pipelines

10.3.6 Spray-in-place pipe (SIPP)

Cementitious materials

Cementitious materials are used in both water and sewer applications primarily for their most basic characteristic – cost effectiveness in corrosion protection. In water applications, cement mortar lining is most common and serves two main functions: the alkalinity of the cement inhibits corrosion (especially in the case of a cast iron pipe), and the relatively smooth internal surface reduces hydraulic roughness and improves flow characteristics. It should be noted that cement mortar lining is also applied to many new cast iron and ductile iron water pipes to inhibit corrosion. In this application, the lining does not fulfill a structural function; rather it reduces the rate at which the pipe will deteriorate. This lining is not appropriate for pipes that leak, or where corrosion has reduced the wall thickness significantly. It is also important to apply sufficient thickness of mortar to create the alkaline environment at the mortar–iron interface for iron water pipes. As with steel reinforcement in concrete structures, inadequate cover of the metal will allow the onset of corrosion, which will cause the mortar to crack and spall.

In sewer applications, cementitious linings comprise various base materials, such as Portland cement and calcium aluminate. These materials are used to line underground structures to prevent infiltration and rebuild structures deteriorated because of age and erosion from I/I. Calcium aluminate cements are frequently used to provide some sulfide resistance in renewing sewer structures where mild corrosion is evident. Cementitious materials are also frequently used as a basecoat material for polymer linings in more corrosive sewer environments.

Polymeric (epoxy, polyurethane and polyurea) materials

Solvent-free, 100% solids epoxy, polyurethane, and polyurea coatings have unique advantages for successful renewal and corrosion protection. These thermoset coatings are essentially inert plastics when cured and they are therefore resistant to most corrosive elements found in these environments. They have longer life expectancies when properly applied by professionally trained applicators using specialized spray equipment to achieve optimal performance as defined by the product manufacturers. Because these products contain no solvents, they do not require an evaporative process to cure and emit no destructive volatile organic compounds (VOCs) found in many applications used in above ground coatings. Polymer formulations with 100% solids are safer in confined spaces and can also be formulated for thicker, structure enhancing applications. Such products now dominate the underground infrastructure protection industry because many can cure and bond to concrete, brick, steel, and cast iron in damp underground environments, while providing necessary protection from the aggressive sewer environment.

Compared to cementitious products, most polymers deliver superior chemical resistance in the most severe corrosive environments, although manufactures should be consulted for specific recommendations. Polymers can be formulated to be structural or non-structural, as they can have extraordinary long-term strengths, flexibility, and elongations. Most polymers used in underground applications have been formulated for ultra high-build thicknesses of 40 to 250 mils single-coat capability, towing to the jagged and deep profiles found on new and deteriorated concrete surfaces. Polymer linings are often used to topcoat cement mortar linings in severe duty environments.

Polyurea/polyurethane hybrid

Polyurea/polyurethane hybrid formulations can be defined as the result of a chemical reaction between an isocyanate and a mixture of polyol and amine reactants. These formulations generally provide an 'intermediate' polyurea that displays many of the same properties as a polyurea. However, hybrid formulations can also display some of the negative problems associated with polyurethane chemistry. In coatings formulation, hybrids generally contain a polyether/polyester polyol and a primary amine resulting in a chemical backbone with amine and hydroxyl functionality. Table 10.7 presents a comparison between polyurethane, epoxy and polyurea. Table 10.8 presents the main characteristics of spray-in-place pipe. Figure 10.11 illustrates a schematic diagram of spray-in-place lining.

Table 10.7 Polyurethane, epoxy and polyurea comparison

Polyurethane	Epoxy	Polyurea
Forms a weak bond when applied directly to concrete. Usually used as a topcoat with an epoxy-based primer.	Epoxy forms a very strong bond to concrete.	Polyurea forms an extremely strong bond to concrete.
Has been in use for 20+ years	Has been in use for 65+ years.	Has been in use for 30+ years
Must be top-coated every 4–7 years.	With the exception of cracking, epoxy lasts indefinitely if applied properly in the right environment.	Lasts indefinitely if applied properly – even in challenging environments.
Polyurethane has elongation properties similar to polyurea.	Epoxy cannot handle expanding and contracting concrete, or heavy impact.	Polyurea has elongation properties more than 50 times that of epoxy – making it highly resistant to cracking.
Cures in hours.	Minimum of 18 hours.	Cures in minutes to an hour.
Highly resistant to oil and most contaminants	Most epoxies absorb oil and other common contaminants, making stains impossible to clean.	Highly resistant to oil and most contaminants.
Moderate VOC levels	Epoxy has high levels of VOCs when compared to polyurea.	Most forms of polyurea have zero or extremely low VOCs.

Table 10.8 Main characteristics of spray-in-place pipe (SIPP) (Najafi, 2005)

Method	Diameter range in. (mm)	Max inst. ft (m)	Liner material	Applications
Spray-in-place pipe (SIPP)	Dependent on the type of product and manufacturer	N/A	Portland cement, epoxy, polyurethane, polyurea	Pressure and gravity pipelines

10.3.7 Thermoformed pipe (ThP)

'Thermoformed' is a terminology used in North America for pipes that have a reduced cross-section by folding for insertion and that are subsequently heated in order to thermoform them to conform to the existing pipe dimensions. This type of trenchless pipeline renewal uses a new folded PVC or

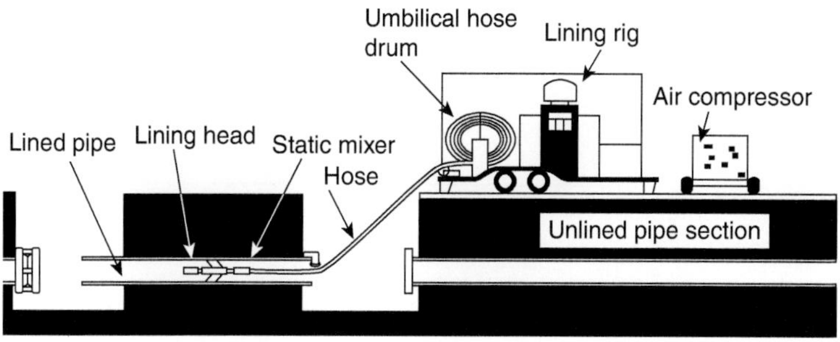

10.11 Spray-in-place pipe lining (courtesy of 3M Water Infrastructure).

10.12 Thermoformed pipe.

Table 10.9 Main characteristics of thermoformed pipe (Najafi, 2005)

Method	Diameter range in. (mm)	Max inst. ft (m)	Liner material	Applications
Thermoformed pipe (ThP)	4–30 (100–800)	700 (210)	PE, PVC	Gravity and pressure pipelines

PE pipe that is expanded by thermoforming to fit tightly inside the existing pipe. Both PVC and PE used for this method have a long performance history in pipe applications that verifies not only their structural capacity, but also other important long-term performance characteristics such as chemical and abrasion resistance. The main characteristics of thermoformed pipe are presented in Table 10.9. Figure 10.12 illustrates the thermoformed pipe.

Table 10.10 Main characteristics of sewer manhole renewal methods (Najafi, 2005)

Method	Diameter range in. (mm)	Max inst. ft (m)	Liner material	Applications
Manhole renewal	Any	N/A	Spray-on lining, PVC, CIPP	Sewer manholes

10.3.8 Sewer manhole renewals (SMR)

Sewer manholes require renewal to prevent surface water inflow and groundwater infiltration, to repair structural damage and to protect surfaces from the damage of corrosive substances. When renewal methods will not solve the problems cost-effectively, manhole replacement should be considered. Selection of a particular renewal method should consider the type of problems, physical characteristics of the structure, location, condition, age and type of original construction. The extent of successful manhole renewal experiences and cost should also be considered. The sewer manhole renewal can be divided into the following methods: cementitious, cast-in-place, cured-in-place, and profile PVC. Table 10.10 presents the main characteristics of sewer manhole renewal methods. Figure 10.13 shows the manhole lining process.

10.4 Selection of renewal method

There are many factors that need to be identified prior to pipe renewal, such as pipe history, cause of failure, fluid quality, leakage, hydraulic capacity, and capability of specific renewal method to address identified problems and defects. Table 10.11 presents the renewal methods and application to service life extension of pipes.

Without a deep and complete understanding of the overall conditions of the problem, it is very difficult to select appropriate renewal measures. The selection of appropriate pipe renewal can be summarized in a two-step process such as (i) identification of deteriorated pipe and definition of the problem, and (ii) selection of the renewal method. These steps are described below:

- Step (i): Identification of deteriorated pipe is the first step in the selection of a particular renewal method. The availability of various closed-circuit inspection (CCTV) equipment and other inspection methods, as well as computer models for flow analysis and water quality computation, geographic information system (GIS), global positioning system

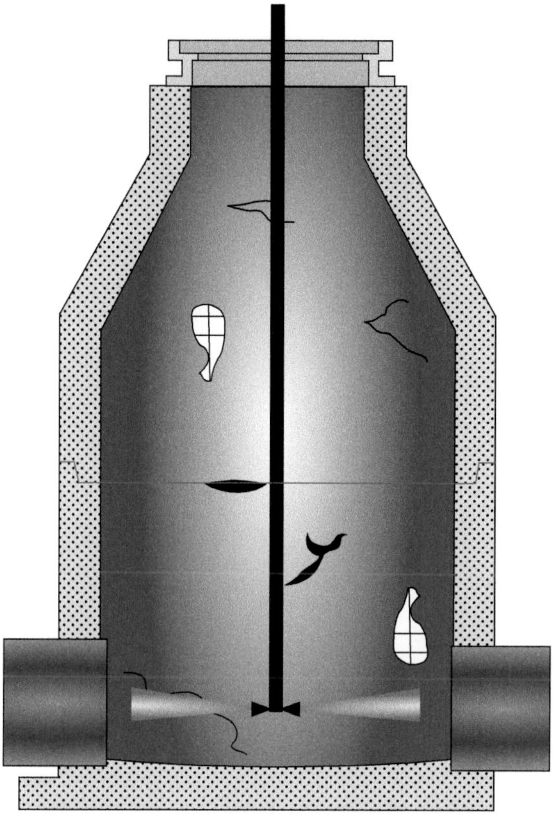

10.13 Manhole lining process (courtesy of Perma Liner Industries).

(GPS) and other software, can help pipeline owners to collect, maintain and analyze data on a system-wide basis. Some of the factors that must be evaluated that will govern the selection of renewal method include age, existing pipe material, history of previous maintenance or repairs, leaks, number and problems with joints, bends, tees, pipe size and length, and number of service laterals. Other pipeline problems include:

- *Hydraulic capacity problems:* Hydraulic problem is failure originating from a decrease in pipe flow capacity or increase in demand. Infiltration and inflow increases the demand as the numbers of structural defects grow and demand from service connections increase. A decrease in capacity is the result of a reduction in diameter due to corrosion, sediment, tree roots and/or grease.
- *Water quality problems:* Water in contact with plastic pipes, surface coatings or other materials can be affected by migration of components that make the water quality unacceptable with respect to aesthetic effects or health (Hem and Aquateam, 2002). Volatile

Table 10.11 Renewal methods and application to service life extension of pipes (Najafi, 2010)

Renewal method	Joint problems	Corrosion	Cracks/holes	Inflow, infiltration, and exfiltration	Structural problems	Inadequate hydraulic capacity	Extension of service life of existing pipe
CIPP	Marginal	Yes	Yes	Yes	Yes	Marginal	Yes
Sliplining	Marginal	Yes	Yes	Yes	Yes	No	Yes
Close-fit pipe	Marginal	Yes	Yes	Yes	No	Marginal	Yes
Thermoformed pipe	Marginal	Yes	Yes	Yes	Yes	Marginal	Yes
SIPP	No	Yes	Marginal	Yes	Marginal	No	Marginal
Modified sliplining	Marginal	Yes	Yes	Yes	Yes	No	Yes
Inline replacement	Yes	Yes	Yes	Yes	Yes	Yes	Yes

Note: Marginal means special preparation (such as a point repair) may be required before installation of the renewal method and/or the renewal solution may be effective dependent on the specific product (manufacturer), quality of application, and project conditions.

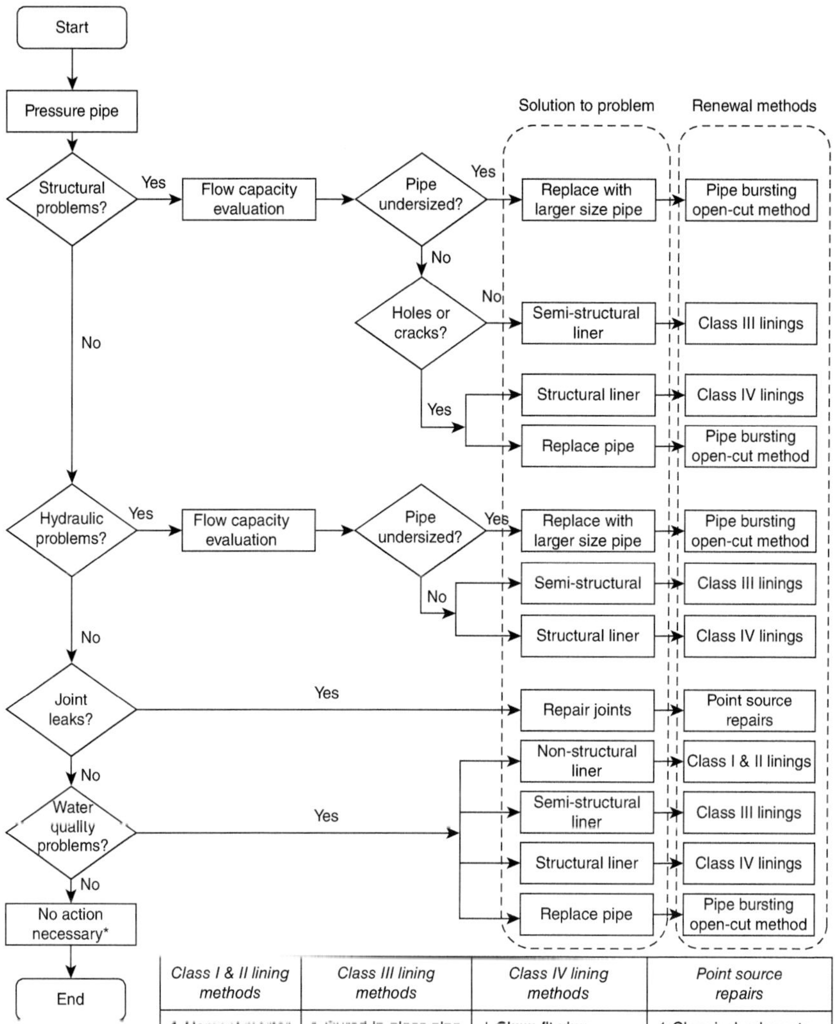

Class I & II lining methods	*Class III lining methods*	*Class IV lining methods*	*Point source repairs*
1. Cement mortar lining 2. Epoxy lining	1. Cured-in-place pipe 2. Polyurea lining 3. Polyurethane lining	1. Close-fit pipe 2. Cured-in-place pipe 3. Sliplining 4. Thermoformed pipe	1. Chemical gel grouts, cement based grouts 2. Expoxy resins, cement mortar 3. Internal seal 4. Point CIPP

*Continue to monitor based on your asset management plan and pipe material and condition requirements.

10.14 Decision support system for pressure pipe renewal methods (Najafi, 2010).

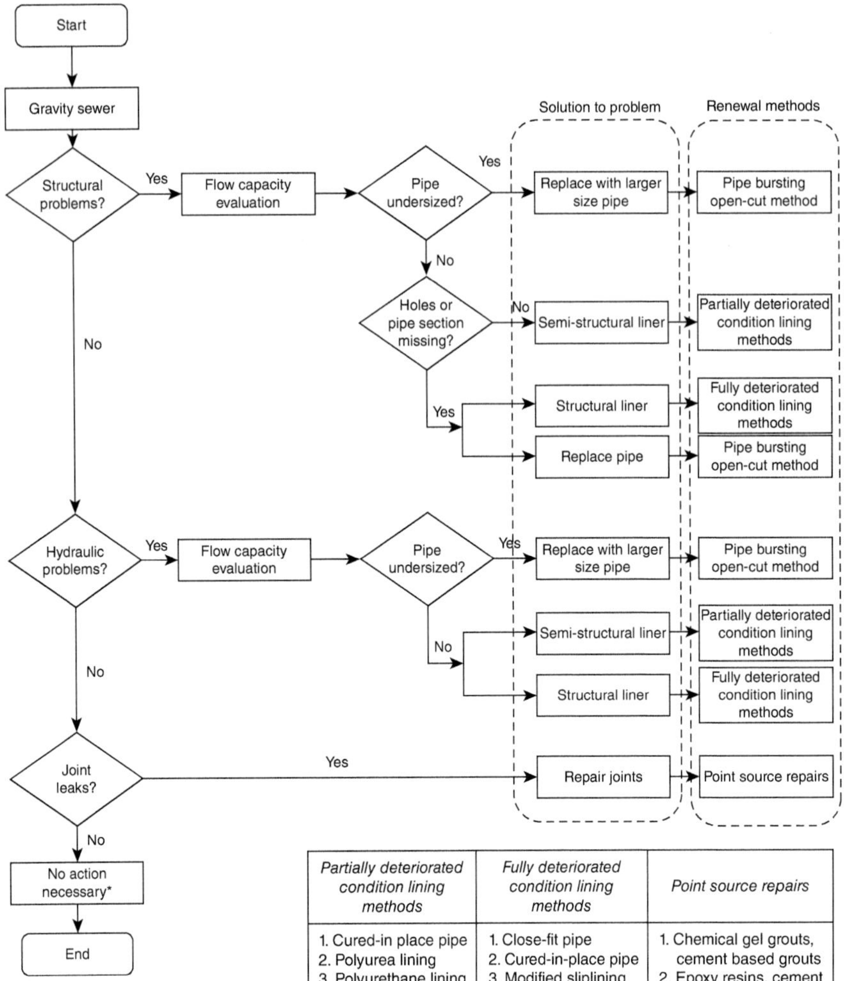

Partially deteriorated condition lining methods	*Fully deteriorated condition lining methods*	*Point source repairs*
1. Cured-in place pipe 2. Polyurea lining 3. Polyurethane lining	1. Close-fit pipe 2. Cured-in-place pipe 3. Modified sliplining 4. Sliplining 5. Thermoformed pipe	1. Chemical gel grouts, cement based grouts 2. Epoxy resins, cement mortar 3. Internal seal 4. Point CIPP

*Continue to monitor based on your asset management plan and pipe material and condition requirements.

10.15 Decision support system for gravity sewer pipe renewal methods (Najafi, 2010).

organic compounds (VOCs) leach from defective or cracked pipes and linings. These VOCs can give taste and odor to the drinking water, and can cause the water to fail the acceptable limits.

- *Structural problems:* Structural defect failure mechanisms include cracks, fractures and breaks in the pipe material that are caused by change in internal or external pressure. Tree roots, broken or cracked pipes, missing or broken cleanout caps, or undersized gravity pipes can cause blockages. Infiltration and inflow has great impact on pipe capacity and is caused when groundwater enters the pipe system through pipe defects and leaking joints. Many surface collapse failures are associated with degraded but functioning pipes that fail owing to a large one-time event (Delleur, 1989; Water Research Centre, 1986). Smaller cyclical dynamic loads include load transfer from above ground activities, such as routine truck, machinery, and bus or train traffic, or in ground movements, such as those caused by expansive soils or frost heave (Najafi, 2005).

- Step (ii): Once the pipe to be renewed is selected, the flow diagrams in Figures 10.14 and 10.15 can be used to suggest appropriate renewal methods.

10.5 Current design concepts for pipeline renewals

The objective of pipe renewal design involves a set of equations considering factors such as groundwater, soil, traffic loads, and other loading conditions. As for any structural design, there are many variables and parameters used in the design of the pipe wall thickness that provide the required performance requirements to resist the loads. There is no single design equation that can be used for all the different conditions that must be taken into account for the proper design of a lining, so it is necessary to divide these conditions into different groups. For both gravity flow and internal pressure, design equations have been categorized into partially and fully deteriorated on the basis of ASTM F1216. These piping condition are defined below.

10.5.1 Partially deteriorated pipe

A partially deteriorated existing pipe is defined as one where the pipe is cracked or corroded, but has no missing pipe sections. The soil–pipe system in this case is capable of supporting all the soil, groundwater and surface load. Lined underground pipe is designed to support external hydrostatic loads due to groundwater as well as to withstand the internal pressure in spanning across specific holes in the original pipe wall.

10.5.2 Fully deteriorated pipe

In fully deteriorated condition the existing pipe is not structurally sound and cannot support soil and live loads, or is expected to reach this condition over the design life of the renewed pipe. This condition is evident when sections of the original pipe are missing, the pipe has lost its original shape (possibly collapsed), or the pipe has corroded due to the effects of the fluid, atmosphere, soil, or applied loads.

10.6 Emerging design concepts for pipeline renewal systems

A conventionally buried pipe is structurally designed to 'carry' or 'transfer' all service loads, which include the weight of soil backfill placed over it, hydrostatic pressures, vacuum, internal working pressures, and loads applied at the ground surface. Current methods for structural design of flexible gravity pipe liners defined by Appendix X1 of ASTM F1216-09, generally consider two external load cases for dimensioning of the liner pipe. The first is sustained hydrostatic pressure due to groundwater acting in the annular space between the liner pipe and the existing pipe. The second load case supposes that earth and traffic loads will, in due course, be transferred from the existing pipe–soil structure to the liner pipe. This excessive conservatism is particularly true in the treatment of the 'fully deteriorated pipe condition' by ASTM F1216-09, for two reasons.

First, the soil load reaching the liner pipe is overestimated by treating the liner as if it had been directly buried in a trench. The more correct liner pipe condition, even in circumstances where the existing pipe may continue to deteriorate after renewal, is a tunnel lining situation (Schrock *et al.*, 1997). Second, the formula used to describe liner response to transferred soil load, and hence calculate the required wall thickness, has been incorrectly modified from an already conservative theory for direct-bury applications that entails further irrational safety factors (Grumbel, 1998).

The basic design concept proposed at the ASCE Pipeline Division web site (www.pipelinedivision.org) is to develop buckling pressure or safe water head charts for each renewal technique that already incorporates its characteristic imperfections, but allow variation of the relevant system imperfections according to the condition of the pipe to be renewed. The most appropriate form of chart may vary according to the type of renewal technique and/or liner material (ASCE, 2007). The reader is encouraged to refer to the above publication for more information.

10.7 Long-term testing

Concern with the design life of structures has become more of an issue recently due to early deterioration of some pipe materials, as well as unexpected and excessive repair and maintenance costs. At present, there are limited established design methodologies for pressure pipe lining systems. The existing ASTM Standard F1216-09 provides some information on design methods and procedures for the renewal of gravity and pressure pipes, using a cured-in-place pipe lining method. Some short-term tests have been conducted to evaluate the key mechanical properties of the lining and its constituents. The data obtained from these tests can be used for defining the properties of the liner during numerical analyses. For long-term testing and prediction of structural failure of flexible pipes when subjected to an external hydrostatic pressure, there are various alternatives available. The most widely used approach involves modification of the Timoshenko equation covering the buckling of unconfined pipe (Timoshenko and Gere, 1961).

Additionally, ASTM D2990-01 (2001), *Test Methods for Tensile, Compressive and Flexural Creep and Creep-rupture of Plastics* provides a test of simply supported beam samples of plastic where 'creep modulus' is measured over 10000 hours with extrapolating long-term values. These data are used to define an apparent long-term flexural modulus of elasticity for use in the buckling equations for plastic pipes. ASTM D1598-02 (2002), ASTM D2992-06 (2006) and ASTM D2837-08 (2008) are used to test pipe samples that are subjected to a varying internal pressure to fail the lining pipes in 10000 hours. A graph of hoop stress with respect to time is plotted on the log scale and extrapolated to determine long-term hoop stress on the pipe and the lining. Table 10.12 presents the expected useful life of various renewal methods in service life extension of the pipe, based on the various publications indicated.

Table 10.12 Expected useful life of various pipe renewal methods

Method	Material used	Expected useful life[†]	Reference
Spray-in-place pipe	Cement mortar	>50 years	Deb *et al.* (1990)
	Epoxy resin	>75 years	Watson (1998)
	Polyurea	>50 years	3 M Water Infrastructure
Cured-in-place pipe	Thermoset resin/ fabric composite	>50 years	TTC (1994)
Sliplining	Polyethylene	50 years	Silbert *et al.* (2002)
Thermoformed	PE & PVC	100 years	Najafi, M, 2005
Close-fit pipe	PE & PVC	>50 years	Selvakumar *et al.* (2002)
Modified sliplining	HDPE, PE & PVC	>50 years	Silbert *et al.* (2002)
Pipe bursting	PE, PVC, HDPE & GRP*	>50 years	Silbert *et al.* (2002)
Pipe removal	PE, PVC, HDPE & GRP*	>50 years	Silbert *et al.* (2002)

* Glass Reinforced Plastic (GRP) has a useful life in excess of 100 Years (Silbert *et al.* (2002).

[†]Expected useful life is a loose term that depends on many factors, such as quality of design and installation, liner pipe material properties, pipe loadings and pipe environmental conditions, type of application, fluid properties, and existing pipe condition and level of its deterioration.

10.8 Conclusions

With the continued aging of existing pipelines and underground infrastructure, the need for renewal of these pipelines is also increasing. The needs for improving quality of life, reduction of inconvenience to society, and less impact on the environment make trenchless renewal methods a sustainable choice for pipeline owners and public agencies. The main benefits of trenchless renewal methods are not only being cost-effective but also providing environmentally friendly and sustainable construction operations. Proper selection of the renewal method reduces the possibility of pipeline failures and the massive costs of emergency repairs. Understanding the causes of pipeline failures helps in the selection of an appropriate renewal method and in obtaining extended service life of the pipe. Once a renewal method has been determined and executed, an assessment of its effectiveness needs to be performed. This will determine the suitability of that particular method for future applications.

This chapter has presented an overview of pipeline deterioration mechanisms and a summary of different trenchless renewal methods, with factors to consider in selecting a specific renewal method. These methods have become methods of choice over traditional open-cut pipeline installations and replacements. Trenchless renewal methods (TRM) stop leaks, resist corrosion and abrasion, and install a new pipe in place of the existing and deteriorated pipe, providing a new design life. An overview of current and emerging design concepts has also been presented.

10.9 Nomenclature

AMSA	Association of Metropolitan Sewerage Agencies
ASCE	American Society of Civil Engineers
ASTM	American Society for Testing and Materials
AWWA	American Water Works Association
CCTV	Closed circuit television
CFP	Close-fit pipe
CIPP	Cured-in-place pipe
CUIRE	Center for Underground Infrastructure Research and Education
DD	Drawdown
DI	Ductile iron
F&F	Fold and formed
GIS	Geographic information system
GPS	Global positioning system
GRP	Glassfiber reinforced polyester
HAB	Horizontal auger boring

HDD	Horizontal directional drilling
HDPE	High density polyethylene
ID	Inside diameter
ILR	In-line replacement
MDPE	Medium density polyethylene
MSL	Modified sliplining
OD	Outer diameter
PCCP	Prestressed concrete cylinder pipe
PE	Polyethylene
PL	Panel lining
PP	Polypropylene
PVC	Polyvinyl chloride
QA	Quality assurance
QC	Quality control
RCP	Reinforced concrete pipe
RD	Rolldown
SIPP	Spray-in-place pipe
SL	Sliplining
SMR	Sewer manhole renewal
SWP	Spiral wound pipe
ThP	Thermoformed pipe
TRM	Trenchless renewal system
TT	Trenchless technology
UV	Ultraviolet
VOC	Volatile organic compound

10.10 References

ASCE (2007). *Emerging Concepts for the Design of Pipeline Renewal Systems*, ASCE-Pipeline Division, Pipeline Infrastructure Task Committee, available at: www.pipelinedivision.org.

ASSOCIATION OF METROPOLITAN SEWERAGE AGENCIES (2002). *Managing Public Infrastructure Assets to Minimize Cost and Maximize Performance*, AMSA.

ASTM F1216 (2009). *Standard Practice for Rehabilitation of Existing Pipelines and Conduits by the Inversion and Curing of a Resin-Impregnated Tube*, West Conshohocken, USA.

ASTM D 2990 (2001). *Standard Test Methods for Tensile, Compressive, and Flexural Creep and Creep-rupture of Plastics*, American Society for Testing and Materials, West Conshohocken, Pa, USA.

ASTM D 1598 (2002). *Standard Test Method for Time-to-Failure of Plastic Pipe Under Constant Internal Pressure*, American Society for Testing and Materials, West Conshohocken, Pa, USA.

ASTM D 2992 (2006). *Standard Practice for Obtaining Hydrostatic or Pressure Design Basis for 'Fiberglass' (Glass-Fiber-Reinforced Thermosetting-Resin) Pipe and Fittings*, American Society for Testing and Materials, West Conshohocken, Pa, USA.

ASTM D 2837 (2008). *Standard Test Method for Obtaining Hydrostatic Design Basis for Thermoplastic Pipe Materials or Pressure Design Basis for Thermoplastic Pipe Products*, American Society for Testing and Materials, West Conshohocken, Pa, USA.

AWWA (2002). *Decision Support System for Distribution System Piping Renewal*, Denver, Co, USA.

DAVIES, J.P., CLARKE, B.A., WHITER, J.T., and CUNNINGHAM, R.J. (2001). 'Factors Influencing the Structural Deterioration and Collapse of Rigid Sewer Pipes,' *Urban Water* 3, 73–89.

DEB, A.K., SNYDER, J.K., CHELIUS, J.J., and O'DAY, D.K. (1990). *Assessment of Existing and Developing Water Main Rehabilitation Practices*. AWWA Research Foundation, Denver, CO, USA.

DELLEUR, J. (1989). 'Evaluation criteria for sewerage failure and rehabilitation.' *Journal of Water Resources Planning and Management*, 115 (3), 278–298.

GRUMBEL, J.E. (1998). 'Structural Design of Pipe Linings 1998 – Review of Principles, Practice and Current Developments Worldwide'. Paper presented at the *4th ASTT Conference*, Brisbane, Australia, August.

HEM, L.J. and AQUATEAM, A.S. (2002). 'Potential Water Quality Deterioration of Drinking Water Caused by Leakage of Organic Compounds from Materials in Contact with the Water', *Proceedings 20th No-Dig Conference*, Copenhagen May 28–31, 2002.

MCKIM, R.A., KATHULA, V.S., and SINHA, S.K. (2000). Integrating Digitized Imaging and Stochastic Modeling for Pipeline Management. In: *Proceedings of the 2000 6th Environmental Engineering Specialty Conference*, CSCE, London, Ontario, Canada, pp. 351–360.

NAJAFI, M. (2005). *Trenchless Technology: Pipeline and Utility Design, Construction, and Renewal*, McGraw-Hill, New York.

NAJAFI, M. (2010). *Trenchless Technology Piping: Installation and Inspection*, McGraw-Hill, New York.

NAJAFI, M. (1994). *Trenchless Pipeline Rehabilitation: State-of-the-Art Review*, Trenchless Technology Center, Ruston, Louisiana, USA.

NIOSH, NATIONAL INSTITUTE FOR OCCUPATIONAL SAFETY AND HEALTH (1977). *Criteria for a Recommended Standard Occupational Exposure to Hydrogen Sulfide*. Cincinnati, Ohio: National Institute for Occupational Safety and Health. DHEW Publication No. NIOSH 77–158.

O'DAY, D.K., WEISSS, R., CHIAVARI, S., and BLAIR, D. (1986). *Water Main Evaluation for Rehabilitation/Replacement*, American Water Works Association Research Foundation (90509), Denver, Colorado, USA.

POLYUREA DEVELOPMENT ASSOCIATION (PDA), available at www.pda-online.org (accessed 07/15/09)

SCHROCK, B.J. and GRUMBEL, J. (1997). 'Pipeline Renewal', Conference Paper, Workshop 1-1, *North American No-Dig '97, NASTT North American Conference and Exhibition of Trenchless Technology*, April 18–21, 1997, Seattle, Washington, USA.

SELVAKUMAR, A., CLARK, R.M., and SIVAGANESAN, M. (2002). *Cost for Water Supply Distribution System Rehabilitation*. U.S. Environmental Protection Agency, Washington, DC, USA.

SERPENTE, R.F. (1993). Understanding the Models of Failure for Sewers. *Proceedings of International Conference on Pipeline Division*. ASCE, New York, pp. 86–100.

SILBERT, L.K., DEB, A.K., and SCHOSER, H. (2002). *Decision Support System for Distribution System Piping Renewal*, American Waters Works Association, Denver, CO, USA.

TIMOSHENKO, S.P. and GERE, J.M. (1961). *Theory of Elastic Stability*, 2nd ed., McGraw-Hill, New York, NY.

TRENCHLESS TECHNOLOGY CENTER (TTC). (1994). *Long-term Structural Behavior of Pipeline Rehabilitation Systems*. TTC Technical Report #302.

WATER RESEARCH CENTRE (1986). *Sewerage Rehabilitation Book*, 2nd ed., Water Research Centre/Water Authorities Association, United Kingdom.

WATSON, C. (1998). 'Rehabilitation of Potable Water Mains – The Changing Policy of Northumbrian Water'. Proceedings of International Conference on Rehabilitation Technology for Water Industry, Lille, France.

Index